S0-BDZ-483

About Island Press

Island Press is the only nonprofit organization in the United States whose principal purpose is the publication of books on environmental issues and natural resource management. We provide solutions-oriented information to professionals, public officials, business and community leaders, and concerned citizens who are shaping responses to environmental problems.

Since 1984, Island Press has been the leading provider of timely and practical books that take a multidisciplinary approach to critical environmental concerns. Our growing list of titles reflects our commitment to bringing the best of an expanding body of literature to the environmental community throughout North America and the world.

Support for Island Press is provided by the Agua Fund, The Geraldine R. Dodge Foundation, Doris Duke Charitable Foundation, The Ford Foundation, The William and Flora Hewlett Foundation, The Joyce Foundation, Kendeda Sustainability Fund of the Tides Foundation, The Forrest & Frances Lattner Foundation, The Henry Luce Foundation, The John D. and Catherine T. MacArthur Foundation, The Marisla Foundation, The Andrew W. Mellon Foundation, Gordon and Betty Moore Foundation, The Curtis and Edith Munson Foundation, Oak Foundation, The Overbrook Foundation, The David and Lucile Packard Foundation, Wallace Global Fund, The Winslow Foundation, and other generous donors.

Conservation of Rare or Little-Known Species

Conservation *of* Rare *or* Little-Known Species

Biological, Social, and Economic Considerations

Edited by
Martin G. Raphael
Randy Molina

Foreword by
Nancy Molina

Washington · Covelo · London

Library of Congress Cataloging-in-Publication Data

Conservation of rare or little-known species : biological, social, and economic considerations / edited by Martin G. Raphael and Randy Molina ; foreword by Nancy Molina.

p. cm.

Includes bibliographical references and index.

ISBN 978-1-59726-165-4 (hardcover : alk. paper) — ISBN 978-1-59726-166-1 (pbk. : alk. paper) 1. Biodiversity conservation. I. Raphael, Martin G. (Martin George) II. Molina, R.

QH75.C6814 2007

333.95'16—dc22

2007026191

British Cataloguing-in-Publication Data available.

Printed on recycled, acid-free paper 

Manufactured in the United States of America

10 9 8 7 6 5 4 3 2 1

Contents

Foreword

Like many works of its kind, this book grew out of a need to synthesize scientific information and thinking, and to apply it to real-world problems for which there were no easy or obvious solutions. Throughout the 1990s and early 2000s, the federal agencies tasked with implementing the Northwest Forest Plan in the Pacific Northwest were challenged by the need to inventory and conserve populations of well over 300 species that were believed to be associated with old-growth forests, and whose status was possibly imperiled. The vast majority of those taxa were either exceedingly rare or were so poorly understood that their status could not be determined with any degree of confidence. The costs and technological challenges of carrying out a conservation program for such a large number of organisms (and moreover doing so in an environment with significant uncertainty and political controversy) were staggering.

At the time, biodiversity conservation science relevant to these problems basically boiled down to a debate about what has come to be termed "coarse filter" versus "fine-filter" approaches, and many questions were being raised that today remain unanswered. Examples of attempts to integrate the two approaches, with an adequate track record upon which to judge success, were (and remain) hard to find. The Northwest Forest Plan agencies decided to convene a symposium (Innovations in Species Conservation: Integrative Approaches to Address Rarity and Risk. September 30–October 1, 2003, Portland, Oregon, http:/outreach.cof.orst.edu/isc/index.htm) that would air the biological, ecological, and social issues associated specifically with the conservation of large numbers of rare or little-known taxa from across the spectrum of life forms. As a follow-up to the symposium, the authors of this book were asked to further synthesize relevant information from their respective fields, and craft a framework of concepts and information that could be used both to ease the Northwest Forest Plan situation and to more generally advance the state-of-the-art of biodiversity conservation. Following are some of the questions they were asked to address:

- What are some alternative approaches to conservation of rare or little-known species? What are their goals, and what is the likelihood they will be successful?
- How do different groups of constituents in society feel about these approaches?
- What are the economic implications?
- What are the legal and policy requirements associated with different conservation approaches?
- What constraints are imposed on land management and natural resource use by the various approaches?

As a result of grappling with these and other questions, the authors of this book are able to present many elements of the solution set for conservation of rare or little-known species, especially within a broader context of sustainable ecosystem management. This is a unique work in its focus on the rare or little-known components of biodiversity, and in the blending of the biological and social disciplines in envisioning how to care for those components. This book should provide a valuable resource for anyone—land manager, policy-maker, scientist, or member of the public—who cares about the conservation of organisms that are cryptic, often overlooked, and problematic to protect.

NANCY MOLINA
December 2006

Acknowledgments

This book was stimulated by presentations during a 2-day symposium, "Innovations in Species Conservation," held in Portland, Oregon, in 2003. We are grateful to the sponsors of the symposium—the USDA Forest Service, USDI Geological Survey, USDI Fish and Wildlife Service, USDI Bureau of Land Management, Oregon State University, the Nature Conservancy, and the Society for Conservation Biology.

We thank the Pacific Northwest Research Station, the Pacific Northwest Region, and the U.S. Geological Survey for funding to support author participation in this project. We thank Nancy Molina and James F. Quinn for their reviews of the entire draft, and three anonymous reviewers for comments on early versions of several draft chapters. We also thank the editorial staff of Island Press, especially Barbara Dean and Erin Johnson, for their guidance and support.

1

Introduction

Martin G. Raphael and Bruce G. Marcot

Conservation and restoration of biological diversity are often-stated objectives of land management. Doubtless, most natural resource managers will say they are achieving those objectives. But how can this be determined? The term "biological diversity" has been given many definitions (see Baydack and Campa 1999 for a listing of 19 alternatives), and most of these emphasize the variety of life and its processes. Much of this variety pertains to various aspects of species—their presence, distribution, occurrence within ecological communities, gene pool diversity, ecological functions, and other parameters—about which we know little, if anything for most species.

What's in This Book

The authors of this book describe a variety of approaches that move natural resource management toward the goal of achieving biological conservation, particularly species conservation. Natural resource managers usually focus on conservation practices that address only the larger and better-known taxa. This book focuses on the rare or little-known taxa, particularly species.

What is meant by "rare" or "little known"? Figure 1.1 illustrates the set of species the authors are addressing. Consider a large planning area and the assemblage of species that occur there (denoted as set A). Among those species are those that are rare (set B; as described in chap. 3) and those that are little known (set D; as described in chap. 4). One might wish to focus on

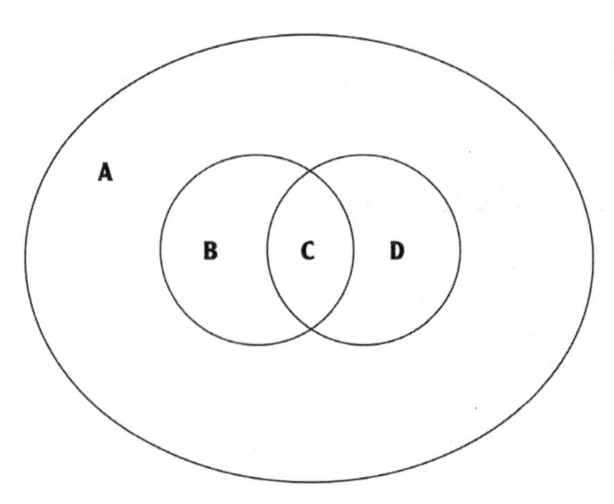

Figure 1.1. The species "universe." Set A represents all species in a given planning area. Set B represents all species in A that are rare. Set D represents all species in A that are little known. The area of overlap between sets B and D (labeled as C) is the set of species that are both rare and little known. The union of sets B and D is the set of species that are rare or little known.

the set of rare species, which might be at risk due to their small population size and increased vulnerability to disturbance. But one must not ignore the little-known species, because many of these, if we did know more about them, also might prove to be at risk because they are rare or because they are also vulnerable to changes in their environment. Conceptually, it is the combined sets of species B and D that define the group of species we call rare or little known. And perhaps the more vulnerable subset of these species are those that are both rare *and* little known (species set C; discussed in chap. 5).

The authors of this book evaluate how well various conservation approaches provide for rare or little-known species, either directly or indirectly. A few conservation strategies are fairly well aimed at providing for such species, but most focus on other aspects of biological diversity and conserve such species at best opportunistically or incidentally. For these strategies, the bottom line for conserving such species is largely that "it depends" on how the strategies are specified and implemented. The authors discuss the legal, biological, sociological, political, administrative, and economic dimensions by which conservation strategies can be gauged, to help managers determine which strategy or combination of strategies would best meet their goals and objectives. There are no fixed, single, or easy answers.

This book focuses primarily on terrestrial ecosystems and rare or little-known species associated with those systems. However, rarity concepts surely extend to aquatic, riparian, and wetland systems, and the authors provide some examples from these ecosystems when relevant to overall objectives. It is outside the scope of this book, however, to provide detailed discussions of river (fish) and marine issues.

This work was also prompted by the need to provide information on alternative approaches that could complement single-species assessment and management. Taking a strictly species-specific approach has proven difficult and expensive, particularly the collection of basic inventory data (Molina et al. 2006). Alternative ways of managing multiple species, including those rare or little-known, as well as ecological systems under scientifically credible approaches might help complement species-specific approaches currently in use.

How can rare or little-known species be managed effectively? There are many possible approaches to conceptualizing, assessing, managing, and monitoring rare or little-known species. In this book, the authors provide a classification of such approaches, offer a summary of the theoretical and conceptual foundations of each approach, evaluate each approach's efficacy in conserving rare or little-known species, and review how each has been used in assessments, management plans, and monitoring activities. The goal is to give land managers access to this diverse literature and to provide them with the basic information they need to select those approaches that best suit their conservation objectives and ecological context.

The book suggests an overall procedure by which management approaches can be identified (fig. 1.2). This entails first describing the legal, social, economic, administrative, political, and ecological requirements and considerations for the management question (discussed here and in chap. 2). The next step is to articulate the overall social, economic, and ecological management goals (chap. 2) and then to identify the rare or little-known species to address (chaps. 3, 4, and 5). Then one conducts a risk assessment that identifies key threats to the species and system factors and possible management sideboards (chaps. 6, 7, and 8). From this assessment, one can identify the management approach or combination of approaches that best addresses the specific risk factors (chaps. 6, 7, and 8), and then evaluate the approaches for their social, economic, ecological, and administrative application (chaps. 8, 9, 10, and 11). A final selection of approaches is made (chap. 12), and a program to monitor and evaluate success of meeting the goals is implemented (chaps. 11 and 12). Results of monitoring can

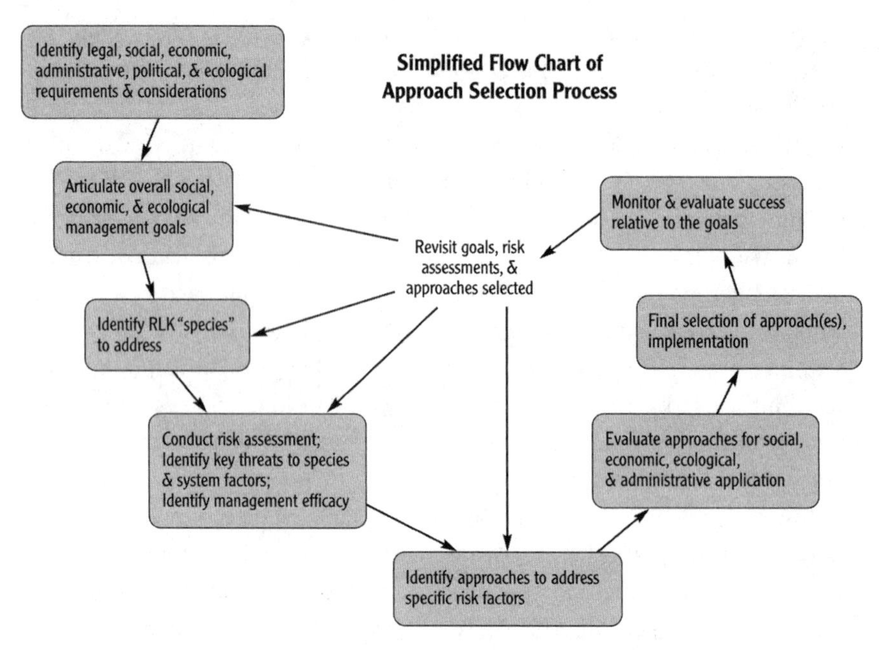

Figure 1.2. Flow chart of the process for selecting and testing alternatives to conservation of rare or little-known species as discussed in this book. A fuller version of this process with explanation of each step is presented in chapter 12.

be used to revisit the goals, the species selected, the risk assessment, and the set of approaches selected. In the end, our objective is to provide a thorough scientific evaluation of approaches and management options for conserving rare or little-known terrestrial species.

Why Should Land Managers Care about Conserving Rare or Little-Known Species?

Land managers should be concerned about rare or little-known species for ethical, ecological, and legal reasons.

Rare or Little-Known Species Contribute to Biodiversity

The reasons may mirror the many practical and ethical reasons given for conserving nature itself—preserving potential future sources of drugs and pharmaceuticals, protecting native pollinators of our food plants, "saving all the pieces" to ensure that the system works, acknowledging the "existence value" of other life forms on Earth, and so on. Bolger (2001) pointed out that, although we do not fully understand how biological diversity provides for ecological functions in ecosystems, assuring some level of redundancy in species and their functions, and protecting species that have particularly salient functions ("keystones"), are probably important for maintaining system stability. Until we learn more, Bolger noted, it is prudent to protect all species lest we sacrifice some unknown critical function. This book provides examples of critical functions and the ecological roles of rare or little-known species.

Conservation science is concerned, in part, with anticipating how natural or human-caused disturbance affects the pattern of commonness and rarity among the biota of a given area (Lubchenco et al. 1991; also see chap. 3). The concept of rarity has several definitions in common usage (see chap. 3), but in the lexicon of conservation biology, "rarity" is most simply defined based on the distribution and abundance of a species (Gaston 1994). However, for many taxa we lack basic information on distribution and abundance (Brown and Roughgarden 1990; Flather and Sieg 2000) from which to judge rarity and basic information on life history and ecology. The authors here refer to these species collectively as "little known." Although it is most certainly true that many little-known species are also rare, there are many examples of little-known species that are likely quite common, such as soil and forest canopy arthropods. As Flather and Sieg discuss in chapter 3, there are different kinds of rarity that may have differing kinds of management implications. Something could be (1) locally common but endemic or highly restricted in spatial distribution, (2) widespread but rare everywhere, (3) a locally rare population of a species that is common elsewhere (i.e., at the edge of the range or an outlying population in a metapopulation, or (4) genetically or behaviorally distinctive.

Rare or Little-Known Species Can Play Key Ecological Roles

Rare or little-known species help provide for productive and diverse ecosystems from which humans can sustainably use renewable natural resources. This is particularly true with native organisms whose roles may be to stave off unwanted invasions by nonnative species; some rare or little-known native organisms provide such ecological services. Rare or little-known species may contribute significantly to the maintenance of ecosystem function (Andrén et al. 1995; Borrvall et al. 2000; Cottingham et al. 2001; Lyons and Schwartz 2001; Lyons et al. 2005), thereby helping to provide the goods and services that humans derive from ecosystems (Chapin et al. 1998).

An example is the set of soil microspiders ("micryphantids" of the family Araneae) that collectively occur in abundance, that consist of many poorly known, undescribed, or possibly rare species, and that are likely important invertebrate predators at a very fine scale. Overall, soil microspiders contribute to the balance of soil food webs and maintenance of nutrient pools, and thus to productivity of soils that in turn provide ecosystem services of growing food crops of commercial interest to people (Mansour and Heimbach 1993). In another example, empirical studies have demonstrated how rare broadleaved plant species can control the abundance of dominant grasses by governing ecosystem processes (Boeken and Shachak 2006). In a review of how biodiversity can affect ecosystem functioning, Hooper et al. (2005) noted in general that even relatively rare species such as keystone predators can strongly influence energy and material flows through an ecosystem.

Likewise, referring to invertebrates, Black (2002, 3) wrote:

> Though endangered invertebrates are unlikely to determine the fate of a large ecological system, as a group they may have a large effect. Often, endangered invertebrates are specialists that perform vital ecosystem functions such as pollination or the recycling of nutrients. . . . Endangered species also may play a linchpin role in small, specialized systems such as caves, oceanic islands, or some pollinator–plant relationships.
>
> . . . [S]ome endangered species might also provide useful products—such as new defenses against diseases or tools for studying various ecosystem or organismal processes—as well as direct material benefits.

Such beneficial ecological roles are played by organisms that are tiny and hidden in soil, forest canopies, and other places difficult to study. Wilson (1987) tells us it is the "little things that run the world." Most of these little things, and some of the big things, are effectively invisible to science but may play major roles in keeping ecosystems healthy and diverse. Wilson argues that some invertebrates (e.g., ants and termites) can have enormous abundances and impacts on overall community energetics. He estimates that the global biomass of ants alone substantially exceeds that of all vertebrates combined, yet the number of management-oriented ant and termite experts and studies is tiny. Patrick (1997, 17) noted that "terrestrial ecosystems are dependent on a high diversity of macro- and microscopic organisms" and their ecological functions.

In some cases, increasing rarity of a species, such as caused by anthropogenic stressors, might sacrifice the ecological roles of such species. McConkey and Drake (2006) discovered this with pteropodid fruit bats (flying foxes) on islands of the South Pacific Ocean, where the bats' role in seed dispersal declined as the bats became scarcer. In such cases, a rare species that otherwise plays an important ecological role, such as dispersal of large seeds by fruit bats, may reach a threshold of "functional extinction" before it reaches numerical extinction.

Such ecological significance of rare or little-known species may not be evident except in changing systems or systems under stress. On this, Andrén et al. (1995, 141) wrote: "It is possible that the major importance of biodiversity for ecosystem processes is not apparent under relatively stable conditions, but that diversity is imperative for an ecosystem's response to stress or major environmental changes, such as climatic change, without any loss of ecosystem function. Perhaps rare species become important when conditions change."

One example is the rare polylepis tree (*Polylepis weberbaueri*) of family Rosaceae, which occurs at 2000 to 4500 m elevation in the Andes Mountains of South America, including in Ecuador. Much of the high-elevation woodlands of the altiplano plateau of Ecuador have been eliminated for agriculture, leaving only remnant pockets of this rare polylepis woodland. These pockets are now increasingly critical habitat for a variety of high-elevation birds, including giant conebill (*Oreomanes fraseri*), rufous antpitta (*Grallaria rufula*), and red-rumped bush-tyrant (*Cnemarchus erythropygius*).

Rare or Little-known Species Can Have Evolutionary Significance

Over evolutionary time, species that are rare and that are closely tied to specific environments may become the basis for unexpected adaptive radiation when environmental conditions change. Regional climate changes likely spurred great speciation events in the Amazon (antbirds), coastal

Australia (treecreepers), and central Africa (many primates, squirrels, and small carnivores). Also, rare founder populations on oceanic islands and continental habitat islands can become the basis for evolution of locally endemic taxa. For instance, the Hawaiian Islands contain local avian endemics, including subspecies such as the Hawaiian owl (*Asio flammeus sandwichensis*), species such as the Hawaiian crow (*Corvus hawaiiensis*), genera such as the Hawaiian goose (*Branta sandvicensis*), and even subfamilies such as Drepanidinae (Hawaiian honeycreepers); the origins of such local endemics were likely rare immigrant founders.

Another reason why rare or little-known species may be of conservation interest pertains to extinction risk. All other things being equal, rare species are more apt to be lost from the regional or local species pool than are common species (Pimm et al. 1988; Johnson 1998).

Rare or Little-Known Species Are Conserved by Some Legal Regulations

Land managers may also care about conserving biodiversity because they must adhere to pertinent legal or regulatory provisions. However, such provisions in many countries do not target rare or little-known species per se, but rather address conservation of listed threatened or endangered species or more general biodiversity goals (see chap. 2). Examples include the United States' Endangered Species Act, Canada's Species at Risk Act, and Australia's Environment Protection and Biodiversity Conservation Act of 1999 under which individual states and territories provided further specifications, such as New South Wales' Threatened Species Conservation Amendment Bill of 2006.

Also relevant to rare species conservation in Australia are the Regional Forest Agreements (RFAs), which establish processes and consequences of the recent extensive redistribution of public forest lands into many new conservation reserves. This redistribution was based largely on modeled habitat requirements of threatened species and the public desire to represent all vegetation communities within the Australian conservation reserve system. The RFAs also specify management prescriptions for all threatened species in commercial forestry operations. The RFAs are based on a collage of state and commonwealth legislation, such as the Forestry and National Park Estate Act of 1998 (Rod Kavanagh, pers. comm.).

In another example, India's National Wildlife Protection Act of 1972 covers major fauna (vertebrates and some insects) and very few plant species, but it does not cover rare, little-known, or endangered species per se. The 2006 amendment creates a National Tiger Conservation Authority that specifically calls for conservation of the tiger as a flagship and endangered species. India's Biodiversity Act of 2002 does address conservation of endangered species likely on the verge of extinction. Also, India's Environmental Protection Act defines and provides protection for sensitive sites, and like most other countries, India follows regulations under the Convention on International Trade in Endangered Species of Wild Fauna and Flora (CITES) to prohibit illegal international trafficking of listed species (Ashish Kumar, P. K. Mathur, pers. comm.).

In some other countries there is either no specific law or regulation comparable to these acts, or legislation is less directly related to protection of rare or little-known species. For instance, Argentina has no endangered species act but instead has National and Provincial Natural Monument Laws to protect some particular endangered species (Ana Trejo, pers. comm.). The People's Republic of China has no national endangered species act; instead, there are lists of wild fauna and flora afforded "special state protection" that were promulgated and approved by the Chinese State Council in 1988. This provides some measure of protection of biota when proposing activities that might arguably impact them. China also has a new Environmental Impact Assessment (EIA) law that requires an EIA for a government "plan" (analogous to a "programmatic EIS" in the United States), but this requirement has yet to be fleshed out in specific regulations. The EIA law also requires an EIA for any construction project, and this dimension has been fleshed out in regulations, although there seems to be little guidance on rare and little-known species (G. Gordon Davis, pers. comm.).

Rare or Little-Known Species Can Be Targets for Conservation in Planning

Finally, as motivated by the foregoing concerns, some legal or regulatory provisions (e.g., the U.S. Endangered Species Act; the Canadian Species at Risk Act) require land-managing agencies to explicitly consider rare or little-known species in their resource planning activities. The basic problem with managing for rare or little-known species is that rarity itself imposes risks to viable, persistent populations, and lack of knowledge imposes uncertainties regarding the necessary conditions for such species. Moreover, the number of species that qualify as rare or little known can be quite large, further complicating the management problem.

For example, in the interior Columbia basin, U.S., Marcot et al. (1997) estimated that 22% of all bird species were either "irregular" or "rare" in occurrence in that region (fig. 1.3). Also, of a total 43,825 estimated species of fungi, lichens, bryophytes, vascular plants, mollusks, arthropods, and vertebrates expected to occur in that region, they found 60% were little known and had insufficient information for scientists to judge their occurrence in the planning area. Given inherent extinction risks, information constraints, and the sheer number of species that may be classified as rare or little known, it is easy to understand why the management issue can be overwhelming. If so many species are unknown and undescribed, or at best poorly studied, how do we know we are conserving biological diversity? How do we go about setting management practices to achieve the conservation of biological diversity?

Knowledge Is Increasing

Scientific knowledge on rare or little-known species has been increasing, but disproportionately more so for taxa that are easier to locate and study. As an example, the number of papers published each decade in all journals of the Ecological Society of America (ESA) since its inception in the 1940s has increased overall, with the greatest recent increases since the 1980s being studies on bryophytes, bacteria, fungi, and ferns (fig. 1.4a), vascular plants (fig. 1.4b), and invertebrates, fishes, and birds (fig. 1.4c).

However, the *proportions* of all ESA papers published on only vascular plants, fishes, and birds have increased; whereas the proportions published on all other taxa have not. This may be explained in part by the emergence of other, taxa-specific publications (particularly on invertebrates, fungi, lichens, and bryophytes) that might have drawn potential papers away from ESA journals. But it appears that scientific knowledge of many of the other taxonomic groups of rare or little-known taxa is not keeping pace with that on vascular plants and vertebrates. A consequence of this imbalance will be a widening gap in knowledge between these taxonomic groups. In addition, taxon experts, especially systematists, are not being trained and hired by agencies or universities/museums at even a replacement rate (Wheeler and Cracraft 1997).

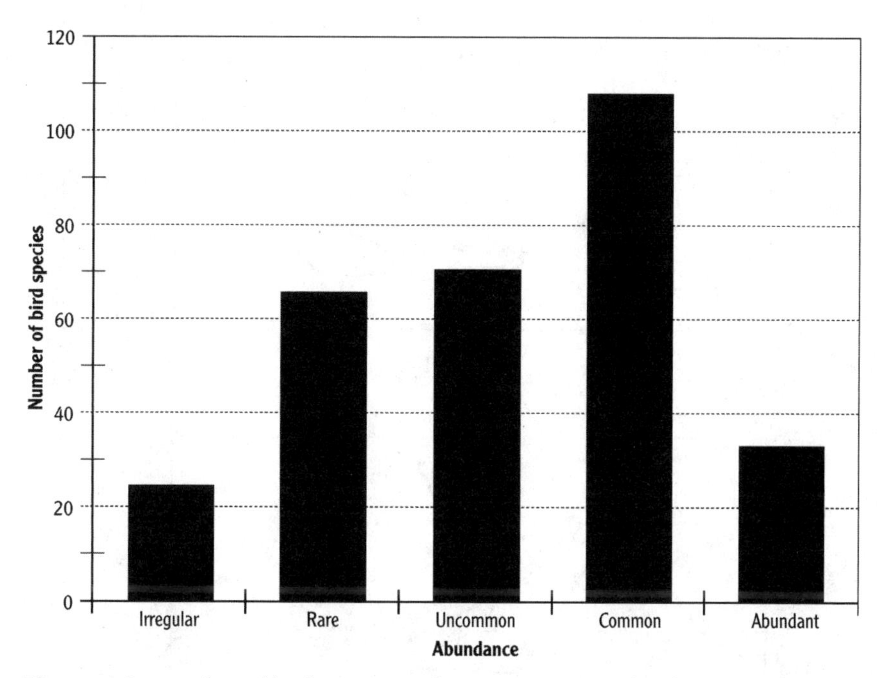

Figure 1.3. Number of bird species in the interior U.S. Columbia River basin by abundance class. *Source*: Marcot et al. 1998.

Another way to view amount of, and change in, knowledge is on a per capita (per species) basis. Dividing the total number of ESA publications (fig. 1.4d) by the globally known number of species per taxon (see chap. 4) reveals that most publications per capita are on known vertebrates and known bacteria and fewest on known fungi and known invertebrates. When considering high-end estimates of numbers of species (described plus unknown), fewest publications per taxon are on invertebrates, bacteria, and fungi. These taxa are in need of much basic taxonomic, biological, and ecological work.

Conclusion

Rare or little-known species are, by their nature of being scarce or poorly understood, often not explicitly included in conservation and natural resource planning. Although still incomplete, there is growing scientific evidence that many rare or little-known species may play key ecological roles in the function, structure, and composition of some ecological com-

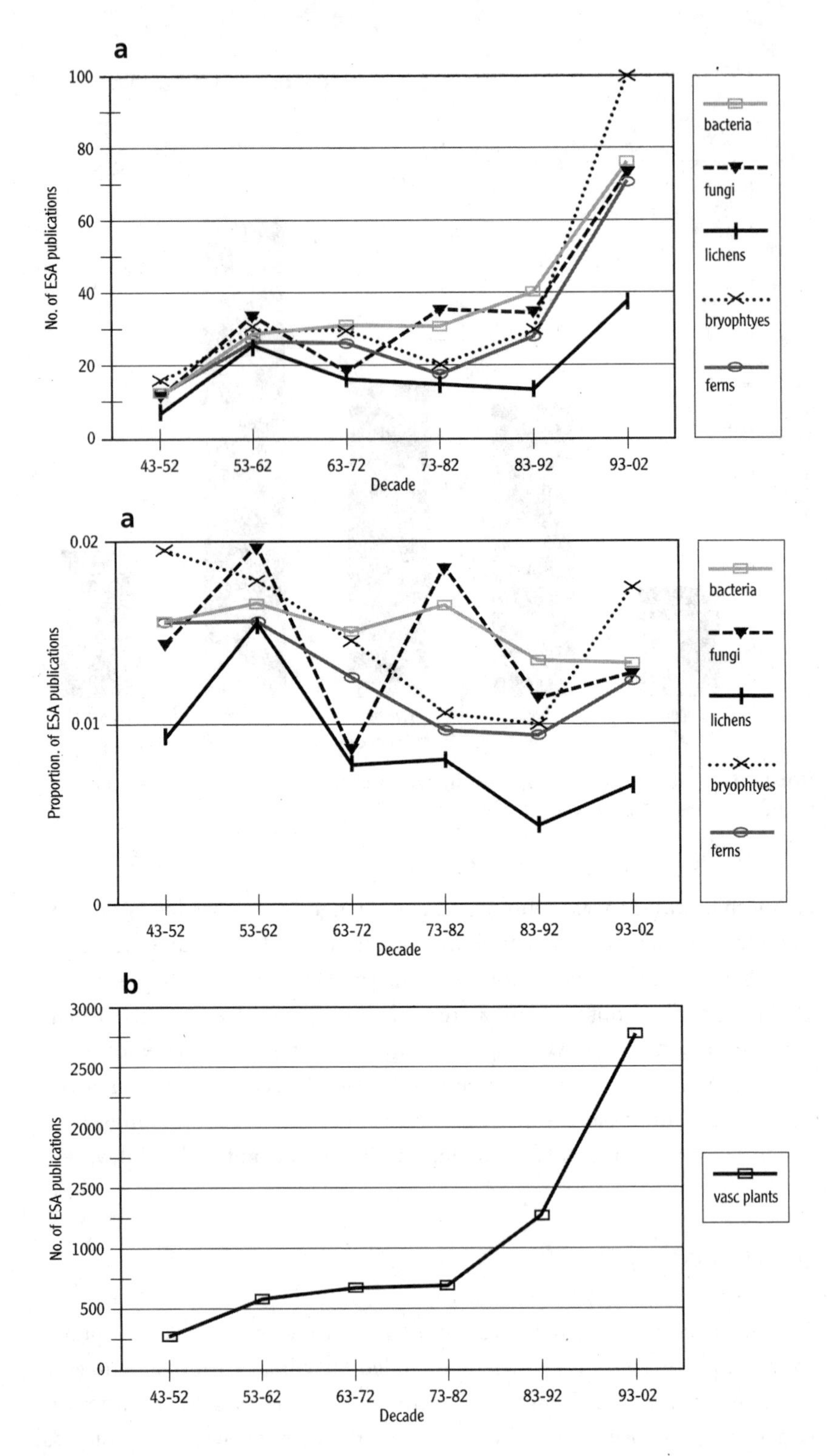

Figure 1.4. *Continues*

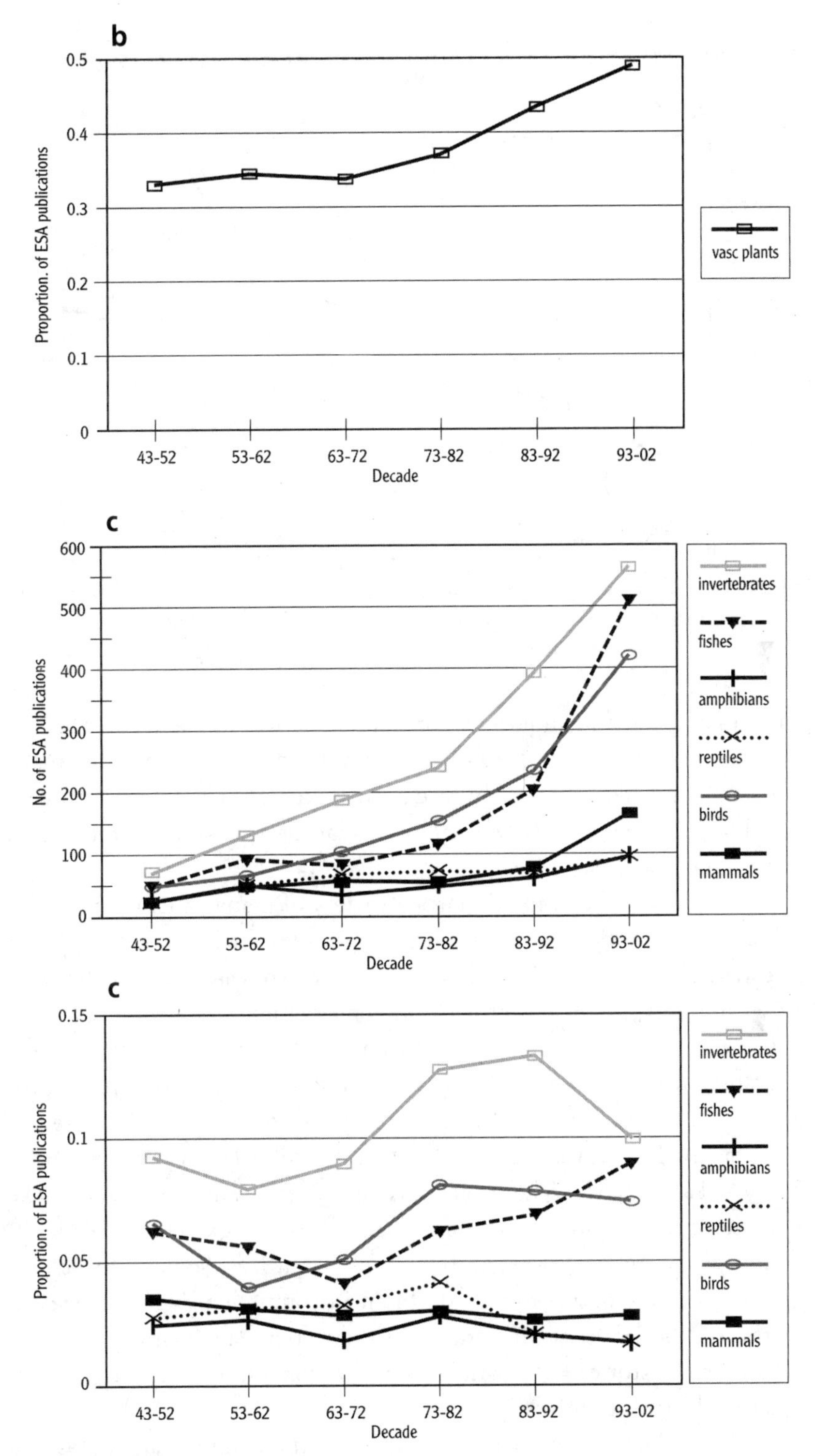

Figure 1.4. *Continued*

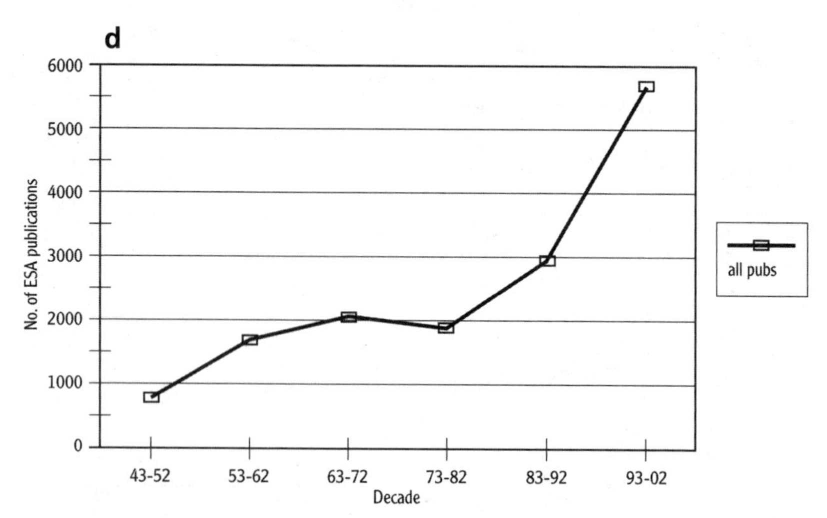

Figure 1.4. Number and proportion of publications by the Ecological Society of America on various taxonomic groups (a–c) and on all topics (d).

munities and could be important for long-term evolutionary potential. To conserve and restore the breadth of natural biodiversity, resource managers could include attention on rare or little-known species and their ecological functions, as well as on the more usual charismatic or better-known species and processes.

This book describes and evaluates a variety of management approaches and practical considerations for conserving rare or little-known species in terrestrial ecosystems. This focus is unique for two reasons. First, although there are many books and articles on conservation of rare species, most deal with conserving large and well-known species. However, in most ecosystems the vast majority of species consists of rare or little-known species that have received paltry or incomplete coverage in the conservation biology literature (e.g., fungi, lichens, mollusks, bryophytes, non-showy soil arthropods). These taxa have many unique attributes that make their detection, identification, and quantification extremely difficult. Given the significant ecosystem functions that at least some of these taxa perform, it is important to include them in comprehensive conservation efforts. Yet the inherent uncertainty in their poorly understood ecologies and natural histories makes their conservation on a species-by-species approach difficult to infeasible.

A major focus of the book thus addresses their unique attributes and provides detailed descriptions of conservation approaches that can be considered and evaluated for effectiveness in meeting specific conservation goals. Second, the book emphasizes the practical considerations that land managers face in developing and implementing conservation strategies to deal with rare or little-known species. These include integration of social, economic, and biological goals as well as practical matters of costs, personnel, and dealing with uncertainty and risks in decision making.

References

Andrén, O., J. Bengtsson, and M. Clarholm. 1995. Biodiversity and species redundancy among litter decomposers. Pp. 141–51 in *The significance and regulation of soil biodiversity*, ed. H. P. Collins, G. P. Robertson, and M. J. Klug. Dordrecht: Kluwer.

Baydack, R. K., and H. Campa III. 1999. Setting the context. Pp. 3–16 in *Practical approaches to the conservation of biological diversity*, ed. R. K Baydack, H. Campa III, and J. B. Haufler. Washington, DC: Island Press.

Black, S. H. 2002. Director's message. *Xerces News* (Summer): 3–4.

Boeken, B., and M. Shachak. 2006. Linking community and ecosystem processes: The role of minor species. *Ecosystems* 9 (1): 119–27.

Bolger, T. 2001. The functional value of species biodiversity: A review. *Biology and Environment: Proceedings of the Royal Irish Academy* 101B (3): 199–224.

Borrvall, C., B. Ebenman, and T. Jonsson. 2000. Biodiversity lessens the risk of cascading extinction in model food webs. *Ecology Letters* 3:131–36.

Brown, J. H., and J. Roughgarden. 1990. Ecology for a changing earth. *Ecological Society Bulletin* 71:173–88.

Chapin, F. S., O. E. Sala, I. C. Burke, J. P. Grime, D. U. Hooper, W. K. Lauenroth, A. Lombard, et al. 1998. Ecosystem consequences of changing biodiversity. *BioScience* 48:45–52.

Cottingham, K. L., B. L. Brown, and J. T. Lennon. 2001. Biodiversity may regulate the temporal variability of ecological systems. *Ecology Letters* 4:72–85.

Flather, C. H., and C. H. Sieg. 2000. Applicability of Montreal Process Criterion 1—conservation of biological diversity—to rangeland sustainability. *International Journal of Sustainable Development and World Ecology* 7:81–96.

Gaston, K. J. 1994. *Rarity*. London, UK: Chapman and Hall.

Hooper, D. U., F. S. Chapin, J. J. Ewel, A. Hector, P. Inchausti, S. Lavorel, J. H. Lawton, et al. 2005. Effects of biodiversity on ecosystem functioning: A consensus of current knowledge. *Ecological Monographs* 75 (1): 3–35.

Johnson, C. N. 1998. Species extinction and the relationship between distribution and abundance. *Nature* 394:272–74.

Lubchenco, J., A. M. Olson, L. B. Brubaker, S. R. Carpenter, M. M. Holland, S. P. Hubbell, S. A. Levin, et al. 1991. The sustainable biosphere initiative: An ecological research agenda. *Ecology* 72:318–25.

Lyons, K. G., and M. W. Schwartz. 2001. Rare species loss alters ecosystem function: Invasion resistance. *Ecology Letters* 4:358–65.

Lyons, K. G., C. A. Brigham, B. H. Traut, and M. W. Schwartz. 2005. Rare species and ecosystem functioning. *Conservation Biology* 19 (4): 1019–24.

Mansour, F., and U. Heimbach. 1993. Evaluation of lycosid, micryphantid and linyphiid spiders as predators of *Rhopalosiphum padi* (Hom.: Aphididae) and their functional response to prey density: Laboratory experiments. *Entomophaga* 38:79–87.

Marcot, B. G., M. A. Castellano, J. A. Christy, L. K. Croft, J. F. Lehmkuhl, R. H. Naney, K. Nelson, et al. 1997. Terrestrial ecology assessment. Pp. 1497–1713 in *An assessment of ecosystem components in the interior Columbia basin and portions of the Klamath and Great basins,* ed. T. M. Quigley and S. J. Arbelbide. Vol. 3. USDA Forest Service General Technical Report PNW-GTR-405. USDA Forest Service Pacific Northwest Research Station, Portland, OR.

McConkey, K. R., and D. R. Drake. 2006. Flying foxes cease to function as seed dispersers long before they become rare. *Ecology* 87 (2): 271–76.

Molina, R., B. G. Marcot, and R. Lesher. 2006. Protecting rare, old-growth forest associated species under the Survey and Manage guidelines of the Northwest Forest Plan. *Conservation Biology* 20 (2): 306–18.

Patrick, R. 1997. Biodiversity: Why is it important? Pp. 15–24 in *Biodiversity II: Understanding and protecting our biological resources,* ed. M. L. Reaka-Kudla, D. E. Wilson, and E. O. Wilson. Washington, DC: Joseph Henry Press.

Pimm, S. L., H. L. Jones, and J. Diamond. 1988. On the risk of extinction. *American Naturalist* 132:757–85.

Wheeler, Q. D., and J. Cracraft. 1997. Taxonomic preparedness: Are we ready to meet the biodiversity challenge? Pp. 435–46 in *Biodiversity II: Understanding and protecting our biological resources,* ed. M. L. Reaka-Kudla, D. E. Wilson, and E. O. Wilson. Washington, DC: Joseph Henry Press.

Wilson, E. O. 1987. The little things that run the world (the importance and conservation of invertebrates). *Conservation Biology* 1 (4): 344–46.

2

Conservation Goals and Objectives

Martin G. Raphael, Richard L. Johnson, John D. Peine, and Cindy S. Swanson

This book explores and evaluates various approaches to the conservation of rare or little-known (RLK) species. However, the objective to conserve RLK species cannot be considered in a vacuum, but, rather, within an overall set of objectives for ecological, social, and economic sustainability. Therefore, it is important to know how various approaches to the conservation of RLK species will affect not only those species but also the other elements of ecological, social, and economic sustainability. For such an evaluation, one must first articulate a suite of objectives under which land managers operate. This chapter uses experiences from federal land management in the United States to create a context for setting these goals and objectives.

Sustainability has been called an overarching goal of the management of federal lands (Committee of Scientists 1999). Sustainability generally consists of interdependent ecological, social, and economic elements, which, taken together, ensure that an area's capacity to provide present and future generations with needed ecosystem benefits is undiminished over time (Lélé and Norgaard 1996). This definition implies that (1) management of natural resources will not degrade those systems being utilized (Lubchenco et al. 1991), and (2) the current generation of humans will leave an equitable share of resources for future generations (Meyer and Helfman 1993). Ecological sustainability provides a foundation upon which the other elements depend. If ecological systems fail, social and economic elements of sustainability will fail (Committee of Scientists 1999). All three elements are, however, interdependent, and thus we have listed goals that address each of these. Before discussing these goals, we illustrate the legal foundation for the conservation of RLK species (using laws in the United States as examples) and then discuss the broader context for conservation of biological diversity.

Legal Foundation for Conservation Objectives

U.S. law provides a variety of mechanisms to protect RLK species; these mechanisms address the ecosystems in which the species live and mechanisms to protect the species themselves. Preserved federal lands, such as national parks, wilderness areas, and national wildlife refuges, play a central role in preserving ecosystems and the species that reside within them. The national parks are governed under the National Park Service Organic Act (16 U.S.C. §§ 1-18f). Congress intended for the Park Service to be responsible for protecting the integrity of national parks from threatening activities; thus preservation of ecosystems has been a primary goal of park management (Keiter et al. 1999). Wilderness areas are governed under the Wilderness Act of 1964 (16 U.S.C. §§ 1131-1136). Under this act, wilderness areas are to be managed to preserve natural conditions while providing access for recreation; other commercial activities are generally prohibited. Conservation of biological diversity within wilderness areas is not an explicit goal, but preservation of natural ecosystems within wilderness areas is an important contribution toward national biodiversity objectives.

The National Wildlife Refuge System Administration Act (16 U.S.C. §§ 668dd-668ee) governs the system of national wildlife refuges. Under the 1997 amendments to this act, the mission of these refuges was clarified to stress conservation, management, and restoration of native fish, wildlife, and plant resources and their habitats (Keiter et al. 1999). The Multiple Use-Sustained Yield Act of 1960 (16 U.S.C. §§ 528-31) requires the U.S. Department of Agriculture (USDA) Forest Service to manage lands for multiple uses without impairing their productivity. Further, due consideration shall be given to the relative values (economic and intrinsic) of the various resources and should conform to changing needs and conditions of resources and related services. This act made clear for the first time that wildlife and fish resource management was a valid purpose for administering the national forests (Wilkinson and Anderson 1987). This direction to recognize the importance of wildlife within forest ecosystems was further strengthened with passage of the National Forest Management Act (NFMA) (16 U.S.C. §§ 1601-14) and the Federal Land Policy Management Act (FLPMA) (43 U.S.C. §§ 1701-1783). These acts are the principal statutes governing management of lands administered by the Forest Service (NFMA) and the Bureau of Land Management (FLPMA). The NFMA directs the Forest Service to "provide for diversity of plant and animal communities based on the suitability and capability of the specific land area in order to meet overall multiple-use objectives" (16 U.S.C. § 1604(g)(3)(B)). The FLPMA, like the NFMA, requires the Bureau of Land Management to manage its lands for multiple uses and calls for management that will not impair the land and quality of the environment. The FLPMA does not contain an explicit biodiversity provision as does the NFMA, but it does emphasize ecological sustainability and thus contains a legal basis to maintain a diverse array of ecosystems over time. The National Environmental Policy Act (NEPA) (42 U.S.C. § 4321) requires federal agencies to take a hard look at the environmental consequences of planned actions before an agency decision to commit resources is made. Under NEPA, an Environmental Impact Statement must be prepared that discloses the ecological effects of planned actions; this offers a framework for pursuit of ecosystem management goals.

The legal basis for species-level conservation is embodied in the Endangered Species Act (16 U.S.C. § 1531), which gives conservation of species a higher priority than any other activity. The act applies only to species that are listed as threatened (likely to become endangered within the foreseeable future) or endangered (in danger of extinction throughout all or a significant portion of their range). Under the act, federal agencies must avoid jeopardizing the continued existence of a species or its critical habitat. Conservation of rare or little-known species is mandated under this act but only if those species are formally listed, a process that often requires substantial scientific knowledge about the species, its habitat requirements, and its population trend. Such knowledge is often lacking for RLK species.

The planning regulations under the NFMA (36 C.F.R. § 219.19) provide a broader mandate for species conservation on national forest lands and have recently been revised. Under the previous regulations, rules stated that "fish and wildlife habitat shall be managed to maintain viable populations of existing native and desired non-native vertebrate species in the planning area." The rules also required the agency to insure that viable populations are "well distributed in the planning area" and to provide habitat that "must be well distributed so that reproductive individuals can interact with others in the planning area" (36 C.F.R. § 219.19). These regulations made distribution of wildlife habitat a controlling factor in forest planning (Wilkinson and Anderson 1987). The regulations further indicated that forest plans should preserve and enhance diversity, taking account of "natural" diversity (that which might have existed if the forest had not been managed) and existing diversity. This broader diversity mandate, coupled with provisions to select management indicator species that can be selected from any taxon, has been interpreted to apply viability standards to a wide variety of plant and animal taxa, including nonvascular plants and invertebrates (e.g., FEMAT 1993).

NFMA also provides a framework from which economic values of RLK species should be addressed either quantitatively or qualitatively by requiring an analysis of present and anticipated uses, demands for, and supply of the renewable resources. Guidelines for developing land management plans should include consideration of economic and environmental aspects associated with fish and wildlife. These economic values include commodity and nonmarket considerations in the context of a benefit–cost analysis.

In January 2005, the Forest Service issued a new set of planning regulations (*Federal Register* 70 (3): 1023–61). Under these new regulations, "the overall goal of the ecological element of sustainability is to provide a framework to contribute to sustaining native ecological systems by providing ecological conditions to support diversity of native plant and animal species in the plan area" (§ 219.10(b)). These regulations, unlike the viability standards of the previous regulations, do not provide a measurable or enforceable standard by which to determine whether the broad goals of the NFMA are met (Noon et al. 2005).

There are many other statutes at the state and local level that provide a legal foundation for conservation goals. Those already cited are offered as examples of the primary foundation for the legal authority for conservation of terrestrial ecosystems and species at the national level. With these in mind, we can turn to a discussion of some of these goals.

Biological Diversity Goals

As delineated in the Montréal Process (http://www.mpci.org/home_e.html), the key indicators of biological diversity are species diversity and ecosystem diversity. The Montréal Process was developed by the Working Group on Criteria and Indicators for the Conservation and Sustainable Management of Temperate and Boreal Forests. The Working Group was formed in Geneva, Switzerland, in June 1994 to develop and implement internationally agreed upon criteria and indicators for the conservation and sustainable management of temperate and boreal forests. Understanding species diversity naturally requires consideration of species, but is it necessary to evaluate the status of individual species? or can land managers somehow rely on the amounts and types of ecosystems to assess diversity? The ecological literature contains an active debate on the relative values of ecosystem-based and species-based approaches to the conservation of biological diversity. One forum, "Preserving Biodiversity" (*Ecological Applications* 1993 3 (2): 202–20), and subsequent letters to the editor (*Ecological Applications* 1994 4 (2): 205–9), contain a lively discussion that highlights why both approaches are necessary. As stated by Wilcove, "The distinction between 'single-species management' and 'ecosystem management' is a false dichotomy; both are part of a continuum of steps necessary to protect biodiversity (Wilcove 1994). Ecosystem-based approaches are needed because there are too many species to handle on a species-by-species approach: such an approach will exhaust available time, financial resources, society's patience, and scientific knowledge (Franklin 1993). However, as pointed out by Wilcove (1994), an ecosystem plan that fails to ensure a viable population of, say, marbled murrelets (*Brachyramphus marmoratus*) or northern spotted owls (*Strix occidentalis caurina*) would be considered a failure. Needs of individual species, whether species listed under the Endangered Species Act or other legislation, or species afforded special status in their role as keystone or umbrella species, are a vital part of any conservation strategy.

Conservation of biological diversity is at the heart of ecological sustainability (Lindenmayer and Franklin 2002). What does it mean to conserve biological diversity? Does it mean to retain representatives of all extant living organisms? Does it mean to retain these representative organisms in their current state? or to mimic a past or desired future state? Does it mean to retain extant species and to restore populations of those species that are no longer present or that are in danger of local extinction? Does it mean to retain that subset of species that are thought to be key to maintaining system functionality? Is it sufficient to retain representative species (presence)? or is it desired to maintain species in their current abundance or at population levels that will ensure population viability? How do land managers account for the dynamic nature of populations and metapopulations (McCullough 1996), including annual or longer population fluctuations? These questions are meant to convey the notion that the phrase "conservation of biological diversity" does not, in itself, provide sufficient direction to land managers, so they set operational objectives.

Viability assessments, as conducted in support of NFMA planning regulations, are an example of procedures to evaluate the consequences of management actions on wildlife species. The intent has been to document how land management decisions might influence the numbers and distribution of species of interest: managers are interested in wildlife *outputs* given some management decision (Andelman et al. 2001). An example of this approach is the recent viability assessment conducted to evaluate effects of proposed land management alternatives on habitat conditions for fish and wildlife under the Interior Columbia Basin Ecosystem Management Project (Raphael et al. 2001; Rieman et al. 2001). Results of this assessment indicated, for example, that habitat conditions for terrestrial species associated with sagebrush habitats were not likely to improve or may decline relative to current conditions under any of the alternatives considered. Using models to relate populations to habitat conditions, the team suggested that populations of some of the species were likely to experience local extirpation because of loss of habitat (Raphael et al. 2001).

Given these general considerations, we recognize that there are a variety of complementary objectives that may be used to evaluate the effectiveness of alternative approaches to conservation of biological diversity. Land managers must evaluate the approaches against all of the objectives to gain a full picture of their effectiveness. Some approaches may do a great job of conserving RLK species but make it more difficult to accomplish other objectives. Conversely, some approaches may make a smaller contribution to conservation of RLK species but be more effective in accomplishing other objectives. In chapter 8, Holthausen and Sieg assess the utility of alternative approaches to the conservation of RLK species against a list of these objectives.

Species Diversity

Land managers can promote conservation of RLK species at a variety of hierarchical levels, including conservation- or restoration-focused viability of RLK species, viability of RLK species as part of a broader suite of all species in the planning area, or conservation of all species in natural patterns of abundance.

Maintain or Restore Viable Populations of Rare or Little-Known Species

This objective, in itself, does not fully meet the broader objective of conserving biological diversity. Rather, it maintains a focus on the subset of RLK species in the planning area. The information needed in support of this objective would be a list of RLK species known to occur in the area; an assessment of their current distribution, abundance, and population trend; and an evaluation of potential threats to persistence of each species due to planned management actions, natural disturbances, or stochastic events. The evaluation of approaches would focus on the degree to which any approach might provide for conditions to support viable populations of RLK species. An inherent dilemma in applying this objective is that, because these species are rare or little known, there will typically be insufficient information to assess their presence, status, and trend.

Maintain or Restore Viable Populations of All Species

This objective focuses on the extent to which a conservation approach provides ecological conditions to support viable populations of all species expected to occur within a planning area, including RLK species. As noted earlier, we recognize that there are too many species to handle with a species-by-species approach: such an approach will exhaust available time, financial resources, society's patience, and scientific knowledge. But the intent is to provide conditions that are capable of supporting all species, even if specific attention is not placed on every one of them. The information necessary to evaluate this objective can take two forms. The first form would be similar to maintaining viable populations of RLK species, except it would be extended to a wider array of species (a list of selected species known to occur in the area, an assessment of their current status, and an evaluation of potential threats to persistence of each species due to planned management actions, natural disturbances, long-term environmental changes, or stochastic events). Species can be selected on the basis of their perceived vulnerability to disturbance. Species that are legally classified as threatened, rare, vulnerable, or endangered and species at risk of not maintaining viable breeding populations, as determined by scientific assessment, would be the highest priority for evaluation.

The second form would focus more specifically on the characteristics of a collection of species within the planning area. For example, a simple count of species is a common and easily understood measure of species diversity (Gaston 1996; Purvis and Hector 2000). Because a general sign of ecosystem stress is a reduction in the variety of organisms found in a given locale (Rapport et al. 1985, Loreau et al. 2001), species counts have a long history of use in assessing ecosystem status (Magurran 1988; Reid et al. 1993). One problem with a simple count of species is its insensitivity to changes in species composition. Native species can become extinct (locally or globally) and new (perhaps exotic) species can colonize and become established in the species pool. Because a count of species does not capture changes in species composition, measures of faunal integrity (Karr 1990) or community completeness (Cam et al. 2000) have been recommended as more informative measures of species diversity. Under the notion of integrity, the observed species pool is compared to the expected species pool as a means of evaluating whether species diversity is being conserved.

Maintain or Restore Viable Populations of All Species in Natural Patterns of Abundance

This objective is similar to maintaining viable populations of all species, but in this case the objective is to maintain the full assemblage of species with a composition, abundance, and distribution similar to that which might be expected in the absence of human disturbance (Noss and Cooperrider 1994). Defining this "natural" pattern of abundance is a major challenge because few data exist on patterns of abundance in unaltered landscapes, and virtually all landscapes have a long history of native human influence. What, then, should be the target upon which to compare current conditions? Species populations and ecosystems are dynamic, and one cannot expect patterns of abundance under one moment in time to persist. Therefore, it is probably more appropriate to consider the range of variability to set population targets. Patterns of abundance that fall within this range might then meet the standard, whereas populations that are either more or less abundant than the limits of the range of variability would be outside the range and would not be defined as meeting the natural condition.

Genetic Diversity

An objective for genetic diversity is to maintain natural genetic variation. The geographic ranges of species are continuously responding to phenomena such as glaciation, vegetation migration, climate fluctuation, interspecific interaction, or human alteration of habitats. Species that currently occupy only a small portion of their former range might have lost some of their genetic variation. They are at risk of losing much of their remaining variability due to natural (e.g., fires, hurricanes or typhoons, diseases) or human-caused events (e.g., road development, reservoirs) that can decimate local populations. This erosion in genetic variation results in the species being less able to adapt to changes in its environment brought on by humans, climatic change, or the invasion of exotic species. The result is a higher risk of species extinction. The ecosystem of which the species is a part will itself become less resilient to change. This objective, then, is another indicator of the resilience of species to environmental change.

Population size is a crucial parameter in determining the amount of genetic variability that can be maintained in a population (Lande and Barrowclough 1987). Genetic variability, in turn, influences the likelihood of species persistence because genetic variation is necessary for evolutionary adaptation to a changing environment. Maintaining genetic variability is thus a crucial aspect of biological diversity conservation.

Population viability analyses sometimes address effects of environmental stressors on population genetics and their implications for population size and distribution. Often of concern are several situations related to population genetics that could ultimately affect the numerical viability of a population or species:

- Genetic bottlenecks whereby low numbers of breeding individuals cause a diminution of allelic diversity (e.g., Brookes et al. 1997; Bouzat et al. 1998; Beebee and Rowe 2001)
- Inbreeding depression whereby increased relatedness between breeding individuals results in dominant or homozygous expression of deleterious alleles and erosion of allelic diversity (Keller and Waller 2002) (a variant of this, outbreeding depression, is seldom a concern in population viability analysis, but in some instances should be included; Fenster and Galloway 2000; Edmands and Timmerman 2003)

- Genetic drift, wherein modal genotypes change and allelic diversity decreases over generations in small isolated populations due to sampling effects (Nei and Tajima 1981; Hitchings and Beebee 1998)

Genetic bottlenecks, inbreeding depression, and genetic drift are distinct processes (they are sometimes confused), but all of them could be caused by severing or reducing a population into smaller isolates, and by other factors, such as appearance of behavioral or reproductive isolating mechanisms within a species, that reduce the effective size and distribution of a breeding population (Wright 1938; Kimura and Crow 1963; Beissinger and McCullough 2002).

Beyond these concerns for population numerical viability are additional situations that may reduce the longer-term evolutionary potential of species (Soulé 1980; Posadas et al. 2001). That is, a species may be demographically and numerically viable—abundant enough to be self-sustaining in a broad array of habitats and geographic locations, to withstand normal variations in resource availability and environmental factors—but may still lose some evolutionary potential. There are several additional concerns related to reduction of evolutionary potential (that would not affect likelihood of continuance of the overall population):

- Loss of small, isolated populations (e.g., Pettersson 1985)
- Loss of peripheral areas of a species' distributional range (e.g., Lammi et al. 1999)
- Loss of a portion of a population's genome that would confer adaptability to future changes in environmental conditions (e.g., Daley 1992)
- Loss of genetic variation within polymorphic populations or within sibling (cryptic) species, superspecies, or subspecies complexes (e.g., Avise 1995; McElroy et al. 1997)

Small, isolated populations and peripheries of species' ranges are often locations where "evolutionary experiments" occur through differential selection pressures causing character divergence from the modal genotypes of the species (Sayama et al. 2003). Such selection pressures may be precursors to the evolution of new subspecies or full species (e.g., Baker et al. 2003), and they may also result in preadaptations that could prove to be of adaptive significance in the face of later environmental changes.

In addition to the species focus, genetic concerns can also address lower taxonomic levels, particularly variability within and among subspecies, races, and morphs within a species, which provide specific kinds of variability by which a species' lineage can adapt to changes in environmental conditions (Wiens 1999). Thus management objectives for maintaining genetic diversity may pertain not just to numerical viability of a species but also to the diversity within and among isolated populations, demes, morphs, subspecies, races, and distributional peripheries, all of which can contribute to ensuring a species' longer-term evolutionary potential beyond shorter-term numerical viability.

Finally, evolutionary history may also provide information relevant to setting conservation priorities (Mace et al. 2003). Preservation of the processes that have generated patterns of biodiversity may be as important as preservation of the pattern itself, and new insights from evolutionary history inform this new approach. The challenge for land managers is to find practical ways to recognize and manage these relevant genetic characteristics. In this case, solutions to the inherent difficulties may depend on future advances in genetic research and technology transfer.

Ecosystem Diversity

Ecosystem diversity provides a "coarse filter" approach to the conservation of biological diversity (Nature Conservancy 1982). The inherent assumption of this approach is that biodiversity depends on diverse habitat conditions. The action or requirement that flows from this assumption is to maintain or restore diverse habitat conditions to conserve or restore biodiversity.

Provide for Native Ecosystem Types and Structural Stages within Their Natural Range of Variation

Under this objective, ecosystem diversity would be measured by the extent and distribution of ecosystem types (such as forest cover types) and structural stages. *Structural stages* are the different structural and compositional phases of forests and grasslands that occur over time following disturbances that kill, remove, or reduce vegetation in patches of various sizes. These structural stages are intended to be the major developmental or seral stages that occur within a particular environment. A management goal might be to create an array of structural stages that result in the representation of those that would be expected over time. Management may also focus on the representation of particular elements within these systems, such as small wetlands, rock outcrops, dead and downed wood, and other microhabitats that support particular organisms. Managing for the range of natural variability does not mean simply managing for diverse habitat conditions. Rather, it is an attempt to restore a system to the range of conditions that it experienced prior to "major" anthropogenic disturbance (Aplet and Keeton 1999; Schwartz 1999).

Maintain or Restore Ecological Processes

An objective to maintain or restore ecological processes embodies a dynamic pattern–process view of natural resource management (Spies et al. 2006). It implies that managers consider the components of ecosystems (e.g., species, vegetation communities, physical attributes) and the formative processes (e.g., disturbance regimes, trophic structures, soil and nutrient cycling, interspecific interactions) that lead to the expression of characteristic mixes of species and vegetation across the landscape. Under this objective, managers would focus on ecological processes as well as the distribution and abundance of structures. We note, however, that there are few cases where processes can be managed directly. Such an objective usually means influencing processes indirectly through manipulating structures.

The inherent assumption is that an ecosystem that retains the expected ecological processes would have the necessary ecosystem components and critical processes to provide for the habitat requirements of associated species, but the specific species composition may be less important than the retention of functional processes and functional roles of organisms. Although one might expect that a large proportion of native species richness is required to sustain ecosystem function, there is no clear consensus in the scientific literature. On the one hand, Schwartz et al. (2000) found little support for the hypothesis that there is a strong dependence of ecosystem function on the full complement of diversity within sites. In contrast, Naeem et al. (1999) concluded that ecosystem functioning is decreased as the number of species in a community decreases and that this is particularly pronounced at lower levels of diversity.

Manage Landscapes to Provide for Resiliency of Species

Resilience is a measure of the capability of species (or ecosystems) to recover from disturbance. No natural community is static; communities change over many spatial and temporal scales. Organisms respond to daily fluctuation in temperature and precipitation, to seasonal changes, and to annual variation. Fire, windthrow, flooding, and other acute disturbances change the local composition and structure of ecosystems. Long-term change, such as the chronic effects of global warming, exert still other changes on ecosystems. Recent studies suggest that ecosystem resistance to environmental perturbations may be lessened as biodiversity is reduced (Naeem et al. 1999). For example, loss of native dry shrubland plant communities in the Columbia basin due to years of heavy grazing by domestic livestock and concomitant invasion and dominance by nonnative weedy plant species has resulted in an altered state of the ecosystem: changes in soil and plant communities can no longer recover historical conditions within a foreseeable time (Hemstrom et al. 2001). This objective, then, is meant to assess the degree to which landscapes are capable of recovery following disturbance and the degree to which expected structures, functions, and processes are sustained over time.

Social Goals

The Montréal Process, described earlier, alludes to the relevancy of social dimensions of forest conservation such as social, cultural, and spiritual needs and values. Sustaining viable populations of RLK species will depend on society's commitment to maintaining and adjusting as needed conservation strategies over the long term—a tall order. It is one thing to support the principal of environmental stewardship in general but quite another to support a specific conservation strategy for an RLK species that might directly impact individuals, their families, and their jobs.

The concept of sustainability of RLK species embodies compromise among human values that are often in conflict. Environmental stewardship, aesthetics, protection of sacred places, and low-impact utilization of renewable natural resources represent values likely to support adequate habitat protection to sustain RLK species (Kellert 2002). Human welfare such as health, housing, employment, lifestyle, and commerce are values that can sometimes compete for the ever-growing demand for limited open space (Naidoo and Adamowicz 2001). Exurban sprawl and its related infrastructure are expanding dramatically, driven by population growth and an expanding economy converting open space to development at a rapid pace (Davis et al. 1994). Habitat critical for sustaining populations of RLK species often occurs over a mix of public and privately owned lands, a critical challenge to the conservation of RLK species (Theobold et al. 2004). As a consequence, exploring alternative conservation strategies to sustain RLK species based on biological requirements should be balanced with other priorities of society.

There is a growing trend to apply nontraditional social strategies to meet the challenge of conservation of RLK species via collaboration (Wondolleck and Yaffee 2000). An emerging trend in resource management is to operate in a broader geographic and temporal scale in a nontraditional manner. Risk taking is inevitable in dealing with complexity, uncertainty, and change associated with the conservation of RLK species. This book focuses on the evaluation of alternative strategies to sustain populations of RLK species. The central role of social considerations is to facilitate collaboration among stakeholders.

The challenging social considerations goal is to establish a long-term commitment among all sectors of society to maintain sustainable populations of RLK species. The following objectives provide the building blocks to achieve this ambitious goal.

Ensure Equitable Inclusiveness

We recognize the importance of ensuring equitable inclusiveness among stakeholder participation in the planning, analysis, and decision-making process that evaluates alternative conservation strategies to sustain RLK species. For instance, participation can be restrained by a lack of convenient access to the planning process or a lack of technical knowledge of how to navigate an online survey or Web site. Public meetings are often taken over by aggressive speakers with extreme positions on issues rather than remaining collegial discussions of shared and divergent values (Suskind and McCreary 1985; Nedovic-Budic 1998). The importance of achieving this objective is documented in several case examples of RLK species conservation included in chapter 9. Stakeholder involvement should encompass all steps in the process from identifying shared and divergent values at play and how they relate to alternative conservation strategies under consideration.

Maintain a Holistic and Nonpartisan Social Perspective

A holistic perspective of alternative RLK species conservation strategies under consideration will likely result in identifying more possibilities to mitigate conflicts. Possibilities include the range of social values, social structure and welfare, cultural heritage and values, natural resource utilization, institutional dynamics and risk taking, political influence, and the social dimensions of decision making. Wondolleck and Yaffee (2000) encourage reducing the relevancy of a technocratic model to a more democratic model for evaluation and decision making.

Maximize an Array of Social and Economic Benefits

Focusing on human benefits associated with a conservation strategy without compromising the RLK species is an important building block for achieving the conservation goal. Conservation strategies that focus on this objective, particularly on privately owned land, are more likely to succeed in the long run. Understanding the array of values among all stakeholders as portrayed in chapters 9 and 10 is a precursor to building collaboration leading to long-term commitment. The degree to which this objective is met is an indicator of the potential for a successful decision-making process.

Establish and Maintain an Institutional Framework

Establishing an institutional framework is a considerable challenge that should include long-term commitments to institutional administration and funding sources. Collaboration between resource managers and university scientists strengthens the potential for longevity. This critical social consideration is illustrated in several case examples of successful RLK species conservation practices in chapter 9. A sustainable institutional framework is just as critical as the biological components associated with a successful conservation strategy for RLK species.

Economic Goals

Economics is a decision science that studies how a society allocates its scarce resources. The general objective of environmental economics is to estimate the economic consequences, track financial transactions, and measure the benefits and costs of alternative conservation policies within a framework of constrained market capitalism.

Use a Framework of Constrained Market Capitalism for Conserving Rare or Little-Known Species

Aggregate economic effects, benefits, and costs of environmental goods and services can most easily be measured in price-setting markets. A constrained market economic objective is to create new ownership conventions and regulations that provide economic market incentives for conserving RLK species and sustainable ecosystems. RLK species preservation that enhances rather than impedes market opportunities is more likely to meet both conservation and economic objectives.

Improve Macroeconomic Models to More Accurately Include the Real Economic Effects of Rare or Little-Known Species Conservation

When conventional measures of gross domestic product (GDP) rise, the quality and quantity of public sector natural resources, such as biological diversity and RLK species conservation, usually fall because the consequences of impacts on RLK species are not captured in the price associated with GDP transactions. A goal for RLK species conservation is to add environmental accounts to more standard input–output models of economic activity so that GDP-type estimates better reflect real changes in the stock of sustainable natural environments and the flow of environmental services.

Human well-being ultimately depends on ecosystem functions, such as nutrient cycling, which provide indispensable services to people. Ecosystem services include clean air, clean water, wildlife, recreational opportunities, farmland, and food supply. Even though a species may be rare, obscure, or small in numbers, it can importantly affect the long-term sustainability of an ecosystem and future economic values and resource use options. Beyond its value as a product, an RLK species may also be a valuable input in producing diverse ecosystem services. Scarcity is inherent for most RLK species, and this implies high incremental values for preserving each individual in the population.

Estimate Nonmarket Values for Rare or Little-Known Species

The primary objective of economic analysis in the public sector is to estimate the public's value for sustaining nonmarket goods and services. These values include current demand to use and enjoy resources, the future option of viewing rare species or functioning ecosystems, bequeathing that same opportunity to future generations, and simply knowing individual species and naturally diverse ecosystems exist, even if there is no expectation of using or viewing them now or in the future. Because the measurement costs for complex, poorly understood environmental processes are high, RLK species might also be used as relatively inexpensive indicators for damages to ecosystems. Human beings are environmental creatures; they are part of and entirely dependent on the environment. The value to society of knowing that their environment, including their food supply, is not toxic also needs to be considered and measured in an economic setting to ensure efficient and effective allocation of resources. When basic elements that are essential to long-term human survival are ignored in resource allocation decisions because they are not marketed, markets and governments jointly fail to allocate resources to the best and most productive use.

Economic analysis that incorporates nonmarket values and environmental regulation is needed to compare RLK species conservation with other societal goals that compete for scarce public resources. Although dollar values do not convey the public's preferences for all social and cultural values, they do provide one commensurate measure for comparing broad public purposes. Legal, regulatory, and collaborative processes evolve to convey and balance these dollar-weighted social preferences.

Integrate Economic Analysis in Rare and Little-Known Species Conservation

A general purpose of this book is to provide administrators and managers with a set of tools for use in conserving RLK species. Economic analysis provides estimates about the economic impacts, benefits, and costs of pursuing alternative biological/ecological, economic, and social objectives. Costs of conducting economic analysis vary according to spatial and temporal complexity and extent, magnitude and complexity of the embedded economy, social and management preferences toward risks of extinction, and the resilience of the underlying ecosystem. Management choices about the most appropriate mix of ecological criteria or methods should be made at the same time economic methods are chosen. The best set of economic methods should be uniquely matched with the best ecological criteria or methods for each RLK species study. Economics is just one set of tools that should be considered when making management decisions related to allocation of scarce resources. The economic tools are discussed in chapter 10. Other values need to be expressed through methods already described under Social Goals in this chapter, and in chapter 9. Peine (chap. 9) also describes innovative ways not only for better inclusion of social values beyond economics, but also broad decision processes for integrating ecological, economic, and social considerations.

Conclusion

Setting goals and objectives is an essential step in conservation planning. As described in this chapter, conservation of biological diversity in general, and of RLK species in particular, is motivated in part by applicable laws and regulations. Conservation is also rooted in scientific knowledge of the roles of ecosystems and the species they support in supplying goods and services to people as well as knowledge of the factors that help sustain biological systems. As land managers seek to implement conservation actions, progress is best judged against ecological, economic, and social criteria. These criteria will be effective if they are grounded in well-articulated goals and objectives. We have presented examples of how such goals can be expressed for conservation of species and ecosystems. We have also presented social goals and objectives focused on engaging equitable involvement by all stakeholders in the decision-making process, assessment of all social considerations of concern, and defining shared values on which to build consensus on how best to establish a sustainable conservation strategy. Social considerations should be reflected in all aspects of the assessment process. Finally, we have also described economic goals. Conservation of RLK species and ecosystems will best meet economic goals when it efficiently balances society's preferences for using associated natural resources now and in the future. Nonmarket and market net benefits, alongside macroeconomic impacts, can uniquely make RLK species and ecosystem conservation commensurate with other public investment objectives. Therefore economics is an indispensable, but not the only, social criterion needed to decide the structure and magnitude of RLK species and ecosystem conservation.

Success in the conservation of RLK species is thus based on well-articulated ecological, social, and economic goals. In managing lands to achieve conservation of biological diversity, if land managers do not set clear goals then the adage, "no matter where you go, there you are," certainly applies.

Acknowledgments

We thank Nancy Molina and Jim Quinn for constructive comments on an earlier draft. This chapter also benefited from comments by authors of other chapters in this book, and by comments from Steve O'Dell on the legal foundation for species and ecosystem conservation. We thank the Pacific Northwest Research Station and Region 1 of the National Forest System, as well as the U.S. Geological Survey for support during the production of this chapter.

References

Andelman, S. J., S. Beissinger, J. F. Cochrane, L. Gerber, P. Gomez-Priego, C. Groves, J. Haufler, et al. 2001. Scientific standards for conducting viability assessments under the National Forest Management Act: Report and recommendations of the NCEAS working group. National Center for Ecological Analysis and Synthesis, Santa Barbara, CA. Unpublished report.

Aplet, G. H., and W. S. Keeton. 1999. Application of historic range of variability concepts to biodiversity conservation. Pp. 71–86 in *Practical approaches to the conservation of biological diversity*, ed. R. K. Baydack, H. Campa III, and J. B. Haufler. Washington, DC: Island Press.

Avise, J. C. 1995. Mitochondrial DNA polymorphism and a connection between genetics and demography of relevance to conservation. *Conservation Biology* 9:686–90.

Baker, J. M., E. López-Medrano, A. G. Navarro-Sigüenza, O. R. Rojas-Soto, and K. E. Omland. 2003. Recent speciation in the orchard oriole group: Divergence of *Icterus spurius spurius* and *Icterus spurius fuertesi*. *Auk* 120:848–59.

Beebee, T., and G. Rowe. 2001. Application of genetic bottleneck testing to the investigation of amphibian declines: A case study with natterjack toads. *Conservation Biology* 15:266–70.

Beissinger, S. R., and D. R. McCullough. 2002. *Population viability analysis*. Chicago: University of Chicago Press.

Bouzat, J. L., H. A. Lewin, and K. N. Paige. 1998. The ghost of genetic diversity past: Historical DNA analysis of the greater prairie chicken. *American Naturalist* 152:1–6.

Brookes, M. I., Y. A. Graneau, P. King, O. C. Rose, C. D. Thomas, and J. L. B. Mallet. 1997. Genetic analysis of founder bottlenecks in the rare British butterfly *Plebejus argus*. *Conservation Biology* 11:648–61.

Cam, E., J. D. Nichols, J. R. Sauer, J. E. Hines, and C. H. Flather. 2000. Relative species richness and community completeness: Birds and urbanization in the mid-Atlantic states. *Ecological Applications* 10:1196–1210.

Committee of Scientists. 1999. Sustaining the people's lands: Recommendations for stewardship of the National Forests and Grasslands into the next century. Washington, DC: U.S. Department of Agriculture. http://www.fs.fed.us/forum/nepa/rule/cosreport.shtml.

Daley, J. G. 1992. Population reductions and genetic variability in black-tailed prairie dogs. *Journal of Wildlife Management* 56:212–20.

Davis, J. S., A. C. Nelson, and K. J. Dueker. 1994. The new 'burbs: The exurbs and their implications for planning policy. *Journal of the American Planning Association* 60:45–59.

Edmands, S., and C. C. Timmerman. 2003. Modeling factors affecting the severity of outbreeding depression. *Conservation Biology* 17:883–92.

Fenster, C. B., and L. F. Galloway. 2000. Inbreeding and outbreeding depression in natural populations of *Chamaecrista fasciculata* (Fabaceae). *Conservation Biology* 14:1406–12.

FEMAT (Forest Ecosystem Management Assessment Team). 1993. Forest ecosystem management: An ecological, economic, and social assessment. Washington, DC: U.S. Government Printing Office 1993-793-071. Available at: Regional Ecosystem Office, P.O. Box 3623, Portland, OR 97208.

Franklin, J. F. 1993. Preserving biodiversity: Species, ecosystems, and landscapes. *Ecological Applications* 3:202–5.

Gaston, K. J. 1996. Species richness: Measure and measurement. Pp. 77–113 in *Biodiversity: A biology of numbers and difference*, ed. K. J. Gaston. Cambridge, MA: Blackwell Science.

Hemstrom, M. A., J. J. Korol, and W. J. Hann. 2001. Trends in terrestrial plant communities and landscape health indicate the effects of alternative management strategies in the interior Columbia River basin. *Forest Ecology and Management* 153:105–26.

Hitchings, S. P., and T. J. C. Beebee. 1998. Loss of genetic diversity and fitness in common toad (*Bufo bufo*) populations isolated by inimical habitat. *Journal of Environmental Biology* 11:269–83.

Karr, J. R. 1990. Biological integrity and the goal of environmental legislation: Lessons for conservation biology. *Conservation Biology* 4:244–50.

Keiter, R. B., T. Boling, and L. Milkman. 1999. Legal perspectives on ecosystem management: Legitimizing a new federal land management policy. Pp. 9–41 in *Ecological stewardship: A common reference for ecosystem management*, vol. 3, ed. W. T. Sexton, A. J. Malk, R. C. Szaro, and N. C. Johnson. Dordrecht: Elsevier Science.

Keller, L. F., and D. M. Waller. 2002. Inbreeding effects in wild populations. *Trends in Ecology and Evolution* 17:230–41.

Kellert, S. R. 2002. Values, ethics, and spiritual and scientific relations to nature. Pp. 29–48 in *The good in nature and humanity: Connecting science, religion, and spirituality with the natural world*, ed. S. R. Kellert and T. J. Farnham. Washington, DC: Island Press.

Kimura, M., and J. Crow. 1963. The measurement of effective population number. *Evolution* 17:279–88.

Lammi, A., P. Siikamaki, and K. Mustajarvi. 1999. Genetic diversity, population size, and fitness in central and peripheral populations of a rare plant *Lychnis viscaria*. *Conservation Biology* 13:1069–78.

Lande R., and G. F. Barrowclough. 1987. Effective population size, genetic variation, and their use in population management. Pp. 87–123 in *Viable populations for conservation*, ed. M. E. Soulé. Cambridge: Cambridge University Press.

Lélé, S. and R. B. Norgaard. 1996. Sustainability and the scientist's burden. *Conservation Biology* 10:354–65.

Lindenmayer, D. B., and J. F. Franklin. 2002. *Conserving forest biodiversity: A comprehensive multiscaled approach*. Washington, DC: Island Press.

Loreau, M., S. Naeem, P. Inchausti, J. Bengtsson, J. P. Grime, A. Hector, D. U. Hooper, et al. 2001. Biodiversity and ecosystem functioning: Current knowledge and future challenges. *Science* 294:804–8.

Lubchenco, J., A. M. Olson, L. B. Brubaker, S. R. Carpenter, M. M. Holland, S. P. Hubbell, S. A. Levin, et al. 1991. The sustainable biosphere initiative: An ecological research agenda. *Ecology* 72:318–25.

Mace, G. M, J. L. Gittleman, and A. Purvis. 2003. Preserving the tree of life. *Science* 300:1707–9.

Magurran, A. E. 1988. Ecological diversity and its measurement. Princeton: Princeton University Press.

McCullough, D. R. 1996. *Metapopulations and wildlife conservation*. Washington, DC: Island Press.

McElroy, D. M., J. A. Shoemaker, and M. E. Douglas. 1997. Discriminating *Gila robusta* and *Gila cypha*: Risk assessment and the Endangered Species Act. *Ecological Applications* 7:958–67.

Meyer, J. L., and G. S. Helfman. 1993. The ecological basis of sustainability. *Ecological Applications* 3:569–71.

Naeem, S., F. S. Chapin III, R. Costanza, P. R. Ehrlich, F. B. Golley, D. U. Hooper, J. H. Lawton, et al. 1999. Biodiversity and ecosystem functioning: Maintaining natural life support processes. *Issues in Ecology* 4:1–11.

Naidoo, R., and W. I. Adamowicz. 2001. Effects of economic prosperity on numbers of threatened species. *Conservation Biology* 15 (4): 1021–29.

Nature Conservancy. 1982. Natural heritage program operations manual. Arlington, VA: Nature Conservancy.

Nedovic-Budic, Z. 1998. The impact of GIS technology. *Environment and Planning B: Planning and Design* 25:681–92.

Nei, M., and F. Tajima. 1981. Genetic drift and estimation of effective population size. *Genetics* 98:625–40.

Noon, B. R., P. Parenteau, and S. C. Trombulak. 2005. Conservation science, biodiversity, and the 2005 U.S. Forest Service regulations. *Conservation Biology* 19:1359–61.

Noss, R. F., and A. Y. Cooperrider. 1994. *Saving nature's legacy.* Washington, DC: Island Press.

Pettersson, B. 1985. Extinction of an isolated population of the middle spotted woodpecker *Dendrocopos medius* (L.) in Sweden and its relation to general theories on extinction. *Biological Conservation* 32:335–53.

Posadas, P., D. R. Miranda-Esquivel, and J. V. Crisci. 2001. Using phylogenetic diversity measures to set priorities in conservation: An example from southern South America. *Conservation Biology* 15:1325–34.

Purvis, A., and A. Hector. 2000. Getting the measure of biodiversity. *Nature* 405: 212–19.

Raphael, M. G., M. J. Wisdom, M. M. Rowland, R. S. Holthausen, B. C. Wales, B. G. Marcot, and T. D. Rich. 2001. Status and trends of habitats of terrestrial vertebrates in relation to land management in the interior Columbia River basin. *Forest Ecology and Management* 153:63–88.

Rapport, D. J., H. A. Regier, and T. C. Hutchinson. 1985. Ecosystem behavior under stress. *American Naturalist* 125:617–40.

Reid, W. V., J. A. McNeely, D. B. Tunstall, D. A. Bryant, and M. Winograd. 1993. *Biodiversity indicators for policy-makers.* Washington, DC: World Resources Institute.

Rieman, B., J. T. Peterson, J. Clayton, P. Howell, R. Thurow, W. Thompson, and D. Lee. 2001. Evaluation of potential effects of federal land management alternatives on trends of salmonids and their habitats in the interior Columbia River basin. *Forest Ecology and Management* 153:43–62.

Sayama, H., L. Kaufman, and Y. Bar-Yam. 2003. Spontaneous pattern formation and genetic diversity in habitats with irregular geographical features. *Conservation Biology* 17:893–900.

Schwartz, M. W. 1999. Choosing the appropriate scale of reserves for conservation. *Annual Review of Ecology and Systematics* 30:83–108.

Schwartz, M. W., C. A. Brigham, J. D. Hoeksema, K. G. Lyons, M. H. Mills, and P. J. van Mantgem. 2000. Linking biodiversity to ecosystem function: Implications for conservation ecology. *Oecologia* 122:297–305.

Soulé, M. E. 1980. Thresholds for survival: Maintaining fitness and evolutionary potential. Pp. 151–70 in *Conservation biology: An evolutionary–ecological perspective,* ed. M. E. Soulé and B. A. Wilcox. Sunderland, MA: Sinauer Associates.

Spies, T. A., M. A. Hemstrom, A. Youngblood, and S. Hummel. 2006. Conserving old-growth forest diversity in disturbance-prone landscapes. *Conservation Biology* 20:351–62.

Susskind, L., and S. T. McCreary. 1985. Techniques for resolving coastal resource management disputes through mediation. *Journal of the American Planning Association* 51:365–74.

Theobold, D. M., N. T. Hobbs, T. Bearly, J. A. Zack, T. Shenk, and W. E. Riebsame. 2004. Incorporating biological information in local land-use decision making: Designing a system for conservation planning. *Landscape Ecology* 15 (1): 35–45.

Wiens, J. J. 1999. Polymorphism in systematics and comparative biology. *Annual Review of Ecology and Systematics* 30:327–62.

Wilcove, D. 1994. Preserving biodiversity: Species in landscapes: Response. *Ecological Applications* 4:207–8.

Wilkinson, C. F., and H. M. Anderson. 1987. *Land and resource planning in the national forests.* Washington, DC: Island Press.

Wondolleck, J. M., and S. L. Yaffee. 2000. *Making collaboration work.* Washington, DC: Island Press.

Wright, S. 1938. Size of population and breeding structure in relation to evolution. *Science* 87:430–31.

3

Species Rarity: Definition, Causes, and Classification

Curtis H. Flather and Carolyn Hull Sieg

In virtually all ecological communities around the world, most *species* are represented by few individuals, and most *individuals* come from only a few of the most common species. Why this distribution of species abundances is so regularly observed among different taxonomic sets in geographically diverse systems is a question that has received considerable theoretical and empirical investigation (Preston 1948, 1962; Harte et al. 1999; Hubbell 2001). Understanding the mechanisms leading to the pattern of few common and many rare species extends beyond basic interest in how natural communities are assembled. It is also of great practical importance to conservation science since human uses of ecosystems can greatly affect the pattern of commonness and rarity in the biota inhabiting those same ecosystems (Lubchenco et al. 1991).

Because budgets for biodiversity conservation are limited, a common strategy for allocating scarce conservation resources has been to focus on species that are thought to have the highest extinction risk (Sisk et al. 1994; Flather et al. 1998). What is the ecological justification for a conservation paradigm focused on rare species? How does one go about distinguishing rare from nonrare species? And what factors contribute to species rarity? We address these questions so as to provide a foundation for why resource managers need to be concerned about rare species and how they can identify them. We also discuss the implications that various causes of rarity have on the choice, and likely success, of management actions to protect and enhance populations of rare species.

Why Do We Care about Rarity?

Among a set of ecologically similar species, those that are rare will have a greater extinction risk than those that are common (Johnson 1998; Matthies et al. 2004). Small populations are more likely to be impacted by chance demographic and environmental events, such as failure to find a mate or reproduce, diseases, floods, and fires (Boyce 1992). Furthermore, the genetic simplification that often accompanies severe population declines can reduce a species' ability to adapt to changing environmental conditions, lead to higher rates of inbreeding and the expression of deleterious genes, or, conversely, lead to outbreeding depression (Ellstrand and Elam 1993; Lande 1995). For these reasons, conservation science has become preoccupied with identifying at-risk species and focusing conservation efforts on those most likely to be lost from the species pool.

The conservation focus on rare species has been further justified by the potential role that rare species may play in maintaining overall ecosystem functionality. How rare species affect ecosystem processes is actually a variant of a much broader and important ecological question: What is the relationship between biodiversity and ecosystem functioning (Loreau et al. 2001)? Our understanding of the relationship between species richness and ecosystem function is incomplete, and this uncertainty is fueling an ongoing debate among ecologists (Kaiser 2000). One contention, call it the "complementarity hypothesis," is that niche differentiation results in unique resource use such that the loss of any species would reduce ecosystem functions (e.g., productivity, nutrient cycling, or resilience) or stability (e.g., cascading extinctions, ecosystem invasibility) (Chapin et al. 1998; Borrvall et al. 2000; Cottingham et al. 2001; Loreau et al. 2001; van Ruijven et al. 2003). An alternative view, call it the "redundancy hypothesis," is that species functions are substitutable. Therefore, a tenable conservation goal under this view would be to judiciously target an appropriate subset of species that provide for key ecosystem functions (see reviews by Schwartz et al. [2000] and Hector et al. [2001]). Yet a third perspective, call it the "facilitative hypothesis," is that the relationship between species diversity and function is an accelerating curve owing to the increased probability of positive species interactions that manifest as increasing marginal gains in functional response as richness increases (Cardinale et al. 2002).

The essence of the biodiversity–functionality debate can be captured with a simple conceptual graphic of three hypothetical curves (fig. 3.1). If

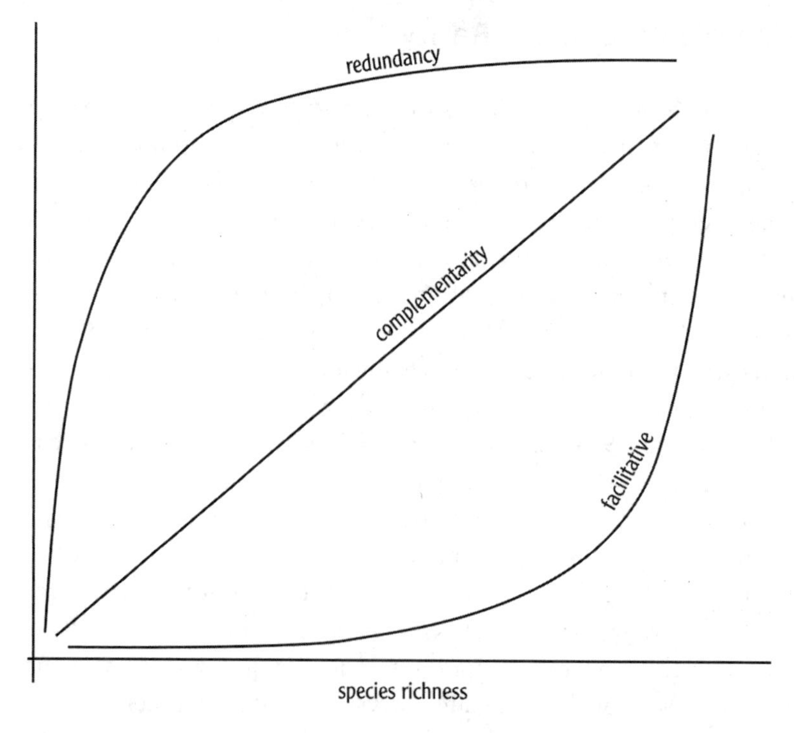

Figure 3.1. Three alternative perspectives on the relationship between biodiversity (species richness) and ecosystem functionality.

ecosystem function (however measured) has a positive linear relationship with species number then there is support for the complementarity hypothesis. If ecosystem function is approached asymptotically with increasing species number, reaching saturation at diversity levels below the full complement of species, then there is support for the redundancy hypothesis. Finally, if functionality is gained at an increasing rate as we move from species-poor to species-rich assemblages, then there is support for the facilitative hypothesis.

Although there appears to be an emerging consensus that ecosystem function is positively related to biodiversity in both terrestrial and aquatic systems (Covich et al. 2004; Balvanera et al. 2006), this qualitative pattern does not distinguish among the functional forms in figure 3.1. Schwartz et al.'s (2000) literature review found that, among observational studies, there was only weak evidence for complementarity, whereas experimental studies more commonly supported redundancy. Evidence in support of facilitative relationships is less common (Cardinale et al. 2002; but see Duarte 2000). Consequently, research results to date have not identified a generally accepted biodiversity–ecosystem relationship. For this reason, attempts to derive practical conservation implications from this research have been controversial and at times contentious, leaving the question unresolved as to what role rare species play in ecosystems.

A number of factors contribute to the ambiguous conservation implications stemming from diversity-function research. First, much of the research is short term and small scale, has focused on a limited set of ecosystem processes, has focused on the diversity–function relationship within a single trophic level, and has not adequately addressed whether diverse regional pools are critical to maintaining local species numbers (Hector et al. 2001; Duffy 2003; Thompson et al. 2005). Second, much of the experimental research has tested these relationships using ecologically unrealistic collections of individuals among species rather than attempting to mimic species-abundance distribution patterns typically observed in natural assemblages (Schwartz et al. 2000). Consequently, the applicability of these experimental findings to nonexperimental systems has been questioned (Symstad et al. 2003). Third, conclusions appear to vary depending upon the kind of ecosystem studied (e.g., forest, grassland, soil, freshwater) and the ecosystem property selected (e.g., primary productivity, nutrient cycling, invasibility) as the functional response variable (Rosenfeld 2002; Balvanera et al. 2006). Evidence in support of each of the three functional forms (fig. 3.1) has been observed within and among studies as the function metric is varied (Duarte 2000). Fourth, scale dependencies (i.e., local versus regional effects) appear to preclude unequivocal expectations for how ecosystem function may behave as species are gained or lost (Chase and Ryberg 2004). Finally, the effect of species on ecosystem function appears to vary depending on their commonness and rarity. For example, Smith and Knapp (2003) found that a threefold reduction in the diversity of rare species had no detectable effect on total above-ground net primary productivity, yet reductions in the abundance of dominant species resulted in immediate and negative impacts on productivity over both years of the study. Conversely, Lyons and Schwartz (2001) found that species assemblages where rare species were removed, reducing overall richness, were more prone to exotic species establishment than were plots where an equivalent biomass of common species was removed.

Given these contrasting results from these pioneering research efforts, a precautionary approach suggests that conserving the full complement of species would be wise until the relationships between biotic structure and ecosystem function are more clearly understood (Rosenfeld 2002; Lyons et al. 2005). Furthermore, ecosystem function is but one argument for the conservation of rare species. There are other, equally legitimate, arguments that derive from legal, ethical, aesthetic, and utilitarian values (see chaps. 9 and 10) that are independent of the functional importance of species (Chapin et al. 1998; Hector et al. 2001; Rosenfeld 2002). Because species abundances are distributed inequitably, and because those that are less abundant are more likely to be lost from regional or local assemblages than common species, a conservation focus on rare species to maintain biodiversity remains justified.

Definitions of Rare Species—How Do We Identify Them?

The concept of rarity has several definitions in common usage, but in the lexicon of conservation biology a species' rarity is most simply based on its distribution and abundance (Gaston 1994). According to Reveal (1981, 42) "rarity is merely the current status of an extant organism which . . . is restricted either in numbers or area to a level that is demonstrably less than the majority of other organisms of comparable taxonomic entities."

An important aspect of this definition is that rarity is a relative, rather than an absolute, concept. Species that are restricted in numbers or spatial occurrence are considered to be rare *relative* to the distribution and abundance of other species making up the pool of interest. Thus it is quite common to see rare species delineated based on some quantile of the frequency distribution of geographic range size, abundance, or both (Gaston 1994). For example, one may choose to define as rare that 10% of species with the lowest abundance estimates. The actual rarity cutpoint selected is a subjective decision, although Gaston (1994, 19) recommends using the 25th quantile because it is practical (for sampling reasons), and it is commonly used in the conservation literature. Such use of quantiles to delineate rare species is restricted to species that are taxonomically (e.g., plants, mammals, passerine birds) or ecologically (e.g., forest interior obligate, serpentine annuals) similar. It is difficult to conceive of how a general threshold of rarity could be applied to a group of species with dissimilar life histories.

One of the weaknesses of the quantile definition of rarity is that a species' status is defined based only on its rank abundance or distribution. Shifts in the species abundance distribution caused by natural or anthropogenic disturbance (e.g., Flather 1996) will not register as an increase or decrease in the number of rare species by the quantile definition. The quantile is a fixed proportion of the species pool, so while the identity of species constituting the "rare" set may change, the number of rare species remains unchanged (assuming a stable species pool size). One way to address this limitation is to define rarity using absolute criteria that focus on the occurrence (e.g., insect species recorded from ≤ 15 10 km survey quadrats from a possible 2862 [Hopkins et al. 2002]) or abundance (e.g., woody tropical plant density < 1 individual/ha [Hubbel and Foster 1986]) of species across some geographic area of interest (Schoener 1987).

Both the relative and the absolute definitions of rarity have been criticized for the lack of an objective ecological justification, and the decision of "where to draw the line" remains a difficult challenge (Magurran 2004, 70). One approach to reduce the arbitrariness of rarity definitions comes from the bioassessment and monitoring literature where reference sites or reference conditions are used to specify the species abundance (or occupancy) distribution expected for pristine or minimally disturbed systems (Reynoldson et al. 1997). Although the definition of rarity among reference sites is still characterized by an arbitrary cutoff, a rarity threshold so defined sets a standard against which to judge whether the degree of rarity is trending toward or away from that expected under the reference conditions.

Three additional issues related to the definition and identification of rare species warrant remark. First, rarity is conditioned on the geographic scale of interest. Certainly a species could be regarded as rare on a local scale (e.g., a management unit within a nature reserve or park), yet common at a regional or global scale. The effects of scale are not trivial since they can greatly affect the number of species that would qualify as rare (e.g., Butchart 2003). There is some evidence that commonness and rarity may be assessed more appropriately if focused on a core set of species within some geographic area of interest (Magurran and Henderson 2003)—removing from consideration those that can be recognized by some criteria as vagrants or other nontarget species.

Second, it should be recognized that the two criteria to judge rarity—geographic range and abundance—are not necessarily independent. One of the more notable macroecological patterns is a positive relationship between a species' range and its local abundance (Brown 1995). Consequently, species with broad geographic ranges tend also to be relatively abundant locally, whereas rare species face a kind of "double jeopardy" (Lawton 1993) whereby their narrow distributions also tend to be characterized by low local abundance. So, although it may be tempting to use distribution and abundance as independent axes to define categories of rarity, perhaps what their interrelationship suggests is that the sets of species identified as rare using a distribution or abundance criterion are likely to be very similar. This has important practical implications since the collection of presence–absence data to quantify species distributions may suffice (He and Gaston 2000) in defining the rare species set, saving the considerable expense associated with estimating species abundance.

Third, sampling artifacts are prevalent among rare species, making their identification as rare problematic (Gaston 1994, 26; McGill 2003). Rare species are often cryptic or furtive or have special life history strategies that can reduce detectability, leading ultimately to substantial underestimates of distribution or abundance if sampling is not done at the appropriate time or place as discussed further in chapters 4 and 5. These detectability issues can cause two kinds of error. Most obviously, underestimates of abundance or range can inflate the number of species considered to be rare. The literature has many examples of a species considered to be rare turning out to be much more abundant or widespread than originally thought (e.g., Espadaler and López-Soria 1991; NavarreteHeredia 1996). A less obvious error can occur when a rare species is not detected at all and is therefore omitted from the rarity list within an area of interest (Green and Young 1993; Venette et al. 2002).

Causes of Rarity

As we reviewed earlier, relatively uncommon species dominate most species assemblages—whether one samples pristine or perturbed systems (McGill 2003). For this reason, there is no ecological justification for treating rarity only as an acquired characteristic of species whose numbers or distributions have been eroded by anthropogenic activities. Rather, the number of rare species varies from place to place because of natural variation in species abundance distributions, or variation in levels of anthropogenic stress. Therefore, the causes of rarity can be lumped into two broad categories: (1) *natural* or *intrinsic* causes defined by a species' inherent biological or ecological characteristics; and (2) *anthropogenic* or *extrinsic* causes defined by harmful human activities that have resulted in limited distribution and abundance, independent of their biology (Pärtel et al. 2005). Although there is a tendency to discuss intrinsic causes of rarity as factors predisposing species to elevated extinction risk that is ultimately governed by human impacts (McKinney 1997), Pärtel et al. (2005) found little overlap between the group of vascular plant species considered intrinsically rare and the group thought to be extrinsically rare. Therefore, a separate discussion of intrinsic and extrinsic causes of rarity seems warranted.

Rarity as an Intrinsic Attribute of Species Assemblages

There is a long and well-known list of inherent biological and ecological attributes that are associated with rare species, and these attributes are often used to classify species into categories of rarity for regulatory or conservation planning purposes (see section on "Classifications of Rare Species and Conservation Priority"). Any assemblage of species is expected to have a relatively high number of species with limited abundances or restricted geographic ranges for no other reason than that the number of individuals and their pattern of occurrence follow a statistical distribution with greater frequencies toward the rare end of the scale. It is important that resource managers acknowledge rarity as an intrinsic property of a suite of species inhabiting any given locale.

Other natural factors that are associated with limited distribution or abundance can be further classified into species traits and ecosystem traits. *Species traits* include those factors affecting basic population vital rates such that species with "slow" life histories (e.g., low growth rates, small litter size, long generation time, few reproductive episodes in a lifetime) may be predisposed to extinction risk (McKinney 1997) and may also be disproportionately represented among species considered to be rare (Pilgrim et al. 2004). Related species traits that may also be associated with rarity include large area requirements, occupying higher trophic levels, complex social structure, high specialization (or low ecological amplitude), low vagility, and large body size (McKinney 1997; Purvis et al.

2000b). *Ecosystem traits* are characteristics of the environments inhabited by species. Some habitats have inherently low carrying capacities (Harper 1977), or suitable habitat may occur only rarely across the landscape (Pärtel et al. 2005), both of which will constrain the observed abundance or occurrence levels of species. Still other environments may be characterized by natural disturbance regimes that act to depress the abundance levels of some species (Boughton and Malvadkar 2002), increasing the likelihood that those species would fall below some defined rarity threshold. Certainly, species traits can interact with ecosystem traits to affect the expression of species rarity. Habitat specificity (species trait), availability of suitable habitat (ecosystem trait), and dispersal capability (species trait) will jointly affect the potential rarity of a species. For example, regional endemics typically have high specificity and are restricted to one or a few sites where their appropriate habitat occurs (Kruckeberg and Rabinowitz 1985).

A final intrinsic factor that may explain some of the variation in observed rarity rates is actually an emergent property of species that share a common taxonomy. Taxon size, or the number of species within a particular taxonomic level (e.g., family, genus), is thought to be a potentially important trait that presages the prevalence of rarity among a set of species. However, conclusions to date are equivocal, with plants and insects showing evidence that diverse taxa have more rare species than expected if rarity occurred randomly among a collection of species (Schwartz and Simberloff 2001; Ulrich 2005), whereas species-poor taxa have disproportionately high rarity among birds and mammals (Russell et al. 1998; Purvis et al. 2000a). An explanation for the observed divergence in the taxon-size effect is being debated, but it may be related to what Kelly (1996) termed the "cost of mutualism." Species whose life histories are linked inextricably, such that the fate of one is conditioned on the fate of the other, may constrain the mutualists to a rarer existence than those not linked in this way. It is noteworthy that such interspecific interactions are prominent among plants and insects. For example, plant species with specialized pollinators or seed dispersers that are declining, or butterflies dependent on endangered larval host plants, are especially vulnerable to these cascading effects (Pilgrim et al. 2004). Indeed, Koh et al. (2004) estimated that there may be as many as 6300 species that are "coendangered" through such symbiotic relationships.

Extrinsic Factors Resulting in Increased Rarity

Human alteration of the environment has become so pervasive that no ecosystem is free of the impacts that Vitousek et al. (1997) attributed to the "growing scale of the human enterprise." Caughley (1994) also observed that human factors are implicated in most post-Pleistocene extinctions. The principal human factors that reduce species abundance have been categorized in various ways—often with metaphorical reference to the "evil quartet" (Diamond 1989, 39) or the "mindless horseman of the environmental apocalypse" (Wilson 1992). Although the factors constituting these lists vary, the common denominators are human land transformations leading to habitat loss and degradation; biotic mixing stemming from the introduction of nonindigenous (exotic) species; direct human exploitation for control, subsistence, or collecting; and pollution through the alteration of biochemical cycles or the introduction of synthetic organic compounds.

Habitat loss and habitat degradation may rank as the most important factors leading to increased rarity—they are certainly the most cited factors contributing to the listing of species as threatened or endangered under the Endangered Species Act (ESA), and imperiled under NatureServe's classification (table 3.1). In the United States, agricultural conversion and its associated land management practices, land conversion for urban and commercial development, and water developments are the top three types of habitat alterations threatening species (Wilcove et al. 2000).

Table 3.1. *The percentage of species listed as threatened or endangered in the United States under the Endangered Species Act or ranked as imperiled under NatureServe's classification whose increased rarity was judged to be affected by five major factors (from Wilcove et al. 2000, 243)*

	Number of Species	Habitat Loss/ Degradation	Exotic Species	Pollution	Exploitation	Disease
		Percent of Species				
All species	1880	85	49	24	17	3
Vertebrates	494	92	47	46	27	11
Invertebrates	331	87	27	45	23	0
Plants	1055	81	57	7	10	1

Habitat loss and degradation also rank as the leading causes of mammal extinctions in Mexico (Ceballos and Navarro 1991), are implicated in 70% of the vertebrates considered to be imperiled in China (Li and Wilcove 2005), and in 84% of 488 endangered species in Canada (Venter et al. 2006).

After habitat loss and degradation, interactions with exotic species are considered the next most important cause of species imperilment in the United States (Flather et al. 1994; Wilcove et al. 2000). Exotic species affect nearly 50% of imperiled or federally listed species (see table 3.1), and are particularly harmful to the native biota inhabiting island systems (Simberloff 1995). Exotic organisms can contribute to the rarity of species through a number of mechanisms, including predation, pathogenesis, competition, hybridization, and alteration of disturbance regimes (Crooks 2002). For many rare species, the spread of exotic species can further reduce the odds of recovery. Such is the case with four species of endangered fish in the greatly altered lower Colorado River ecosystem, where predation by nonnative fish precludes their recruitment (Minckley et al. 2003). Declines in some mammal species in Mexico have been linked with the introduction of cats, pigs, goats, and rats (Ceballos and Navarro 1991). The accidental or intentional introduction of nonnative species is increasingly being recognized as contributing to the decline of species worldwide (Pimentel et al. 2000). However, the prevalence of exotic species introductions as a factor contributing to species rarity does vary greatly among studies. In an analysis of the global Red List of Threatened Species maintained by the International Union for the Conservation of Nature and Natural Resources (IUCN), Gurevitch and Padilla (2004) found that only 6% of imperiled taxa listed exotic species as either a direct or an indirect factor contributing to their decline, and Li and Wilcove (2005) found that the threat of alien species was cited in only 3% of imperiled Chinese vertebrates.

Pollution and human exploitation are the next most important factors listed as contributing factors to imperiled species in the United States. Some pesticides, such as DDT, were considered the primary cause of increased rarity among a number of bird species, but have now been banned in the United States. Unfortunately, some pesticides prohibited in the United States are still widely used in other countries, such as Mexico, and may be responsible in declines of insectivorous bats (Ceballos and Navarro 1991). Other forms of pollution, including siltation and agricultural amendments, continue to be an important cause of rarity, especially among aquatic species in the United States (Wilcove et al. 2000). The rapidity with which pollution agents can decimate species is well illustrated by recent population collapses among several Old World vultures across the Indian subcontinent—collapses that were ultimately traced to birds feeding on livestock carcasses treated with a common antiinflammatory drug (Green et al. 2004). Direct exploitation by humans is often highlighted as an important factor contributing to rarity. However, overexploitation is blamed for the listing of only 17% of threatened and endangered or imperiled species in the United States (see table 3.1). Even fewer species (7.6%) are on the global IUCN Red List because exploitation is considered an important factor in their population decline (Gurevitch and Padilla 2004). However, as is the case with most of these factors, the relative importance of exploitation does vary among taxa and locale. In Canada, overexploitation contributed to declines of 32% of endangered species (Venter et al. 2006); in Estonia, an estimated 31% of plants are threatened by collection (Pärtel et al. 2005); and overexploitation is cited as a contributing factor to increased rarity of 78% of imperiled Chinese vertebrates (Li and Wilcove 2005).

Management Implications

Understanding the causes of rarity is fundamental to developing strategies to reduce extinction threats associated with species rarity. Indeed, the particular causes of rarity may dictate the suite of management approaches that will be most successful in species recovery efforts (see chap. 8). Moreover, if the likelihood of success is a criterion used to set conservation priorities (see Mace and Lande 1991), then an understanding of rarity could also help identify which species are most likely to respond to management efforts.

For instance, species that are naturally rare, and those that have been so over evolutionarily significant periods of time, may have life history characteristics adapted to their rarity. Species that are intrinsically rare may not warrant management directed at intensive "population recovery" efforts, for there may be very little practical opportunity for accomplishing such a conservation objective. For these species, extensive efforts to protect their habitat (see "Locations of Target Species at Risk" in chap. 6)

Table 3.2. *Conservation priorities based on population trajectories and causes of species rarity*

Population Trajectory	Cause of Rarity	
	Intrinsic (natural)	Extrinsic (anthropogenic)
Increasing	No immediate conservation concern	Population recovering, conservation priority conditioned on deviation from historical occupancy or abundance
Stable	No immediate conservation concern	Population maintaining, conservation priority conditioned on deviation from historical occupancy or abundance
Decreasing	High conservation priority but prospects for recovery may be limited	High conservation priority but prospects for recovery may be great

may be sufficient to ensure their persistence. Certainly the "double jeopardy" (Lawton 1993) associated with small populations and restricted distributions makes extinction risk a concern for naturally rare species, but it may be that conservation priorities can only be set after considering extrinsic factors that may be further eroding the population or distribution of intrinsically rare species (table 3.2).

Conversely, species that have become rare rather recently due to human land transformation activities or direct exploitation may not have life history characteristics that are adapted to low numbers and may actually be more threatened with extinction than intrinsically rare species. However, extrinsically rare species may, paradoxically, be more responsive to management actions designed to ameliorating the anthropogenic threats (see table 3.2; see also Barrett and Kohn 1991; Gaston and Kunin 1997). Among the rare flora of Estonia, Pärtel et al. (2005) estimated that nearly 50% (of 301 species) would benefit if land management on grassland, agriculture, and forestry lands shifted from intensive to more traditional extensive land management (see "Maintaining Disturbance Regimes" in chap. 7). An additional 18% of species threatened by collecting would benefit from upgrading legal regulations and public education. Such examples provide hope that rare species with declining population trajectories driven largely by extrinsic factors have high prospects for recovery if shifts in public values are sufficiently strong to alter our resource management behavior and the way we derive goods and services from ecosystems (see chaps. 9 and 10).

Rarity and Threat

Species considered threatened with extinction will more than likely also be considered rare. The converse is not necessarily true. A species may qualify as rare but may not be considered at risk of extinction (Gaston 1994). This may seem to contradict an earlier statement that rare species are more likely to become extinct than common species. However, we are not comparing rare with common species here. Rather, if we restrict our comparison to those species determined to be rare, not all will share the same probability of persistence—which is to say that the risk of extinction will vary. Consequently, one element considered in establishing conservation priorities is to focus on the subset of rare species that are under the greatest threat or most vulnerable to extinction (Mace and Lande 1991).

One of the early reviews of the concept of threat was completed by Munton (1987) and many of the criteria for evaluating threat are the same criteria that have been used to define rarity—a confounding of terminology that is common in the literature. However, Munton's (1987) review highlights the role of population dynamics in evaluating threat. Rare species that have been, in the recent past, declining in abundance or occupancy are more threatened than rare species with stable or increasing trends. Furthermore, the degree of threat assigned to a species can also be affected by the predicted trends in distribution or abundance in response to various human impacts. These early, sometimes characterized as subjective, efforts to evaluate threat based on population dynamics actually foreshadowed the emergence of population viability analysis as a standard approach to persistence probability estimation (Boyce 1992). Unfortunately, the substantial data requirements for formal viability assessments will limit the number of species for which a viability-based threat assessment can be completed (see "Conservation of Individual Species Based on the Concepts of Population Viability" in chap. 6). However, there is emerging evidence that categorizations of threat can be predicted using a combination of intrinsic life history traits and estimated population variances from temporal monitoring programs—a much more limited and feasible set of data requirements when compared to a typical population viability analysis. Fagan et al. (2001) were able to rate the vulnerability of more than 750 species into three broad extinction threat categories. They were also able to show for mammals that body size, age at first reproduction, and average number of offspring correctly predicted the extinction threat category for 83% (60 of 72) of species.

Until further research can verify whether the "shortcut" approach examined by Fagan et al. (2001) has broad applicability, uncertainty (i.e., data-poor species) will continue to plague efforts to evaluate threats to rare species. Obviously, uncertainty affects which species can be evaluated, but it can also affect how we assign species into certain threat categories. Under the precautionary principle, conservationists often pursue a risk-averse strategy such that species may be placed in a higher threat category, or at least placed in a category that acknowledges the uncertainty (e.g., suspected threatened), to guard against treating a species as relatively secure when it is in fact at risk. Given the lack of knowledge about the status of species in most taxonomic groups, this strategy has the potential to designate a large number of species as threatened with extinction due to their "little known" status (Mace 1994; and see chap. 4). The risk-averse strategy for judging extinction threat is a double-edged sword. On the one hand it guards against an undesirable and irreversible outcome—species extinction (Prato 2005). On the other hand, inaccuracies in judging threats may erode public support for biodiversity conservation (Roberts and Kitchener 2006).

Classifications of Rare Species and Conservation Priority

A number of strategies have been proposed to classify the members of an assemblage into rarity categories. Perhaps the best known classification strategy for rare species is Rabinowitz's (1981; Rabinowitz et al. 1986). Her classification is based on three attributes: geographic range (wide, narrow), local abundance (somewhere large, everywhere small), and habitat specificity (broad, restricted). This leads to eight classes: one abundant class, where a species has a wide geographic range, is abundant in some places, and has broad habitat requirements; and seven different forms of rarity.

A number of classification strategies have been proposed to assign conservation priorities to rare species. We reviewed a subset of strategies used by nongovernmental organizations and various countries in an attempt to identify factors used in assigning conservation priorities (table 3.3). Our intent in selecting the strategies we reviewed was not to be comprehensive,

Table 3.3. *Examples of strategies and ecological attributes used in assigning species to rarity and conservation priority classes*

Ecological Attribute											
Strategy	Range/ Distribution	Distribution Trend	Occupancy (area or pattern of occurrence)	Abundance	Abundance Trend	Ecological Specialization	Reproductive Potential	Taxonomic Distinctiveness	Fragility	Habitat Condition/ Protection	Threat	Population Viability
Rabinowitz (1981)	X		X	X		X						
Millsap et al. (1990)	X	X	X	X	X	X	X	X			X	
Burke and Humphrey (1987)	X		X	X		X						
Ceballos and Navarro (1991)	X		X				X				X	
Partners in Flight (Hunter et al. 1993; Dunn et al. 1999)	X		X	X	X						X	
Cofré and Marquet (1999)	X		X	X		X	X	X	X		X	
Pärtel et al. (2005)	X		X	X		X				X	X	
Master et al. (2000) NatureServe (2006)	X	X	X	X	X			X	X	X	X	
IUCN (2001, 2005)	X	X	X	X	X					X	X	X
Canada's Species at Risk Act (Irvine et al. 2005)	X			X	X						X	X
Australia's EPBC Regulations 2000 (OLD 2004)	X	X		X	X							
Mexico's SEMARNAP (2002)	X			X	X					X	X	
U.S.'s ESA (Fish and Wildlife Service) *	X			X				X		X	X	

*Criteria associated with determination of threatened and endangered species under the Endanagered Species Act (PL-205, 87 Stat, 884, as amanded).

but rather to demonstrate how different factors contributing to rarity are incorporated in various classification schemes for assigning conservation priorities to rare species. The majority of the classification strategies we reviewed used measures of distribution and abundance to identify rare species, then assigned them into conservation priority classes. This pattern was expected given that distribution and abundance are fundamental to discussions of rarity. Trend information in either abundance or distribution was used in many classification strategies to help identify species that may not currently qualify as rare but may be on a trajectory toward rarity in the future. The pattern of spatial occurrence was used in over half of the classifications reviewed as a means of capturing the fine-scaled pattern of landscape occupancy. The spatial occurrence pattern differs from species distribution in that it considers whether a population occurs somewhat ubiquitously or patchily throughout its geographic range, and can reflect whether a species is a habitat specialist.

In addition to geographic range, abundance, and habitat specialization, other ecological attributes may be considered in assigning conservation priorities to rare species. Intrinsic, or natural, attributes that can place species at higher risk include a low reproductive potential (Millsap et al. 1990; Ceballos and Navarro 1991), taxonomic distinctiveness (Millsap et al. 1990; Cofré and Marquet 1999), and fragility, or a species' sensitivity to perturbations or intrusions of its biological or physical environment (Master et al. 2000). Extrinsic factors, including habitat condition and amount of habitat occurring in protected areas (Pärtel et al. 2005), as well as other anthropogenic threats to the species are considered in nearly all of the strategies we reviewed to assign conservation priorities to rare species. Only two strategies use an assessment of population viability, which incorporates both intrinsic and extrinsic factors in an attempt to estimate extinction probability. Population viability is a function of population size, number, and condition of occurrences and trends in these factors, as well as threats and landscape connectivity (Master et al. 2000).

The strategies we reviewed differ in the classes (or rankings) of rare species. NatureServe (2006) assigns global (G), national (N), and state (S) level ranks that define the spatial scale over which relative imperilment is assessed. This geographic identifier is followed by a whole number between 1 and 5 to indicate the conservation status of plants, animals, and communities (Master et al. 2000; NatureServe 2006). The numbers are defined as 1, critically imperiled; 2, imperiled; 3, vulnerable to extirpation or extinction; 4, apparently secure; and 5, demonstrably widespread, abundant, and secure. A species that is critically imperiled on a rangewide basis would be ranked as G1, whereas a species ranked as S1 would be critically imperiled in a particular state, regardless of its status elsewhere. Criteria used in ranking species include (1) total number and condition of occurrences, (2) population size, (3) range extent and area of occupancy, (4) short- and long-term trends in the foregoing factors, (5) threats, (6) fragility, and (7) number of adequately protected populations (Master et al. 2000).

The IUCN Red List of Threatened Species is based on risk of extinction criteria (IUCN 2001, 2005). The main criteria used in assessing extinction risk include (1) population size, (2) geographic range (both extent of occurrence and area of occupancy), and (3) population size trajectory. However, concepts of threat are also recognized within these criteria, including degraded habitat quality, levels of exploitation, and the effects of introduced taxa, hybridization, pathogens, pollutants, competitors, or parasites (IUCN 2001). Further, the Red List criteria have procedures for recognizing and dealing with three types of uncertainty: natural variability, semantic uncertainty, and measurement error (Akçakaya and Ferson 2001). The recommended approach is precautionary as opposed to evidentiary and provides plausible ranges of parameters used to evaluate the criteria (IUCN 2001).

Canada, Australia, and Mexico use IUCN criteria as a basis for their national strategies for assigning conservation priorities to rare species. Canada's recently passed Species at Risk Act recognizes "endangered species" as "wildlife species facing imminent extirpation or extinction" and "threatened species" as "a wildlife species that is likely to become an endangered species if nothing is done to reverse the factors leading to its extirpation or extinction" (Irvine et al. 2005). Criteria for designating a species as endangered include (1) a declining population size; (2) a small distribution with declining or fluctuating abundance; (3) a small, declining population size; (4) a very small population size; or (5) a probability of extinction in the wild > 20% in 20 years or five generations, whichever is longer (Irvine et al. 2005).

Australia recognizes "critically endangered," "endangered," and "vulnerable" species based on the total number of mature individuals, the degree to which populations of a species have become reduced, the estimated rate at which the number of individuals will continue to decline, the degree to which its geographic distribution is precarious for its survival and how restricted the distribution is, and the probability of the species' extinction in the wild (Office of Legislative Drafting 2004).

Mexico recognizes species that are in danger of becoming extinct, threatened species, and species subject to special protection (SEMARNAP 2002). Species classified as in danger of becoming extinct are characterized by drastic recent reductions in their distribution or population sizes, which have reduced their viability due to factors such as destruction or modification of their habitat, overuse, or diseases. Threatened species are in danger of becoming extinct if factors that negatively affect their population viability by diminishing their population size or destroying their habitat are not addressed. Species subject to special protection are in danger of becoming threatened by factors affecting their population viability and for which special recovery actions are needed to restore their populations, habitat, or associated species necessary for their recovery.

In contrast to strategies based on the IUCN criteria, the U.S. Fish and Wildlife Service identifies "endangered" or "threatened" species for protection under the ESA. "Endangered" refers to a species "in danger of extinction throughout all or a significant portion of its range," and a "threatened" species is "likely to become an endangered species within the foreseeable future throughout all or a significant portion of its range" (Endangered Species Act, 16 U.S.C. §§ 1531-36, 1538-40, Sec. 3(6) and Sec. 3(20)). Criteria used to classify species into these rarity categories include (1) present or threatened destruction, modification, or curtailment of its habitat or range; (2) overutilization for commercial, recreational, scientific, or education purposes; (3) disease or predation; (4) inadequacy of existing regulatory mechanisms; or (5) other natural or human-made factors affecting its continued existence.

Our intent in reviewing these classification strategies was to provide an overview of the kinds of ecological attributes that are considered in assigning species to rarity categories. We have resisted evaluating which of these strategies, by some standard, is "the best" for conserving rare species for a number of reasons. First, the selection of an existing classification strategy, or the decision to develop a new strategy, will depend on the resource manager's conservation objectives (see chap. 2). Moreover, the conservation strategies we review in table 3.3 vary in their data requirements, with some focusing primarily on current distributional characteristics (e.g., Ceballos and Navarro 1991), and others requiring distributional and abundance trend data, or data on demographic rates (e.g., Millsap et al. 1990).

The kinds of data available will certainly affect resource managers' decisions about which classification strategy to implement.

Such considerations notwithstanding, others have evaluated the relative strengths and weaknesses of rarity classification systems. De Grammont and Cuarón (2006) reviewed 25 systems used in North America to categorize threatened species. Based on 15 characteristics that relate to risk categories, criteria, and other system characteristics, they ranked the IUCN (2001) system as having the highest number of desirable characteristics. In particular, the IUCN (2001) system was superior in that it clearly defines categories and criteria, was the only system that considers uncertainty in the assessment, and was applicable at both national and regional levels (de Grammont and Cuarón 2006). Their recommendations for improving the IUCN (2001) assessment approach focused on defining locations quantitatively and removing subjective words such as "typically." Rodrigues et al. (2006) also found the IUCN (2001) assessment protocols to be useful and noted the need for compiling point locality data that will be useful in both identifying priority sites for conservation and rapidly updating species conservation assessments in the future. Other authors (e.g., Eaton et al. 2005) have concurred on the need to remove subjectivity in IUCN (2001) protocols, especially in regard to the persistence potential of species whose status is secure in other regions. Although varying objectives and data availability will affect the choice of which rarity classification schemes can be used, perhaps what these evaluation efforts offer is a rigorously defined goal that resource managers can strive to meet with incremental inventory improvements over time (see chap. 5).

Conclusion

Rarity in natural systems is common. In any given species assemblage, most species will be relatively rare, whereas only a few will be common. Rarity is most often defined by two attributes: a species' distribution and its abundance. Species are considered rare if their area of occupancy or their numbers are small when compared to the other species that are taxonomically or ecologically comparable. Conservation science is concerned with how natural or human-caused changes to ecosystems affect both the number of species considered to be rare, and the population trends of species that are rare. Because species abundances are distributed inequitably, and because those that are less abundant are more likely to be lost from regional or local assemblages than common species, a conservation focus on rare species in order to maintain biodiversity appears justified.

However, given that species might be naturally rare and not considered at risk of extinction, conservation science must also consider the immediacy of the threats to species in order to identify those rare species that have the greatest likelihood of being lost. A number of strategies for classifying rare species have been proposed, with most using information on the current status or trends in the distribution and abundance of a species either to determine if a species qualifies as rare or to categorize the species into a rarity type. Conserving rare species should focus on factors that have resulted in increased rarity: habitat loss and degradation, introduction or invasion of exotic species, pollution, and direct human exploitation. Species that have become rare recently due to extrinsic human activities may not have life history characteristics that are adapted to low numbers and may actually be more threatened with extinction than intrinsically rare species but may also be more responsive to management actions that address the anthropogenic threats.

Although rarity is common among taxonomic groups that are well studied, our current understanding of rarity may in fact be biased. Much of the world's biodiversity remains poorly studied and our ignorance about the full complement of species inhabiting a given locale limits our ability to quantify "who is" and "who is not" rare and this impedes our ability to assess potential impacts of resource management activities on RLK species. This leads to what Molina and Marcot call the "conundrum" of little-known species, the implications of which are the subject of the next chapter.

Acknowledgments

We would like to thank Tom Sisk (Northern Arizona University) for providing some very constructive and insightful comments that improved our presentation of species rarity issues. Suzy Stephens (Rocky Mountain Research Station) is also acknowledged for her electronic graphics support.

References

Akçakaya, H. R., and S. Ferson. 2001. RAMAS Red List: Threatened species classifications under uncertainty. Version 2.0. New York: Applied Biomathematics.

Balvanera, P., A. B. Pfisterer, N. Buchmann, J. S. He, T. Nakashizuka, D. Raffaelli, and

B. Schmid. 2006. Quantifying the evidence for biodiversity effects on ecosystem functioning and services. *Ecology Letters* 9:1146–56.

Barrett, S. C. H., and J. R. Kohn. 1991. Genetic and evolutionary consequences of small population size in plants: Implications for conservation. Pp. 3–30 in *Genetics and conservation of rare plants,* ed. D. A. Falk, and K. E. Holsinger. New York: Oxford University Press.

Borrvall, C., B. Ebenman, and T. Jonsson. 2000. Biodiversity lessens the risk of cascading extinction in model food webs. *Ecology Letters* 3:131–36.

Boughton, D., and U. Malvadkar. 2002. Extinction risk in successional landscapes subject to catastrophic disturbances. *Conservation Ecology* 6 (2): 2. http://www.consecol.org/vol6/iss2/art2.

Boyce, M. S. 1992. Population viability analysis. *Annual Review of Ecology and Systematics* 23:481–506.

Brown, J. H. 1995. *Macroecology*. Chicago: University of Chicago Press.

Burke, R. L., and S. R. Humphrey. 1987. Rarity as a criterion for endangerment in Florida's fauna. *Oryx* 21:97–102.

Butchart, S. 2003. Using the IUCN Red List criteria to assess species with declining populations. *Conservation Biology* 17:1200–1.

Cardinale, B. J., M. A. Palmer, and S. L. Collins. 2002. Species diversity enhances ecosystem functioning through interspecific facilitation. *Nature* 415:426–28.

Caughley, G. 1994. Directions in conservation biology. Journal of Animal Ecology 63:215–44.

Ceballos, G., and D. Navarro L. 1991. Diversity and conservation of Mexican mammals. Pp.167–98 in *Latin American mammalogy: History, biodiversity, and conservation,* ed. M. A. Mares and D. J. Schmidly. Norman: University of Oklahoma Press.

Chapin, F. S., O. E. Sala, I. C. Burke, J. P. Grime, D. U. Hooper, W. K. Lauenroth, A. Lombard, et al. 1998. Ecosystem consequences of changing biodiversity. *BioScience* 48:45–52.

Chase, J. M., and W. A. Ryberg. 2004. Connectivity, scale-dependence, and the productivity–diversity relationship. *Ecology Letters* 7:676–83.

Cofré, H., and P. A. Marquet. 1999. Conservation status, rarity, and geographic priorities for conservation of Chilean mammals: An assessment. *Biological Conservation* 88:53–68.

Cottingham, K. L., B. L. Brown, and J. T. Lennon. 2001. Biodiversity may regulate the temporal variability of ecological systems. *Ecology Letters* 4:72–85.

Covich, A. P., M. C. Austen, F. Bärlocher, E. Chauvet, B. J. Cardinale, C. L. Biles, P. Inchausti, et al. 2004. The role of biodiversity in the functioning of freshwater and marine benthic ecosystems. *BioScience* 54:767–75.

Crooks, J. A. 2002. Characterizing ecosystem-level consequences of biological invasions: The role of ecosystem engineers. *Oikos* 97:153–66.

De Grammont, P. C., and A. D. Cuarón. 2006. An evaluation of threatened species categorization systems used on the American continent. *Conservation Biology* 20:14–27.

Diamond, J. 1989. Overview of recent extinctions. Pp. 37–41 in *Conservation for the twenty-first century,* ed. D. Western and M. C. Pearl. New York: Oxford University Press.

Duarte, C. M. 2000. Marine biodiversity and ecosystem services: An elusive link. *Journal of Experimental Marine Biology and Ecology* 250:117–31.

Dunn, E. H., D. J. T. Hussell, and D. A. Welsh. 1999. Priority-setting tool applied to Canada's landbirds based on concern and responsibility for species. *Conservation Biology* 13:1404–15.

Duffy, J. E. 2003. Biodiversity loss, trophic skew and ecosystem functioning. *Ecology Letters* 6:680–87.

Eaton, M. A., R. D. Gregory, D. G. Noble, J. A. Robinson, J. Hughes, D. Procter, A. F. Brown, and D. W. Gibbons. 2005. Regional IUCN Red Listing: The process as applied to birds in the United Kingdom. *Conservation Biology* 19:1557–70.

Ellstrand, N. C., and D. R. Elam. 1993. Population genetic consequences of small population size: Implications for plant conservation. *Annual Review of Ecology and Systematics* 24:217–42.

Espadaler, X., and L. López-Soria. 1991. Rareness of certain Mediterranean ant species: Fact or artifact? *Insectes Sociaux* 38:365–77.

Fagan, W. F., E. Meir, J. Pendergast, A. Folarin, and P. Karieva. 2001. Characterizing population vulnerability for 758 species. *Ecology Letters* 4:132–38.

Flather, C. H. 1996. Fitting species-accumulation functions and assessing regional land use impacts on avian diversity. *Journal of Biogeography* 23:155–68.

Flather, C. H., L. A. Joyce, and C. A. Bloomgarden. 1994. *Species endangerment patterns in the United States.* Rocky Mountain Research Station General Technical Report RM-241. Fort Collins, CO: USDA Forest Service.

Flather, C. H., M. S. Knowles, and I. A. Kendall. 1998. Threatened and endangered species geography: Characteristics of hot spots in the conterminous United States. *BioScience* 48:365–76.

Gaston, K. J. 1994. *Rarity*. London: Chapman and Hall.

Gaston, K. J., and W. E. Kunin. 1997. Concluding comments. Pp. 262–72 in *The biology of rarity: Causes and consequences of rare-common differences,* ed. W. E. Kunin and K. J. Gaston. London: Chapman and Hall.

Green, R. H., and R. C. Young. 1993. Sampling to detect rare species. *Ecological Applications* 3:351–56.

Green, R. E., I. Newton, S. Shultz, A. A. Cunningham, M. Gilbert, D. J. Pain, and V. Prakash. 2004. Diclofenac poisoning as a cause of vulture population declines across the Indian subcontinent. *Journal of Applied Ecology* 41:793–800.

Gurevitch, J., and D. K. Padilla. 2004. Are invasive species a major cause of extinctions? *Trends in Ecology and Evolution* 19:470–74.

Harper, J. L. 1977. *Population biology of plants.* London: Academic.

Harte, J., A. P. Kinzig, and J. Green. 1999. Self-similarity in the distribution and abundance of species. *Science* 284:334–36.

He, F., and K. J. Gaston. 2000. Estimating species abundance from occurrence. *American Naturalist* 56:553–59.

Hector, A., J. Joshi, S. P. Lawler, E. M. Spehn, and A. Wilby. 2001. Conservation implications of the link between biodiversity and ecosystem functioning. *Oecologia* 129:624–28.

Hopkins, G. W., J. I. Thacker, A. F. G. Dixon, P. Waring, and M. G. Telfer. 2002. Identifying rarity in insects: The importance of host plant range. *Biological Conservation* 105:293–307.

Hubbell, S. P. 2001. *The unified neutral theory of biodiversity and biogeography.* Princeton: Princeton University Press.

Hubbell, S. P., and R. B. Foster. 1986. Commonness and rarity in a neotropical forest:

Implications for tropical tree conservation. Pp. 205–31 in *Conservation biology: The science of scarcity and diversity*, ed. M. E. Soulé. Sunderland, MA: Sinauer.
Hunter, W. C., M. F. Carter, D. N. Pashley, and K. Barker. 1993. The Partners in Flight species prioritization scheme. Pp. 109–19 in *Status and management of neotropical migratory birds*, ed. D. M. Finch and P. W. Stangel. General Technical Report RM-229. Fort Collins, CO: USDA Forest Service, Rocky Mountain Forest and Range Experiment Station.
Irvine, J. R., M. R. Gross, C. C. Wood, L. B. Holtby, N. D. Schubert, and P. G. Amiro. 2005. Canada's Species at Risk Act: An opportunity to protect "endangered" salmon. *Fisheries* (December): 11–19.
IUCN. 2001. IUCN Red list categories and criteria: Version 3.1 IUCN Species Survival Commission. Gland, Switzerland, and Cambridge, UK: IUCN. http://www.redlist.org/info/categories_criteria2001.html.
———, 2005. Guidelines for using the IUCN Red List categories and criteria. April 2005. Standards and Petitions Subcommittee of the IUCN Red List Programme Committee. http//www.redlist.org/webfiles/doc/SSC/RedList/RedListGuidelines.pdf.
Johnson, C. N. 1998. Species extinction and the relationship between distribution and abundance. *Nature* 394:272–74.
Kaiser, J. 2000. Rift over biodiversity divides ecologists. *Science* 289:1282–83.
Kelly, C. K. 1996. Identifying plant functional types using floristic data bases: Ecological correlates of plant range size. *Journal of Vegetation Science* 7:417–24.
Koh, L. P., R. R. Dunn, N. S. Sodhi, R. K. Colwell, H. C. Proctor, and V. S. Smith. 2004. Species coextinctions and the biodiversity crisis. *Science* 305:1632–34.
Kruckeberg, A. R., and D. Rabinowitz. 1985. Biological aspects of endemism in higher plants. *Annual Review of Ecology and Systematics* 16:447–79.
Lande, R. 1995. Mutation and conservation. *Conservation Biology* 9:782–91.
Lawton, J. H. 1993. Range, population abundance and conservation. *Trends in Ecology and Evolution* 8:409–12.
Li, Y. M., and D. S. Wilcove. 2005. Threats to vertebrate species in China and the United States. *BioScience* 55:147–53.
Loreau, M., S. Naeem, P. Inchausti, J. Bengtsson, J. P. Grime, A. Hector, D. U. Hooper, et al. 2001. Biodiversity and ecosystem functioning: Current knowledge and future challenges. *Science* 294:804–8.
Lubchenco, J., A. M. Olson, L. B. Brubaker, S. R. Carpenter, M. M. Holland, S. P. Hubbell, S. A. Levin, et al. 1991. The sustainable biosphere initiative: An ecological research agenda. *Ecology* 72:318–25.
Lyons, K. G., and M. W. Schwartz. 2001. Rare species loss alters ecosystem function–invasion resistance. *Ecology Letters* 4:358–65.
Lyons, K. G., C. A. Brigham, B. H. Traut, and M. W. Schwartz. 2005. Rare species and ecosystem functioning. *Conservation Biology* 19:1019–24.
Mace, G. M. 1994. Classifying threatened species: Means and ends. *Philosophical Transactions of the Royal Society of London, Series B* 344:91–97.
Mace, G. M., and R. Lande. 1991. Assessing extinction threats: toward a reevaluation of IUCN threatened species categories. *Conservation Biology* 5:148–57.
Magurran, A. E. 2004. *Measuring biological diversity*. Malden, MA: Blackwell Science.
Magurran, A. E., and P. A. Henderson. 2003. Explaining the excess of rare species in natural species abundance distributions. *Nature* 422:714–16.

Master, L. L., B. A. Stein, L. S. Kutner, and G. A. Hammerson. 2000. Vanishing assets: Conservation status of U.S. species. Pp. 93–118 in *Precious heritage: The status of biodiversity in the United States,* ed. B. A. Stein, L. S. Kutner, and J. S. Adams. New York: Oxford University Press.

Matthies, D., I. Bräuer, W. Maibom, and T. Tscharntke. 2004. Population size and the risk of local extinction: Empirical evidence from rare plants. *Oikos* 105:481–88.

McGill, B. J. 2003. Does Mother Nature really prefer rare species or are log-left-skewed SADs a sampling artifact? *Ecology Letters* 6:766–73.

McKinney, M. L. 1997. Extinction vulnerability and selectivity: Combining ecological and paleontological views. *Annual Review of Ecology and Systematics* 28:495–516.

Millsap, B. A., J. A. Gore, D. E. Runde, and S. I. Cerulean. 1990. Setting priorities for the conservation of fish and wildlife species in Florida. *Wildlife Monographs* 111:1–57.

Minckley, W. L., P. C. Marsh, J. E. Deacon, T. E. Dowling, P. W. Hedrick, W. J. Mathews, and G. Mueller. 2003. A conservation plan for native fishes of the lower Colorado River. *BioScience* 53:219–34.

Munton, P. 1987. Concepts of threat to the survival of species used in Red Data books and similar compilations. Pp. 72–95 in *The road to extinction: Problems of categorizing the status of taxa threatened with extinction,* ed. R. Fitter and M. Fitter. Gland: IUCN/UNEP.

NatureServe. 2006. Explorer (Version 6.1). Arlington, VA: NatureServe. http://www.natureserve.org/explorer/ranking.htm#globalstatus.

NavarreteHeredia, J. L. 1996. Is the apparent rarity of *Liatongus monstrosus* (Bates) (Coleoptera: Scarabaeidae) real or an artifact of collecting? *Coleopterists Bulletin* 50:216–20.

Office of Legislative Drafting (OLD). 2004. Environment Protection and Biodiversity Conservation Regulations 2000. Statutory Rules 2000 No. 181 as amended made under the Environment Protection and Biodiversity Conservation Act 1999. Attorney-General's Department. Canberra, Australia. 233 pp. http://www.comlaw.gov.au/comlaw/legislation/legislativeinstrumentcompilation1.nsf/0/FAA515B854C46E02CA2570C900200F31?OpenDocument.

Pärtel, M., R. Kalamees, Ü. Reier, E. L. Tuvi, E. Roosaluste, A. Vellak, and M. Zobel. 2005. Grouping and prioritization of vascular plant species for conservation: Combining natural rarity and management need. *Biological Conservation* 123:271–78.

Pilgrim, E. S., M. J. Crawley, and K. Dolphin. 2004. Patterns of rarity in the native British flora. *Biological Conservation* 120:161–70.

Pimentel, D., L. Lach, R. Zuniga, and D. Morrison. 2000. Environmental and economic costs of nonindigenous species in the United States. *BioScience* 50:53–65.

Prato, T. 2005. Accounting for uncertainty in making species protection decisions. *Conservation Biology* 19:806–14.

Preston, F. W. 1948. The commonness, and rarity, of species. *Ecology* 29:254–83.

———. 1962. The canonical distribution of commonness and rarity: part I. *Ecology* 43:185–215.

Purvis, A., P. M. Agapow, J. L. Gittleman, and G. M. Mace. 2000a. Nonrandom extinction and the loss of evolutionary history. *Science* 288:328–30.

Purvis, A., J. L. Gittleman, G. Cowlishaw, and G. M. Mace. 2000b. Predicting extinction risk in declining species. *Proceedings of the Royal Society of London, Series B* 267:1947–52.

Rabinowitz, D. 1981. Seven forms of rarity. Pp. 205–17 in *The biological aspects of rare plant conservation*, ed. H. Syngeg. New York: Wiley.

Rabinowitz, D., S. Cairns, and T. Dillon. 1986. Seven forms of rarity and their frequency in the flora of the British Isles. Pp. 182–204 in *Conservation biology: The science of scarcity and diversity*, ed. M. E. Soulé. Sunderland, MA: Sinauer Associates.

Reveal, J. L. 1981. The concept of rarity and population threats in plant communities. Pp. 41–46 in *Rare plant conservation*, ed. L. E. Morse and M. S. Henefin. Bronx: New York Botanical Garden.

Reynoldson, T. B., R. H. Norris, V. H. Resh, K. E. Day, and D. M. Rosenberg. 1997. The reference condition: A comparison of multimetric and multivariate approaches to assess water-quality impairment using benthic macroinvertebrates. *Journal of the North American Benthological Society* 16:833–52.

Roberts, D. L., and A. C. Kitchener. 2006. Inferring extinction from biological records: Were we too quick to write off Miss Waldron's red colobus monkey (*Piliocolobus badius waldronae*)? *Biological Conservation* 128:285–87.

Rodrigues, A. S. L., J. D. Pilgrim, J. F. Lamoreaux, M. Hoffmann, and T. M. Brooks. 2006. The value of the IUCN Red List for conservation. *Trends in Ecology and Evolution* 21:71–76.

Rosenfeld, J. S. 2002. Functional redundancy in ecology and conservation. *Oikos* 98:156–62.

Russell, G. J., T. M. Brooks, M. M. McKinney, and C. G. Anderson. 1998. Present and future taxonomic selectivity in bird and mammal extinctions. *Conservation Biology* 12:1365–76.

Schoener, T. W. 1987. The geographical distribution of rarity. *Oecologia* 74:161–73.

Schwartz, M. W., C. A. Brigham, J. D. Hoeksema, K. G. Lyons, M. H. Mills, and P. J. van Mantgem. 2000. Linking biodiversity to ecosystem function: Implications for conservation ecology. *Oecologia* 122:297–305.

Schwartz, M. W., and D. Simberloff. 2001. Taxon size predicts rates of rarity in vascular plants. *Ecology Letters* 4:464–69.

SEMARNAP (Secretaría de Medio Ambiente, Recursos Naturales y Pesca). 2002. Norma Oficial Mexicana NOM-059-ECOL-20011, Protección ambiental-Especies nativas de México de flora y fauna silvestre—Categorías de riesgo y especifiaciones para su inclusión, exclusión o cambio–Lista de especies en riesgo. Diario Oficial de la Nación, 6 de Marzo del 2002. México, D.F.

Simberloff, D. 1995. Why do introduced species appear to devastate islands more than mainland areas? *Pacific Science* 49:87–97.

Sisk, T. D., A. E. Launer, K. R. Switky, and P. R. Ehrlich. 1994. Identifying extinction threats: Global analyses of the distribution of biodiversity and the expansion of the human enterprise. *BioScience* 44:592–604.

Smith, M. D., and A. K. Knapp. 2003. Dominant species maintain ecosystem function with nonrandom species loss. *Ecology Letters* 6:509–17.

Symstad, A. J., F. S. Chapin III, D. H. Wall, K. L. Gross, L. F. Huenneke, G. G. Mittleback, D. P. C. Peters, and D. Tilman. 2003. Long-term and large-scale perspectives on the relationship between biodiversity and ecosystem functioning. *BioScience* 53:89–98.

Thompson, K., A. P. Askew, J. P. Grime, N. P. Dunnett, and A. J. Willis. 2005.

Biodiversity, ecosystem function and plant traits in mature and immature plant communities. *Functional Ecology* 19:355–58.
Ulrich, W. 2005. Regional species richness of families and the distribution of abundance and rarity in a local community of forest Hymenoptera. *Acta Oecologica* 28:71–76.
Van Ruijven, J., G. B. De Deyn, and F. Berendse. 2003. Diversity reduces invasibility in experimental plant communities: The role of plant species. *Ecology Letters* 6:910–18.
Venette, R. C., R. D. Moon, and W. D. Hutchinson. 2002. Strategies and statistics of sampling for rare individuals. *Annual Review of Entomology* 47:143–74.
Venter, O., N. N. Brodeur, L. Nemiroff, B. Belland, I. J. Dolinsek, and J. W. A. Grant. 2006. Threats to endangered species in Canada. *BioScience* 56:903–10.
Vitousek, P. M., H. A. Mooney, J. Lubchenco, J. M. Melillo. 1997. Human domination of Earth's ecosystems. *Science* 277:494–99.
Wilcove, D. S., D. Rothstein, J. Dubow, A. Phillips, and E. Losos. 2000. Leading threats to biodiversity: What's imperiling U.S. species. Pp. 239–54 in *Precious heritage: The status of biodiversity in the United States*. ed. B. A. Stein, L. S. Kutner, and J. S. Adams. New York: Oxford University Press.
Wilson, E. O. 1992. *The diversity of life*. Cambridge, MA: Belknap.

4

Definitions and Attributes of Little-Known Species

Randy Molina and Bruce G. Marcot

The number of species worldwide has been estimated as 5 to 30 million (Wilson 1988). Only about 1.4 to 1.6 million of those species have been formally described, and even that number is relatively uncertain given the vagaries of systematic convention, geographic variation in species traits, and high levels of taxonomic synonymy (Stork 1997). Most of the global biodiversity, at least as reflected in species number, is unknown to science. Global patterns of estimated known and unknown number of species (fig. 4.1) suggest that the greatest unknowns among macro- and mesoscopic species occur with arthropods, fungi, and mollusks, but the gaps between known and unknown are likely even greater with microscopic species such as soil bacteria. Although much of this phantom biodiversity occurs in the species-rich tropics, there is a lack of comprehensive descriptions of many taxonomic groups in temperate areas as well.

This global pattern is repeated regionally. For example, in the inland West of the United States, of the taxa included in a major regional assessment (the Interior Columbia Basin Ecosystem Management Project of USDA Forest Service and USDI Bureau of Land Management), the greatest disparity between known and estimated numbers of species occurred with arthropods, fungi, and mollusks (fig. 4.2) over an area of 58,470,000 ha.

Before including little-known species in conservation programs, particularly from species-rich taxa such as fungi and arthropods, it is important to understand the inherent difficulty in gathering new information to determine their taxonomic and conservation status. For example, many species in these rich but little-known taxa are extremely difficult to detect

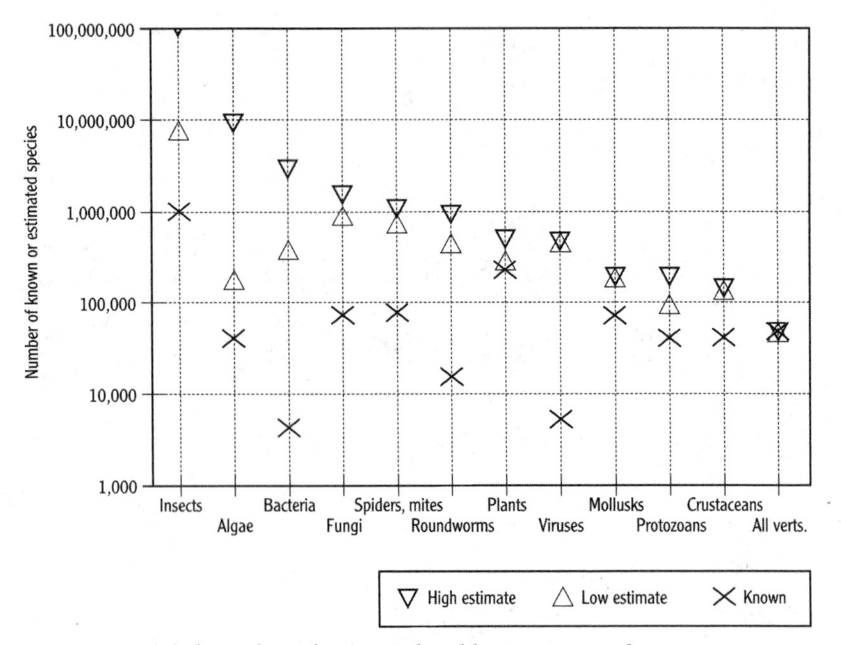

Figure 4.1. Global number of estimated and known species by taxonomic group. *Source*: Wilson 1988, Marcot et al. 1997.

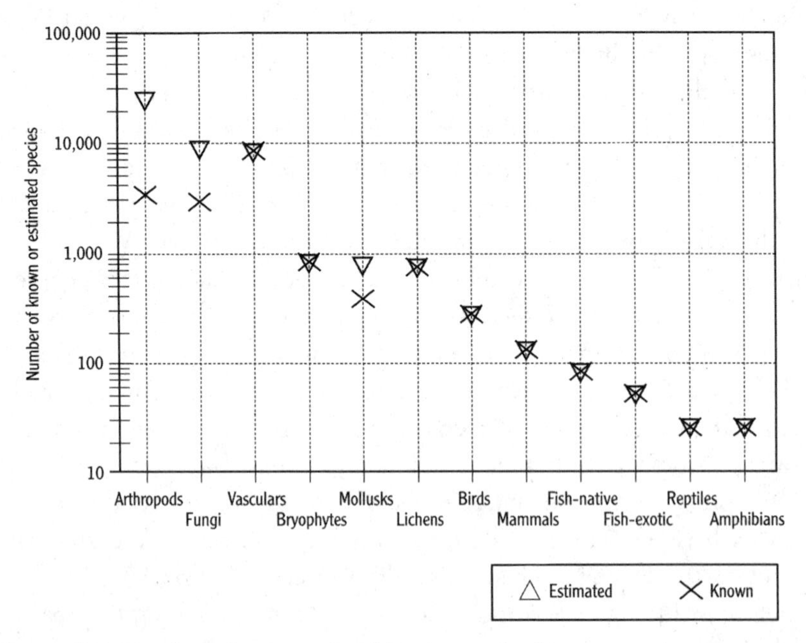

Figure 4.2. Number of estimated and known species by taxonomic group, as evaluated for the interior U.S. Columbia River basin. Values were derived from multiple taxa experts and compilations of species lists. Viruses, algae, phytoplankton, zooplankton, and most aquatic arthropods are not included. Richness of microfungi, bacteria, protozoa, and nematode species is largely unknown but may run into tens or hundreds of thousands of species. *Source*: Marcot et al. 1998.

due to their inconspicuous nature (figs. 4.3, 4.4). They are often hidden (e.g., buried in substrate) or so small that locating them is extremely difficult or impractical. Some species are simply too small to be readily detected in the field. Others have reproductive structures so minuscule (e.g., soil microarthropods) or diagnostic structures so obscure (some bryophytes) that microscopic examination and special expertise are required for identification. Some lichens cannot be reliably identified to

Figure 4.3. Examples of lichen and moss species in forests of the U.S. Pacific Northwest with detectability or identification difficulties, which may lend to their being little known in distribution or ecology. (a) Microscopic characteristics (60×): cord moss (*Atrichum selwynii*). The family, Polytrichaceae, is characterized by capsules, shown here, with 64 blunt peristome teeth, determined with a hand lens; the genus is among the smaller mosses in this family; the similar genus *Timmia* never has lamellae (upright ridges on the upper leaf surface), also requiring a hand lens for identification. (b) Microchemical tests and microscopic characteristics (60×): pixie cup lichen (*Cladonia fimbriata*). Identified by the medulla (loosely packed fungal hyphae below the photosynthetic zone) being K– (potassium hydroxide, negative results), P+ (p-phenylenediamine, positive results of red color), and UV– (no fluorescence under ultraviolet light); and told from *C. chlorophaea* and similar species by its powdery soredia (algal cells in fungal filaments) inside and outside the cups, and by cup shape and structure. (c) Hidden occurrence: podetium (stalk) of a *Cladonia* lichen growing within a clump of *Pohlia cruda* moss on a tree trunk. (d) Microchemical tests and mesoscopic characteristics (60×): beaded bone lichen (*Hypogymnia enteromorpha*). Distinguished from the similar *H. apinnata* by its short, budded side lobes and being P+ and KC+ (potassium hydroxide and sodium hypochlorite test, positive results). (All photos by Bruce G. Marcot.) Source: McCune and Geiser 1997; Vitt et al. 1988.

Figure 4.4. Examples of animal species with detectability or identification difficulties, which may lead to their being little known in distribution or ecology. (a) Microspiders (family Micryphantidae), full-grown adults shown on a U.S. penny (60×). Microspiders are numerous in abundance and species and play key predaceous ecological roles in soil food webs but are seldom seen, occurring in litter, duff, and soil layers. Many species may be undescribed. (b) Soil springtails or snowfleas (*Hypogastrura* sp.; family Hypogastruruidae) (200×). Occur in great numbers in the upper soil and litter layers of forests, may be important decomposers of soil fungi, lichens, bacteria, and decaying plant matter, and creators of topsoil. Some species may be undescribed and difficult to study, although their general presence can indicate healthy and productive soils. (c) Soil mite (*Pergamasus* sp.; family Parasitidae) (60×). Usually < 1 mm (0.04 in.) long, they are difficult to identify to species. Many species are likely undescribed but are numerous and key to soil health through their chewing, decomposition, and herbivory functions. (d) Soil springtail (*Ptenothrix [Dicyrtoma] maculosa*; family Dicyrtomidae) (60×). Tiny soil predators of mites and other invertebrates; difficult to identify to species by nonexperts. (e) Immature Pacific (coastal) giant salamander (*Dicamptodon tenebrosus*; family Dicamptodontidae) shown with a U.S. quarter. Adults more often found, but immatures seldom so, being tiny and cryptic or hiding at the bottom of streams; little is written of habitat ecology and environmental physiology of immatures. (f) Land snail (*Limicolaria subconica*; family Achatinidae), Congo River Basin, Africa. Hidden beneath leaves of shrubs, can easily escape detection. (All photos by Bruce G. Marcot. Photos (a) through (e) from U.S. Pacific Northwest.)

species without chemical tests. Many species are hidden in the forest floor, and extensive movement or sampling of substrates is required to find individuals. Hypogeous fungi (truffles), for example, fruit several centimeters below the forest floor and duff layer, and many mollusks or salamanders live on the undersides of leaves and woody debris or in talus. Others, such as arboreal arthropods, mammals, and lichens, live high in the canopies where extensive climbing or branch clipping might be required to sample species. Such attributes challenge our ability to locate and monitor these species.

Many little-known taxa share other life history characteristics, particularly reproductive strategies and seasonality, which also make detection difficult. Mollusks and amphibians are more active and detectable during wet, warm periods than during dry, cold periods, so effective surveys in regions such as the U.S. Pacific Northwest are limited to spring and fall. Multiple site visits and surveys may be needed to detect surface-active individuals, especially if populations are patchy, small in size, or unpredictable in substrate occurrence. Fungi offer one of the greater challenges because of their highly ephemeral nature and eruptive occurrence. Some reproduce (develop fruiting structures) in spring or fall. Timing and occurrence of fall reproduction is sensitive, however, to onset of rainfall (soil moisture) and patterns of temperature change. In many years, fungi will not reproduce, so 5 years of sampling in one location are typically required to determine their presence and diversity (O'Dell et al. 1999; Molina et al 2001). Many little-known species of fungi and other taxa also are highly patchy in distribution, which may be a response to unique microsite habitat requirements and past disturbances. The consequences of difficult detectability are discussed further under sampling considerations in chapter 5.

This chapter briefly explores causes that underlie the "little-known" conundrum and the types of information needed to make science-based decisions on conservation management of little-known species. We discuss information needs in light of the perceived risk to the species and for meeting other broad management objectives, such as sustainable timber harvest. We also reference examples from the Survey and Manage species conservation program that attempted to survey and conserve rare, little-known species of bryophytes, fungi, lichens, mollusks, amphibians, and arthropod groups as part of the Northwest Forest Plan in the Pacific Northwest (Marcot and Molina 2006; Molina et al. 2006). Chapter 5 then builds on combining concepts of rarity (chap. 3) and little-known (this chap.) to discuss the science implications of conserving species that are both rare *and* little-known.

Primary Causes of "Little-Known" Status

Little-known species fall into three categories of knowledge uncertainty: (1) taxonomic uncertainty, which occurs with overwhelming diversity and incomplete taxonomic description (and limited expertise of most observers); (2) distributional uncertainty, which occurs when few or no species inventory data are available; and (3) ecological uncertainty, which occurs with poor ecological understanding. For some species, information may be lacking in one or two of these categories, but for the great bulk of little-known species, lack of information generally stems from all three causes. The categories we provide evolve from the fact that little-known species are inconspicuous and difficult to detect, which leads to a lack of knowledge.

Taxonomic Uncertainty

It is important to understand the taxonomy of species before striving to conserve them. In addition to practical considerations of identification, taxonomic information brings with it an understanding of relatedness of species. This can be of conservation value when related species share a similar life history or ecology. Unfortunately, providing taxonomic clarity for little-known species comes with many challenges. Some taxonomic groups can number in the tens of thousands of species at the regional scale (e.g., soil microarthropods). With such overwhelming numbers of species, it is not surprising that most remain undescribed. At best, there may be a generally poor understanding of their taxonomy and systematics.

Hawksworth (1991), for example, hypothesized that, of the estimated global 1.5 million species of fungi, less than 5% have been described. Fungal classification is also in flux; for example, 10 years later, Hawksworth (2001) upheld his global estimate of 1.5 million species of fungi but clarified that the number of species now known is 74,000 to 120,000. Although his new estimate is somewhat greater than the original 5%, it remains a small fraction of the total. Even this estimate is uncertain because no comprehensive compilation of species is available. The advent of new molecular approaches that better define taxonomic relations between species has drastically altered our understanding of the major relations of fungal phyla and classes (Blackwell and Spatafora 2004; Wayne and Morin 2004).

Even with species that have been described, many can be extremely difficult to identify, and identification keys are often inadequate or nonexistent. In the Pacific Northwest with its rich array of forest macrofungi (mushrooms, truffles, cup fungi, and their allies), there is a dearth of comprehensive regional keys to species identification. General mycological guides (e.g., Arora 1986) are often the only books available with keys to start the identification process. Likewise, identifying many arthropods often relies on sending voucher specimens to those few overburdened entomologists and experts who specialize in particular genera or families; a few examples are the species-rich taxa of staphylinid beetles, micryphantid microspiders, and soil mites and nematodes. There are generally no comprehensive identification keys or "field guides" to help identification of most terrestrial invertebrate species the way there are for plants, birds and other vertebrate groups.

Well-trained taxonomists are vital to basic understanding of biodiversity (Chambers and Bayless 1983; Huber and Langor 2004), but the current decline in the cadre of taxonomic specialists who do fundamental taxonomic research compounds the challenge of describing the vast numbers of unnamed species and developing identification keys (Molina et al. 2001). In this regard, the development of sophisticated molecular-DNA tools to identify and differentiate species has been a two-edged sword. Although this new science has vastly improved our ability to detect recalcitrant taxa (e.g., of soil microorganisms), the attractiveness of this exploding research field has enticed young scientists away from traditional taxonomic fields. In the United States, the National Science Foundation (NSF) has recognized this issue and enhanced funding of its systematics program:

> The National Science Foundation (NSF), in partnership with academic institutions, botanical gardens, freshwater and marine institutes, and natural history museums, seeks to enhance and stimulate taxonomic research and help prepare future generations of experts. NSF announces a special competition, Partnerships for Enhancing Expertise in Taxonomy (PEET), to support competitively reviewed

research projects that target groups of poorly known organisms. This effort is designed to encourage the training of new generations of taxonomists and to translate current expertise into electronic databases and other formats with broad accessibility to the scientific community (http://web.nhm.ku.edu/peet).

Efforts such as the PEET program, if sustained as part of science endeavors, will help to close the taxonomic information gap on little-known species.

Distributional Uncertainty

Basic inventory of species presence, abundance, and distribution is necessary for understanding species status and for monitoring trends. Lack of basic inventory data is at the root of the little-known status for various reasons. Taxonomic groups such as fungi, arthropods, or mollusks may simply not be included in inventory programs because they are seen as low priority compared to more "charismatic" flora and fauna. Their low-priority status reflects biases in scientific study, conservation, and social interests, and to some extent, the expertise of available biologists and taxonomists. Some taxa, such as spiders, have simply not been included in surveys and inventories, so that basic information on occurrence and distribution is lacking and thus the taxa are excluded from conservation planning (Skerl 1999). More typically, inventory programs find the idea of including such difficult and diverse taxa daunting, and they lack resources and administrative support to tackle the problem.

Issues of high diversity and distributional uncertainty are complicated by the simple dearth of knowledgeable biologists who can effectively deal with species such as fungi, mollusks, and arthropods in inventory programs. Most field botanists and wildlife biologists have been trained in vascular plant and animal species and generally have less knowledge to deal with these other taxa. A lack of familiarity with the ecological roles of little-known taxa makes it difficult to draw on available management tools to conduct specific surveys and studies contributing to species conservation. It is simpler to manage diverse macrovegetation conditions as a basis for habitat of bird or fish populations than to manage dispersed microscale habitats and substrates required by many little-known species. However, these species have been receiving increasing attention. For example, Dunn

(2005) emphasized the need to include insects in biodiversity conservation planning and assessments of recent species extinctions.

Use of population attributes to characterize viability and persistence of many groups of little-known taxa may not yet be practical given the current state of knowledge and slow pace of basic research progress on these taxa. But biologists and resource managers should not be completely discouraged from this general lack of inventory information because some progress is being made for these difficult taxa. Examples of such progress follow.

Attempts to conduct all-taxa biodiversity inventories exemplify the ability of professional societies to coordinate efforts in specific locations such as the Great Smoky Mountains National Park. As of 2006, the program had identified 625 new species and recorded 4666 species previously unknown from the park (http://www.dlia.org/atbi/index.shtml). Forest stand–level inventories and studies conducted at the H. J. Andrews Experimental Forest in the Oregon Cascades (a long-term ecological research site), have yielded comprehensive species lists and diversity, abundance, and importance data for fungi (Luoma et al. 1991; Smith et al. 2002) and arthropods (Parsons et al. 1991). Similar small-scale inventories including many little-known taxa have occurred elsewhere. Examples include surveys and inventories of amphibians in Great Smoky Mountains National Park (Dodd 2003); rare, threatened beetles in boreal forests (Martikainen and Kouki 2003); and mites in caves and deep soil (Ducarme et al. 2004). Although these examples are generally exceptions to most biodiversity inventory and survey projects, much could be learned from their successful methods to integrate into other biodiversity conservation inventory and monitoring programs. The all-taxa biotic inventory at the Great Smoky Mountains National Park, for example, conducts extensive outreach to train and educate volunteers to help with field surveys and other aspects of the program.

In another successful example, in the Pacific Northwest, the U.S. Department of Agriculture (USDA) Forest Service and U.S. Department of the Interior (USDI) Bureau of Land Management have developed survey protocols, field guides to species identification, and distribution maps, and have conducted surveys of 400 little-known species of lichens, bryophytes, fungi, mollusks, vascular plants, and amphibians, resulting in nearly 60,000 records over a 10-year program from 1994 to 2004 (Marcot and Molina 2006; Molina et al. 2006). Training of field biologists and use of parataxonomists (nonprofessionals, field-trained in species inventory and identification) as well as contracted taxonomy experts from academia were instrumental in the success of gathering useful and scientifically valid new information on abundance and distribution of these little-known species.

The current electronic information age also allows for better use of information and distribution of expertise. For example, databases from major herbaria and museums are now coming online over the Internet so historical records on species locations can be searched and analyzed. Many professional societies that focus on major groups of little-known taxa have recognized the need to include their taxa in the growing international call for biodiversity conservation. For example, a cadre of mycologists has recently published the first comprehensive treatise on inventory and monitoring methods for fungi (Mueller et al. 2004). Many authors have published on efficient methods for sampling terrestrial invertebrates, such as the use of enclosures and pitfall traps (Moffatt et al. 2004; Borgelt and New 2005; Hansen and New 2005). Such efforts bring scientific consistency to methodologies so that results can be compared among different regions. For example, the Natural Resources Monitoring Partnership (http://biology.usgs.gov/status_trends/nrmp/MonitoringPartnership.htm) has proposals to publish an inventory of monitoring projects and a library of recommended protocols covering major national and international status and trends programs.

Web-based species identification systems are becoming more available, such as for ants (AntWeb; http://www.antweb.org/index.jsp), grasshoppers (Field Guide to Common Western Grasshoppers; http://www.sdvc.uwyo.edu/grasshopper/fieldgde.htm), and butterflies and moths (Butterflies and Moths of North America; http://www.butterfliesandmoths.org/). In many cases, these tools may be the best or only way that biologists can access keys, photo series, and other expertise to identify sample specimens of little-known species.

Ecological Uncertainty

Even if something is known about the taxonomy and distribution of little-known species, there is typically a lack of information on the ecology of the species, including habitat requirements, community dynamics, response to disturbance, key interactions with other species, or ecosystem functions.

This type of information is critical to evaluating the life histories and ecological functions of little-known species in a holistic ecosystem sense, and to designing management approaches that maintain or restore species persistence and function.

The lack of general inventory data is exacerbated by the lack of information on population biology and natural history attributes needed to describe and forecast species presence, abundance, and population dynamics or trends. Knowledge of dispersal capabilities, either of individuals for mating events (animals) or of sexual or asexual propagules (plants and fungi), is usually lacking. Dispersal may be extremely slow and limited for many species (e.g., terrestrial mollusks).

Specifically with fungi, spore production, dispersal, and reproduction events leading to new fungal individuals and populations are not well understood in terrestrial ecosystems. Most population work with fungi has focused on pathogens to explore concepts of epidemiology, and molecular techniques are only now being developed to discern fungal individuals and populations at landscape scales (Dunham et al. 2003, 2006; Kretzer et al. 2003). Effects of natural and anthropogenic fragmentation of habitat, and the capability of species to disperse across unsuitable areas to maintain gene flow, remain largely unknown. Some fungal species depend on invertebrates or vertebrates for dispersal (e.g., Kotter and Farentinos 1984), further complicating analyses of fungus population distribution, abundance, and persistence.

In addition to knowledge on site occurrence, range, and distribution of little-known species, information is needed on the microhabitat and local site conditions they require. Such information is critical to understanding how these species respond to disturbance, and thus how to ameliorate threats and stressors caused by alteration of their habitat. Developing reliable habitat models of rare species is challenging, especially when species are little known and when their habitat consists of microscale features or very fine patch sizes typically not mappable across planning landscapes (e.g., using geographic information systems software). This is complicated by these species' usually sporadic and patchy distributions and dispersal limitations already noted.

Many little-known species in Pacific Northwest forests, for example, closely associate with large, coarse, woody debris and unknown microscale soil attributes or substrates such as rocks, bark of specific tree species, and moss patches. Little of this is spatially mappable from remote sensing information. Also, given dispersal limitations, patchy distribution of populations, and microscale habitat requirements, presence of habitat does not necessarily mean likely presence of predicted species. Understanding occupancy rates in seemingly optimal habitat is important for evaluating the utility of model- and habitat-based management approaches. Scaling-up from microscale habitat needs to the scale at which conservation or land-use planning occurs will remain a major challenge to predict species presence. We discuss the potential use of habitat modeling approaches in chapter 5.

Most organisms have evolved adaptations to changes in their environment over space and time. The science of disturbance ecology at the landscape scale includes understanding patterns of biotic response to immediate impacts of disturbance (natural and anthropogenic) and to long-term effects of land use and global climate change. However, the disturbance ecology of little-known species remains largely unexplored. This is perhaps one of the more critical and practical information needs as resource managers increasingly rely on re-creating natural disturbance regimes to sustain or restore healthy ecosystems.

Determining Information Needs and Setting Priorities

Addressing conservation concerns about little-known species requires gaining information on needs of the species and setting management priorities. For example, at a small site scale (a stand management unit), the most important information may simply be the presence or absence of the species of concern. This is often a criterion for sensitive and special status species programs of the Forest Service and Bureau of Land Management when planning management activities such as forest thinning projects. Identifying and inventorying microscale habitat features where the species is present may also be necessary so that habitat can be managed appropriately. At larger planning scales (e.g., watersheds to regions), information on distributions of species in reserves or concentrations of individuals may be most useful. Given often-limited resources, it is important to collect information at a scale appropriate to the specific needs of the plan area. For example, if information collection is too project-specific, the accuracy of syntheses or of running models at broad geographic scales can be lost.

Case Study: The Survey and Manage Program of the Northwest Forest Plan

The Survey and Manage program of the Northwest Forest Plan provides an example of strategic, regionwide surveys for hundreds of little-known individual species and arthropod species groups that were thought to be rare and likely associated with late-successional or old-growth forests (Molina et al. 2003, 2006). We first provide background information on the overall program and then discuss how resource managers and scientists have worked together to prioritize information needs and gather new information to reduce uncertainty on these species.

In 1994, the Bureau of Land Management and Forest Service adopted standards and guidelines for the management of habitat for late-successional old-growth (LSOG) forest-related species within the range of the northern spotted owl (*Strix occidentalis caurina*) under the Northwest Forest Plan (NWFP; USDA and USDI 1994). The main conservation elements of the NWFP are a system of reserves (with focus on maintenance and restoration of LSOG forests), an aquatic conservation strategy that protects streams and riparian areas, and various standards and guidelines pertaining to seven other land allocation categories, including "matrix" lands on which more intensive resource production and use may occur. The NWFP included mitigation to protect rare and often endemic species associated with LSOG forests. This mitigation is referred to as the Survey and Manage program. Over 400 species and four arthropod guilds, across eight major taxonomic groups, were listed for protection under this program. Initial risk analyses, which identified which species were to be addressed under the Survey and Manage program, were primarily based on expert opinion because so little quantitative information was available on abundance, distribution, population status, habitat associations, and degree of protection provided by reserve land allocations (Meslow et al. 1994; Raphael and Marcot 1994).

The Survey and Manage standards and guidelines (USDA and USDI 1994, 2001) described an adaptive management approach to conservation. Protection of sites where the organisms were known to exist was combined with regionwide surveys (as well as other information-gathering techniques such as herbarium searches and research) designed to provide new information that addressed uncertainties surrounding species viability. As new data were acquired and analyzed, the status of each species within the Survey and Manage program was revisited in formal annual species reviews, and the species management guidelines were revisited and revised if appropriate. The annual species reviews resulted in category changes among Survey and Manage status rankings and sometimes even removal from the Survey and Manage list, demonstrating the dynamic structure of the conservation program and the successful application of the adaptive management approach. Molina et al. (2003) described the process of acquiring and using new information. Details of the mitigation measure are given in the two records of decision (USDA and USDI 1994, 2001). Molina et al. (2006) provided an overview of how the Survey and Manage program evolved and the many implementation challenges encountered.

At the initiation of the Survey and Manage program, important life history information was lacking on nearly all of the Survey and Manage–listed species. For example, distribution data were not available on most species, and many species were known from only a few historic sites. Virtually no population-level data or habitat requirements were known, thus precluding use of population viability analysis or habitat modeling. In fact, natural histories of most species were so poorly understood that it was difficult to document biological threats (e.g., from limited dispersal, limited habitat availability, and habitat fragmentation) or management threats (e.g., from timber harvest and prescribed fire). Because of this lack of knowledge and high degree of uncertainty as well as the lengthy startup time to organize the program of work, the agencies allowed for at least 10 years of regionwide surveys. This allowed time to better assess persistence concerns and develop species-specific management.

Two fundamental goals of the Survey and Manage program drove the information needs assessment for the plan area: (1) to provide for the persistence of well-distributed populations, and (2) to maximize the role of the reserve lands to meet persistence requirements (e.g., provide habitat and connectivity). The first step in identifying information needs was to collect all available information from herbaria and museum records, agency field records, expert opinion, and the scientific literature, and then to synthesize this information for resource managers and biologists. This initial process indicated where significant knowledge gaps occurred. The next stage involved developing key questions and identifying the types of information needed to answer those questions. For example, what is the species distribution in reserve lands? and does the species require specific microhabitat such as large woody debris? Surveys would then be conducted in reserves, and the amount of woody debris would be measured in locations where target species are found.

Molina et al. (2003) described the types of information needs for the strategic survey effort of the Survey and Manage program (box 4.1). Information needs were organized into general categories of rarity, habi-

Box 4.1. Categories and specific information needs for survey and manage species (Molina et al. 2003).

Rarity

- Number of current and historic known sites
- Relative abundance at historic and known sites
- Size, area, diversity, and extent of inhabited sites on the landscape

Habitat

- Known or suspected habitat requirements
- Description of potential suitable habitat at both the micro- and macroscale
- Ecological amplitude

Distribution

- Historical and current distribution of known sites
- Historical and current distribution of potential suitable habitat
- Portion of suitable habitat that is occupied
- Distribution of known sites in reserve land designations

Persistence concerns

- Population trends and status of isolated populations
- Life history traits that might create additional risk
- Dispersal capacity and requirements
- Fragmentation of suitable habitat in relation to historical connectivity
- Successional trends of potential and suitable habitat
- Threats to occupied areas (both natural and anthropogenic)

Management consideration

- Quality of sites in reserve land designations
- Response to disturbance, natural and anthropogenic
- Connectivity of occupied sites needed to maintain stable populations
- Active and passive management needed to maintain or restore suitable habitat at known or potential sites (habitat quality at known sites)

tat, distribution, persistence concerns, and management considerations. Those items listed under rarity, habitat, distribution, and persistence concerns reflected many important information gaps critical to understanding the status and guiding conservation of individual species. Those items listed under management considerations reflect information needed to address key management objectives. For example, information on species response to disturbance was needed to help resource managers develop site management plans that included activities such as forest stand thinning, prescribed burning, or road building. Some management considerations have higher priority than others and thereby guide decisions on what information is most critically needed and will be gathered. Decisions typically blend information needed to address both species and management priorities.

Once operational decisions were made, surveys were strategically planned and implemented to collect the information. For example, the Survey and Manage program conducted surveys at known sites to collect crucial habitat information and regionwide, random-grid surveys to gain information on rarity and general distribution trends in reserve lands. These strategic surveys used an adaptive process wherein the survey methods and results obtained were periodically analyzed for efficiencies and effectiveness in gaining the needed information. New survey data were analyzed in an annual species review process, and changes to species management were made as appropriate. The species list and survey protocols were adjusted as needed. At the completion of this review, a new cycle of information prioritization, survey planning and implementation, and data analysis was undertaken. All planning was documented in an annual implementation guide. This adaptive management approach allowed the Survey and Manage program to make steady progress in meeting the objectives for this unprecedented conservation program for little-known species (see Molina et al. 2003 for more details on this planning process).

Even with a well-conceived process for acquiring new information on little-known species, the Survey and Manage program ran into many implementation challenges that revolved around many of the issues raised at the beginning of this chapter. These included difficulties in detecting species and determining their actual rarity; the complexity of defining species persistence and evaluating how well the plan's systems of reserves protected individual species; the task of training and maintaining a cadre of taxonomy specialists to identify species and provide expert interpretation of survey results; and the impracticality of targeting 400 little-known species over the 9.7 million ha plan area. Chapter 11 provides an overview of implementation challenges of conserving rare or little-known species with further examples from the Survey and Manage program.

Conducting Threat Assessments for Conservation Planning

As noted earlier, resource managers are often most concerned about the effects of various management practices on species of concern, and they desire information to help ameliorate threats to species. Part of increasing our knowledge about the response of little-known species to anthropogenic disturbances may entail first conducting basic threat assessments. A threat assessment determines what those threats are as well as the likelihood of a decline or loss of small populations in the face of human activities. A threat assessment would determine whether it is reasonable to craft management or monitoring actions for conserving little-known species. Further, it would determine the degree to which changes in management actions would be expected to solve problems; that is, the degree to which resource managers can have an impact on those threats (Morrison et al. 1998).

Results of such a threat assessment would classify species into categories, such as: (1) species for which current scientific knowledge and expert understanding are so poor that both threats and the potential impacts of management activities are very uncertain, and (2) species threatened by human activities for which reasonable (and testable) hypotheses can be devised concerning the role of these threats. The latter category can be further divided into species for which changes in specific land management activities can be expected to help reduce threats, and those that would not be so helped. For example, adverse effects on little-known terrestrial forest species from activities that compact soil could be mitigated by using different machinery such as rubber-tired tree loaders or by conducting these activities during other seasons such as when soil is frozen. Species that are influenced by human actions but that would not be aided by changing local resource management activities may include species generally sensitive to air pollution, such as some pendant arboreal lichens that are sensitive to sulfur oxide and nitrogen oxide in the atmosphere. By conducting threat assessments, the resource manager could describe realistic expectations for how and whether changing management activities would help the species.

In summary, given the high diversity and lack of essential distributional and ecological knowledge on many little-known species or taxonomic groups, it is important to set priorities for information needs to meet specific objectives. Otherwise, a program that includes many little-known species can become unwieldy and ineffective and may lose support. For example, in the Survey and Manage program discussed previously, prior to strategically focusing surveys in reserves to examine species persistence, most surveys were conducted in matrix lands before conducting ground-disturbing activities. Consequently, many sites of listed species were found in matrix lands, and resource managers often chose to forgo management activities such as timber harvest to avoid risking harm to the species. These decisions impacted the ability of the agencies to meet other management goals of the Northwest Forest Plan (e.g., timber harvest). Eventually these management frustrations and litigation from the timber industry led to the abandonment of the Survey and Manage program (USDA and USDI 2004).

Ecological and Social Implications of Little-Known Species

Interest in conservation of little-known species may be more than merely esoteric or academic. Many little-known species perform crucial ecosystem functions, including cycling nutrients, fixing nitrogen, aggregating soil, improving soil structure, and acting as links in the food web (see figs. 4.3, 4.4). Indeed, this is perhaps the most important factor in considering the protection and conservation of the functional diversity within these taxa for meeting broad goals of ecosystem management. For example, some rare plants rely on obligate pollinators (Spira 2001) that, in turn, may be poorly known and in decline (Cane and Tepedino 2001). Other rare plants can help stave off invasion of exotic species (Lyons and Schwartz 2001). Many little-known soil invertebrates play major roles as litter decomposers, and their functional redundancy may help maintain soil productivity (Andrén et al. 1995). Ostfeld and LoGiudice (2003) reported that loss of rare as well as common species led to increased incidence of Lyme disease in their modeled ecosystem.

Many little-known taxa enter into symbioses with other plants and animals and thus influence community and ecosystem dynamics. Many soil fungi, for example, form mutualistic symbioses with plant roots termed mycorrhizae. Mycorrhizal fungi—many species of which are poorly known in terms of specific autecology, distribution, or even taxonomy—strongly depend on host photosynthate as their primary energy source; in return, the plants receive much of their nutrient uptake (as well as other benefits) via their mycorrhizal fungi. Mycorrhizal fungi and plants exhibit varying degrees of host–fungus specificity in their natural associations (Molina et al. 1992), but, from a functional perspective, are obligate symbionts.

In another example of symbiosis, many arthropods tightly couple with other organisms (e.g., obligate pollination and dispersal relationships) in functional interdependencies (Buchmann and Nabhan 1996; Shepherd et al. 2003). Some rare species may be useful in bioassessments and may serve as indicators of ecosystem health (Cao et al. 2001; Welsh and Droege 2001). Conservation of these species necessitates an understanding of their interactions with other biota so that these relationships and key ecological functions can also be conserved. Although much might be known in general about these common symbioses and species interactions, they are poorly understood at the species level within specific ecosystems.

Much research remains to be done to determine the specific ecological functions of little-known species, but examples suggest that retaining little-known species may help maintain ecosystem services, biodiversity, and the full range of system functions. Little-known species can also play key social and cultural roles as well. In fact, many little-known species provide vital services to people. Many native peoples throughout tropical areas use invertebrates, for example, as a source of food and protein, and many native plants for medicines. Other little-known species may play key roles in various cultural rites, religions, and rituals and could be considered in habitat management (Bengston 2004). Yet many of these species are poorly studied or are scientifically unknown (Phillips et al. 1994). Recently, concern has been raised about adverse effects of environmental degradation on conservation of both well-known and little-known medicinal plants (Shanley and Luz 2003). In some cases, entire "ethnobiomedical" forest reserves have been delineated in tropical ecosystems to protect these little-known species (Balick et al. 1994).

Conclusion

This chapter reviewed causes and characteristics of little-known species, their ecological and cultural roles, and implications for management. Species can be little known for a number of reasons, each reason implying very different solutions. Some solutions could include further taxonomic research to describe the species systematically, field inventories to determine presence and distribution, or ecological studies to understand habitat associations, life history, and environmental correlates and stressors.

If the resource manager is interested in conservation of little-known species, the first step could be to take stock, produce a list, and compile whatever information is available in literature and from experts, on species names, taxonomy, distribution, abundance, and autecology. In such a list, the resource manager can begin to identify which species may be associated with particular environments of conservation concern, such as old-growth forests or native grasslands, and also which areas of knowledge are most lacking for each species. Also important may be to determine if species are rare and what might cause rarity (see chaps. 3 and 5).

Further, the resource manager could consider some little-known species as part of species groups such as habitat groups and ecological functional guilds (see chap. 6). For example, many aquatic macroinvertebrates could be combined into functional sets of shredders, predators, and decomposers, and terrestrial lichens can be grouped by growth form (crustose, foliose, fruticose) and substrate association (rock, tree bark, mineral soil, etc.). Some researchers and resource managers have used functional groups to include little-known species with others. Such approaches have commonly been used with vascular plants (Smith et al. 1993; Körner 1994), such as using functional plant groups of invasive species (Ramovs and Roberts 2005). By grouping species, management activities could focus on the habitats, substrates, and other attributes of species groups as an initial "coarse filter" step toward conservation strategies.

Still, by definition of little-known species, much will remain unknown and may require further study or testing of effects of management activities. What can the resource manager do in the face of such uncertainties? The most obvious, and possibly least tenable, solution is to cease all adverse, anthropogenic, environment-disturbing activities until further data can be gathered. This has the greatest chance for ensuring successful conservation of little-known species (presuming the species is not actually dependent upon anthropogenic disturbances), but this is seldom feasible or socially desirable. Such was the case with Survey and Manage species under the Northwest Forest Plan cited earlier; timber harvest and other forest management activities needed to proceed in light of a dearth of information on many species.

Short of conducting rigorous inventories and studies, the resource manager could conduct a more immediate threat assessment on individual species or on a species group. This could help identify key stressors or threats to conservation of the species, or at least key areas of uncertainty and scientific unknowns. The resource manager would then be faced with decision making under such uncertainty and choosing a risk attitude (whether one is risk averse, risk neutral, or risk seeking) toward potential effects on the species of interest from environment-disturbing activities. At this point, the resource manager could use well-established methods of decision analysis and risk management to document knowns and unknowns and their decision criteria and procedures (see chaps. 6 and 7).

The Survey and Manage program discussed earlier used an involved process of identifying little-known species based on syntheses of ecological knowledge, and then rated each species, through panels of biologists and resource managers, to determine their appropriate conservation category. Conservation categories reflected degree of rarity, levels of persistence concern, and required survey activities. The program also conducted a thorough information needs analysis for each species (i.e., what critical information was needed to improve managerial success for maintaining the species in the plan area). Following the process to prioritize species and management needs noted previously (see box 4.1), the program strategically designed surveys and research studies to gather essential information, at times using multiple-species approaches for efficiency. Two examples include a regionwide, plot-based, random grid survey (see Molina et al. 2003, 2006 for details) to improve understanding of rarity and distribution for over 200 species of little-known fungi, lichens, bryophytes, and mollusks, and field research on effects of prescribed fire and thinning on soil arthropod guilds and communities (Niwa and Peck 2002; Peck and Niwa 2005).

Another approach that could be used in tandem with threat assessments and risk management is to use better-known surrogates for species conservation. Such surrogates could include macrovegetation conditions at a broad scale and presence of other indicator species or their environmental conditions. However, as is explored in chapter 6, use of surrogates or indicators typically carries high uncertainty as to how effectively they conserve specific little-known species.

For invertebrates, another approach is use of morphospecies groups as units for conservation (Krell 2004). Morphospecies are groups of organisms (and species) that have similar appearances (morphologies) and that occur in the same location, vegetation condition, or substrate type; each morphospecies generally represents multiple taxonomic species. An example is the set of all large, black, terrestrial ants associated with decaying wood on the forest floor. A parataxonomist—typically a biologist with basic training in identifying characteristics of organisms—sorts specimens according to their common features. This approach can be useful when expertise or base scientific information is lacking to identify each taxon to the species level. For example, Barratt et al. (2003) successfully used student researchers to separate coleopteran beetles into morphospecies groups in New Zealand. The students were able to identify a total number of morphospecies within about 10% of the actual number as identified by a taxonomic expert. In another example, Longino and Colwell (1997) successfully used parataxonomists to identify and prepare specimens of ant morphospecies in a Costa Rican rainforest. Derraik et al. (2002) found that parataxonomists varied in their accuracy in identifying arthropod morphospecies groups, but initial training by expert taxonomists would likely improve results. Of course, the morphospecies approach cannot replace basic work in taxonomy, but it can be helpful to locate little-known and even undescribed organisms and to estimate overall species richness.

Other species and systems approaches to conservation of little-known species are discussed and critiqued in chapter 6. In the end, however, no model, indicator, surrogate, or grouping approach can fully substitute for knowledge gained on little-known species from basic field biology and autecology.

References

Andrén, O., J. Bengtsson, and M. Clarholm. 1995. Biodiversity and species redundancy among litter decomposers. Pp. 141–51 in *The significance and regulation of soil biodiversity*, ed. H. P. Collins, G. P. Robertson, and M. J. Klug. Dordrecht: Kluwer.

Arora, D. 1986. *Mushrooms demystified*. Berkeley: Ten Speed Press.

Balick, M. J., R. Arvigo, and L. Romero. 1994. The development of an ethnobiomedical forest reserve in Belize: Its role in the preservation of biological and cultural diversity. *Conservation Biology* 8:316–17.

Barratt, B. I. P., J. G. B. Derraik, C. G. Rufaut, A. J. Goodman, and K. J. M. Dickinson.

2003. Morphospecies as a substitute for Coleoptera species identification, and the value of experience in improving accuracy. *Journal of the Royal Society of New Zealand* 33:583–90.
Bengston, D. N. 2004. Listening to neglected voices: American Indian perspectives on natural resource management. *Journal of Forestry* 102:48–52.
Blackwell, M., and J. W. Spatafora. 2004. Fungi and their allies. Pp. 7–21 in *Biodiversity of Fungi: Standard Methods for Inventory and Monitoring*, ed. G. M. Mueller, G. F. Bills, and M. Foster. New York: Academic.
Borgelt, A., and T. R. New. 2005. Pitfall trapping for ants (Hymenoptera, Formicidae) in mesic Australia: The influence of trap diameter. *Journal of Insect Conservation* 9:219–21.
Buchmann, S. L., and G. P. Nabhan. 1996. *The forgotten pollinators*. Covelo, CA: Island Press.
Cane, J. H., and V. J. Tepedino. 2001. Causes and extent of declines among native North American invertebrate pollinators: Detection, evidence, and consequences. *Conservation Ecology* 5 (1): 1. http://www.consecol.org/vol5/iss1/art1.
Cao, Y., D. P. Larsen, and R. S.-J. Thorne. 2001. Rare species in multivariate analysis for bioassessment: Some considerations. *Journal of the North American Benthological Society* 20:144–53.
Chambers, S. M., and J. W. Bayless. 1983. Systematics, conservation and the measurement of genetic diversity. Pp. 349–63 in *Genetics and conservation: a reference for managing wild animal and plant populations*, ed. C. M. Schonewald-Cox, S. M. Chambers, B. MacBryde, and L. Thomas. Menlo Park, CA: Benjamin/Cummings.
Derraik, J. G. B., G. P. Closs, K. J. M. Dickinson, P. Sirvid, B. I. P. Barratt, and B. H. Patrick. 2002. Arthropod morphospecies versus taxonomic species: A case study with Araneae, Coleoptera, and Lepidoptera. *Conservation Biology* 16:1015–23.
Dodd, C. K., Jr. 2003. *Monitoring amphibians in Great Smoky Mountains National Park*. U.S. Geological Survey Circular 1258. Tallahassee: U.S. Department of the Interior, U.S. Geological Survey.
http://water.usgs.gov/pubs/circ/2003/circ1258/#pdf.
Ducarme, X., G. Wauthy, H. M. Andre, and P. Lebrun. 2004. Survey of mites in caves and deep soil and evolution of mites in these habitats. *Canadian Journal of Zoology* 82:841–50.
Dunham, S. M., A. Kretzer, and M. E. Pfrender. 2003. Characterization of the extent and distribution of Pacific golden chanterelle (*Cantharellus formosus*) individuals using codominant microsatellite loci. *Molecular Ecology* 12:1607–18.
Dunham, S. M., T. E. O'Dell, and R. Molina. 2006. Spatial analysis of within-population microsatellite variability reveals restricted gene flow in the Pacific golden chanterelle (*Cantharellus formosus*). *Mycologia* 98:250–59.
Dunn, R. R. 2005. Modern insect extinctions, the neglected majority. *Conservation Biology* 19:1030–36.
Hansen, J. E., and T. R. New. 2005. Use of barrier pitfall traps to enhance inventory surveys of epigaeic Coleoptera. *Journal of Insect Conservation* 9:131–36.
Hawksworth, D. L., ed. 1991. *The biodiversity of microorganisms and invertebrates, and its role in sustainable agriculture*. Wallingford, UK: CABI Publishing
Hawksworth, D. L. 2001. The magnitude of fungal diversity: The 1.5 million species estimate revisited. *Mycological Research* 105:1422–32.
Huber, J. T., and D. W. Langor. 2004. Systematics: Its role in supporting sustainable forest management. *Forestry Chronicle* 80:451–57.

Körner, C. 1994. Scaling from species to vegetation: The usefulness of functional groups. Pp. 117–40 in *Biodiversity and ecosystem function*, ed. E. D. Schulze and H. A. Mooney. New York: Springer-Verlag.

Kotter, M. M., and R. C. Farentinos. 1984. Tassel-eared squirrels as spore dispersal agents of hypogeous mycorrhizal fungi. *Journal of Mammalogy* 65:684–87.

Krell, F.-T. 2004. Parataxonomy vs. taxonomy in biodiversity studies: Pitfalls and applicability of "morphospecies" sorting. *Biodiversity and Conservation* 13:795–812.

Kretzer, A. M., S. Dunham, R. Molina, and J. W. Spatafora. 2003. Microsatellite markers reveal the below ground distribution of genets in two species of *Rhizopogon* forming tuberculate ectomycorrhizas on Douglas fir. *New Phytologist* 161:313–20.

Longino, J. T., and R. K. Colwell. 1997. Biodiversity assessment using structured inventory: Capturing the ant fauna of a tropical rain forest. *Ecological Applications* 7:1263–77.

Luoma, D. L., R. E. Frenkel, and J. M. Trappe. 1991. Fruiting of hypogeous fungi in Oregon Douglas-fir forests: Seasonal and habitat variation. *Mycologia* 83:335–53.

Lyons, K. G., and M. W. Schwartz. 2001. Rare species loss alters ecosystem function: Invasion resistance. *Ecology Letters* 4:358–65.

Marcot, B. G., L. K. Croft, J. F. Lehmkuhl, R. H. Naney, C. G. Niwa, W. R. Owen, and R. E. Sandquist. 1998. *Macroecology, paleoecology, and ecological integrity of terrestrial species and communities of the interior Columbia River Basin and portions of the Klamath and Great Basins*. General Technical Report PNW-GTR-410. Portland, OR: U.S. Department of Agriculture, Forest Service, Pacific Northwest Research Station.

Marcot, B. G., M. A. Castellano, J. A. Christy, L. K. Croft, J. F. Lehmkuhl, R. H. Naney, K. Nelson, C. G. Niwa, R. E. Rosentreter, R. E. Sandquist, B. C. Wales, and E. Zieroth. 1997. Terrestrial ecology assessment. Pp. 1497-1713 in: T. M. Quigley and S. J. Arbelbide, eds. An assessment of ecosystem components in the interior Columbia Basin and portions of the Klamath and Great Basins. Volume III. USDA Forest Service General Technical Report PNW-GTR-405. USDA Forest Service, Pacific Northwest Research Station, Portland, OR. 1713 pp.

Marcot, B. G., and R. Molina. 2006. Conservation of other species associated with older forest conditions. Pp. 145–80 in *Northwest Forest Plan—the first ten years (1994–2003): A synthesis of monitoring and research results*, ed. R. W. Haynes, B. T. Bormann, D. C. Lee, and J. R. Martin. General Technical Report PNW-GTR-651. Portland, OR: U. S. Department of Agriculture, Forest Service, Pacific Northwest Research Station.

Martikainen, P., and J. Kouki. 2003. Sampling the rarest: Threatened beetles in boreal forest biodiversity inventories. *Biodiversity and Conservation* 12:1815–31.

McCune, B., and L. Geiser. 1997. *Macrolichens of the Pacific Northwest*. Corvallis: Oregon State University Press.

Meslow, E. C., R. S. Holthausen, and D. A. Cleaves. 1994. Assessment of terrestrial species and ecosystems. *Journal of Forestry* 4:24–27.

Moffatt, C., S. McNeill, and A. J. Morton. 2004. Invertebrate community sampling of woodland field layers: Trials of two techniques involving enclosures. *Journal of Insect Conservation* 7:233–45.

Molina, R., H. Massicotte, and J. M. Trappe. 1992. Specificity phenomena in mycorrhizal symbioses: Community-ecological consequences and practical implications. Pp. 357–423 in *Mycorrhizal functioning: An integrative plant–fungal process*, ed. M. F. Allen. New York: Chapman and Hall.

Molina, R., D. Pilz, J. Smith, S. Dunham, T. Dreisbach, T. O'Dell, and M. Castellano. 2001. Conservation and management of forest fungi in the Pacific Northwestern United States: An integrated ecosystem approach. Pp. 19–63 in *Fungal Conservation: Issues and Solutions*, ed., D. Moore, M. M. Nauta, S. E. Evans, and M. Rotheroe. Cambridge: Cambridge University Press.

Molina, R., D. McKenzie, R. Lesher, J. Ford, J. Alegria, and R. Cutler. 2003. Strategic survey framework for the Northwest Forest Plan Survey and Manage Program. General Technical Report PNW-GTR-573. Portland, OR: U.S. Department of Agriculture, Forest Service, Pacific Northwest Research Station.

Molina, R., B. G. Marcot, and R. Lesher. 2006. Protecting rare, old-growth forest-associated species under the Survey and Manage Program guidelines of the Northwest Forest Plan. *Conservation Biology* 20:306–18.

Morrison, M. L., B. G. Marcot, and R. W. Mannan. 1998. *Wildlife–habitat relationships: Concepts and applications*. 2nd ed. Madison: University of Wisconsin Press.

Mueller, G. M., G. Bills, and M. S. Foster. 2004. *Biodiversity of fungi: Inventory and monitoring methods*. San Diego: Elsevier Academic Press.

Niwa, C. G., and R. W. Peck. 2002. Influence of prescribed fire on carabid beetle (Carabidae) and spider (Araneae) assemblages in forest litter in southwestern Oregon. *Environmental Entomology* 31:785–96.

O'Dell, T. E., J. F. Ammirati, and E. G. Schreiner. 1999. Species richness and abundance of ectomycorrhizal basidiomycete sporocarps on a moisture gradient in the *Tsuga heterophylla* zone. *Canadian Journal of Botany* 77:1699–1711.

Ostfeld, R. S., and K. LoGiudice. 2003. Community disassembly, biodiversity loss, and the erosion of an ecosystem service. *Ecology* 84:1421–27.

Parsons, G. L., G. Cassis, A. R. Moldenke, J. D. Lattin, N. H. Anderson, J. C. Miller, P. Hammond, and T. D. Schowalter. 1991. Invertebrates of the H. J. Andrews Experimental Forest, Western Cascade Range, Oregon, V: An annotated list of insects and other arthropods. General Technical Report PNW-GTR-290. Portland, OR: U.S. Department of Agriculture, Forest Service, Pacific Northwest Research Station.

Peck, R. W., and C. G. Niwa. 2005. Longer-term effects of selective thinning on microarthropod communities in a late-successional coniferous forest. *Environmental Entomology* 34:646–55.

Phillips, O., A. H. Gentry, C. Reynel, P. Wilkin, and C. Galvez-Durand B. 1994. Quantitative ethnobotany and Amazonian conservation. *Conservation Biology* 8:225–48.

Ramovs, B. V., and M. R. Roberts. 2005. Response of plant functional groups within plantations and naturally regenerated forests in southern New Brunswick, Canada. *Canadian Journal of Forest Research* 35:1261–76.

Raphael, M. G., and B. G. Marcot. 1994. Species and ecosystem viability: Key questions and issues. *Journal of Forestry* 92:45–47.

Shanley, P., and L. Luz. 2003. The impacts of forest degradation on medicinal plant use and implications for health care in eastern Amazonia. *BioScience* 53:573–84.

Shepherd, M., S. L. Buchmann, M. Vaughan, and S. H. Black. 2003. *Pollinator conservation handbook*. Portland, OR: Xerces Society.

Skerl, K. L. 1999. Spiders in conservation planning: A survey of U.S. natural heritage programs. *Journal of Insect Conservation* 3:341–47.

Smith, J. E., R. Molina, M. M. P. Huso, D. L. Luoma, D. Mckay, M. A. Castellano, T. Lebel, and Y. Valachovic. 2002. Species richness, abundance, and composition of hypogeous and epigeous ectomycorrhizal fungal sporocarps in young, rotation-age, and old-growth stands of Douglas-fir (*Pseudotsuga menziesii*) in the Cascade Range of Oregon, U.S.A. *Canadian Journal of Botany* 80:186–204.

Smith, T. M., H. H. Shugart, F. I. Woodward, and P. J. Burton. 1993. Plant functional types. Pp. 272–92 in *Vegetation dynamics and global change*, ed. S. M. Solomon and H. H. Shugart. New York: Chapman and Hall.

Spira, T. P. 2001. Plant–pollinator interactions: A threatened mutualism with implications for the ecology and management of rare plants. *Natural Areas Journal* 21:78–88.

Stork, N. E. 1997. Measuring global biodiversity and its decline. Pp. 41–68 in *Biodiversity II: Understanding and protecting our biological resources*, ed. M. L. Reaka-Kudla, D. E. Wilson, and E. O. Wilson. Washington, DC: Joseph Henry Press.

USDA (U.S. Department of Agriculture), Forest Service; USDI (U.S. Department of the Interior), Bureau of Land Management. 1994. Record of decision for amendments to Forest Service and Bureau of Land Management planning documents within the range of the northern spotted owl. Portland, OR: U.S. Department of Agriculture, Forest Service, and U.S. Department of Interior, Bureau of Land Management.

———USDA (U.S. Department of Agriculture), Forest Service; USDI (U.S. Department of the Interior), Bureau of Land Management. 2001. Record of decision and standards and guidelines for amendments to the survey and manage, protection buffer, and other mitigations measures standards and guidelines. Portland, OR: U.S. Department of Agriculture, Forest Service, and U.S. Department of Interior, Bureau of Land Management.

———. 2004. Record of decision to remove or modify the survey and manage mitigation measures standards and guidelines. Portland, OR: U.S. Department of Agriculture, Forest Service and U.S. Department of Interior, Bureau of Land Management.

Vitt, D., J. Marsh, and R. Bovey. 1988. *Mosses, lichens and ferns of northwest North America*. Edmonton, Alberta: Lone Pine Publishing.

Wayne, R. K., and P. A. Morin. 2004. Conservation genetics in the new molecular age. *Frontiers in Ecology and the Environment* 2:89–97.

Welsh, H. H., Jr., and S. Droege. 2001. A case for using plethodontid salamanders for monitoring biodiversity and ecosystem integrity of North American forests. *Conservation Biology* 15:558–69.

Wilson, E. O. 1988. The current state of biological diversity. Pp. 3–18 in *Biodiversity*, ed. E. O. Wilson. Washington, DC: National Academy Press.

5

Special Considerations for the Science, Conservation, and Management of Rare or Little-Known Species

Bruce G. Marcot and Randy Molina

Rare or little-known (RLK) species pose special problems to conservation management. It is difficult to detect and study rare species, and they do not lend themselves to experimentation (see chap. 2). Likewise, locations, numbers, and responses of species that are little known are difficult to impossible to predict (see chap. 4). This chapter addresses special considerations for the science, conservation, and management of RLK species, and also the further joint problem of species that may be both rare and little known, which constitutes the conditions of many inconspicuous taxa that contribute to the bulk of biological diversity.

Recall the Venn diagram in figure 1.1 (chap. 1) which depicts the overlap of the sets of species that are rare and those that are little known. We can expand on this overlap by suggesting three categories of each condition, as shown in table 5.1. Rare species (chap. 3) are those that (1) are scarce everywhere and have low density throughout their distributional range, even if total numbers are high; (2) have a small total population, which means that total global numbers are low, whether locally scarce or locally common; or (3) are genetically or behaviorally distinctive, including those that occur locally as a peripheral population, which means they occur on the edge of their overall distributional range, where local peripheral number or density is low even if the total population elsewhere is not (fig. 5.1). Little-known species are those for which we have (1) an incomplete taxonomic description, so that there is uncertainty over their taxo-

Table 5.1. *Categories and examples of rare and little-known species*

		Category of Rare		
		Scarce everywhere	Small total population	Distinctive or peripheral population
Category of Little-Known	Incomplete taxonomic description	Speckled chub, *Macrhybopsis aestivalis* species complex	Brown kiwi, *Apteryx australis*	White salmon pocket gopher, *Thomomys talpoides limosus*
	Limited species inventory	Sharp-tailed snake, *Contia tenuis*	Fuertes's parrot, *Hapalopsittaca fuertesi*	Bonobo, *Pan paniscus,* near Lac Tumba, western DR Congo
	Poor ecological understanding	Club mushroom, *Podostroma alutaceum*	Chevron skink, *Oligosoma homalonotum*	Dalles side-band snail, *Monadenia fidelis minor*

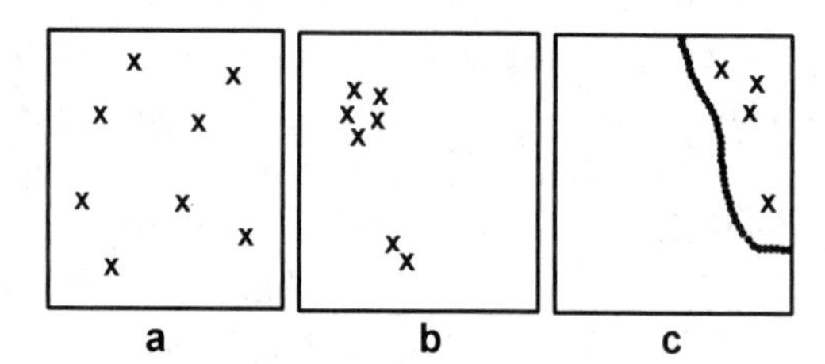

Figure 5.1. Three kinds of species rarity: (a) scarce everywhere, with low density throughout the range; (b) small total population, whether locally scarce or locally common; and (c) genetically or behaviorally distinctive, including peripheral populations on the edge of their distribution range (dotted line). X = occurrence of the species.

nomic status; (2) a limited species inventory, so that we have poor understanding of their basic occurrence and distribution; or (3) a poor ecological understanding, so that we do not know what environmental factors determine their occurrence or distribution.

Some examples illustrate the problems of species that are rare or little known in these various ways (see table 5.1). Among species that are probably scarce everywhere, one species for which we have incomplete taxonomic description is the rare fish speckled chub (*Macrhybopsis aestivalis*). This fish was previously considered a wide-ranging, polytypic species with six subspecies, but genetic and morphometric work has suggested that it may contain at least 10 distinct species, some of which are likely imperiled (Butler and Mayden 2003; Warren et al. 2000). In such cases, the distinct species may warrant separate attention and further inventory to determine their abundance and distribution.

Sharp-tailed snake (*Contia tenuis*), a small colubrid of western North America with limited inventory data (e.g., Leonard and Leonard 1998), has been considered scarce everywhere (but see further discussion later in chapter). Although it is well established taxonomically, its known habit of residing in well decayed downed logs makes it somewhat secretive and difficult to find, and thus it is difficult to determine its true abundance and population trend.

Podostroma alutaceum, a club mushroom, is likely scarce everywhere and is a species for which we have poor ecological understanding. This fungus is widely distributed in coastal and Cascade Range forests from northern Washington south to the San Francisco Bay area but is nearly always solitary in occurrence. Lack of understanding of its ecology means that it cannot be reliably predicted from habitat maps and environmental data.

Continuing with examples, among species with a total small population, one species for which we have an incomplete taxonomic understanding is a New Zealand endemic bird, the brown kiwi (*Apteryx australis*). Although three subspecies of brown kiwi are currently recognized (*A. a. mantelli* on North Island, *A. a. australis* on South Island, and *A. a. lawryi* on Stewart Island), recent genetic work suggests that the species should be split into two distinct species, the brown kiwi with distinct varieties occurring on North Island and near Okarito on South Island, and the tokoeka with distinct varieties near Haast and in Fiordland on South Island and on Stewart Island. The splitting of these species means that the total population of each is smaller than previously thought, which may mean that each has lower probabilities of persistence throughout a smaller range. Smaller populations also mean potentially greater harm to donor populations should removal be done for transplantation or captive breeding.

An example of a species with total small population and adequate taxonomic description but with limited inventory is Fuertes's parrot (*Hapalopsittaca fuertesi*), which, in 2002, was rediscovered after 90 years in the high Andes Mountains of Colombia. The parrot is known to associate with tall mature trees and to feed on berries in epiphyte-laden canopy branches, but surveys are needed to determine if and where more than just the small isolated population still exists. If removal of any of the parrots for captive breeding is to be done, lack of inventory data means great uncertainty as to the degree of potential adverse impact on the donor population. In a similar case, such inventories were also needed, and conducted, on the endangered Puerto Rican parrot before some individuals were removed for captive breeding (Vilella and Garcia 1995). However, in this case, dividing an already tiny wild population may lead to future problems of genetic bottlenecks and loss of genetic diversity (Wilson et al. 1994).

An example of a species with a small overall population for which we have poor ecological understanding is the chevron skink (*Oligosoma homalonotum*), listed by the government of New Zealand as a vulnerable species. Lost for more than 60 years, the chevron skink was first described in 1906, reported again in the 1970s, and rediscovered as a single subadult in 1991 on Little Barrier Island. Less than 250 sightings are known since 1906. It also currently occurs on Great Barrier Island but the apparent 60-year hiatus there was likely due to misidentification in museum records. Although locations and taxonomy are known, and a habitat study was conducted on the species during 1997–2002, little is known of its specific ecology. It was found to inhabit streamsides and damp places, and is likely vulnerable to predation by introduced Norway rats, but its specific habitat requirements are still very poorly known because the lizard is secretive and cryptic. Behavioral studies of a small captive population are ongoing.

Among species with a peripheral population, one taxon for which we have an incomplete taxonomic understanding is the white salmon pocket gopher (also called Columbia River pocket gopher; *Thomomys talpoides limosus*), a rare and locally endemic subspecies of the northern pocket gopher that occurs on the edge of its parent species range in southern Washington. According to the International Union for Conservation of Natural Resources (IUCN 2006), the taxonomy of *Thomomys talpoides* is still uncertain and may consist of several sibling species in the U.S. Pacific Northwest, possibly including the white salmon pocket gopher. The management implication is that, if these incipient sibling species are indeed elevated to species status, there may be conservation interests (or mandates) for some or each of them individually, including the white salmon pocket gopher.

An example of a species occurring as a rare peripheral population for which we have limited inventory is the bonobo or pygmy chimpanzee (*Pan paniscus*) near Lake Tumba in the western Democratic Republic of Congo. Bonobos have only recently been discovered in the old swamp forests of this area, are low in numbers and density there, and are clearly on the edge of their range. Inventories are being conducted under Bonobo Conservation International, but their local distribution, numbers, and population structure are still incompletely known. Discovery of local bonobo troops could mean changes in local forest use practices, such as potential designation of wildlife refuges or development of ecotourism centers from which to visit and view the animals.

And finally, a taxon that occurs as a peripheral population for which we have poor ecological understanding is the locally endemic Dalles sideband snail (*Monadenia fidelis minor*), a terrestrial mollusk found on the east side of Mount Hood in the Cascade Mountains and up the Columbia River Gorge east of the Cascade Mountains crest of northern Oregon. Its parent species is broadly distributed from Canada to California, but the specific ecology of this subspecies is poorly known. Some species, particularly more sessile ones such as mollusks, tend to develop unique habitat and environmental associations on their range peripheries; in some cases this may lead to speciation. Thus understanding the ecology of this species at its peripheral locations and comparing that to elsewhere in its range may be useful for determining whether more locally stringent or different sets of management guidelines might be needed for its overall conservation.

These examples of RLK species illustrate a wide array of conditions, taxa, and challenges for research and conservation. However, some of the attributes they do share include their inconspicuous nature and uncertain taxonomy or ecology, and their scarcity in part or all of their range. We next consider some of the conservation implications of being rare and little known.

Conservation Considerations

Several issues rise when considering conservation of RLK species: issues of scientific value and classification of such species, identification of species and taxa for conservation management, and setting conservation goals and objectives.

Science Issues

A scientific approach to identifying and conserving RLK species could first use the categories within "rare" and "little known" already discussed (see table 5.1). This would help set the stage for identifying what kind of further information (taxonomic, inventory, or ecological) is most critical, and how conditions contributing to rarity would guide conservation.

Further science issues of RLK species pertain to evaluating their persistence and their ecology. Persistence of species that are well known demographically or genetically can be modeled and assessed using population viability analysis (PVA). PVA consists of estimating persistence probabilities (or time to extinction), using quantitative calculations and models of effective population size, influence of genetic bottlenecks, and rate of change in population size. However, with species that are little known, a PVA approach would be fraught with uncertainty, so, at best, a more qualitative population viability evaluation (PVE) may be more in order. A PVE may consist of qualitative or categorical ranking of potential threats and the level of vulnerability of a species to environmental conditions and human activities, and not depend on genetic or demographic data on the population. PVEs have been conducted using expert panels and used to evaluate effects of alternative land management scenarios on many species for which demographic and genetic data are lacking (e.g., Raphael et al. 2001).

Many RLK species may play important ecological roles in their ecosystems (see chap. 4). To the extent possible, the species could be treated in functional groups (see chap. 7) such as soil arthropod decomposers, or in substrate groups such as fungi associated with decaying wood, even if knowledge of individual species is poor. Other species and system-level approaches to representing or indicating RLK species are discussed in chapters 6 and 7. Other science issues of RLK species also pertain to problems of sampling, detectability, and modeling, and depicting and dealing with uncertainty in management (see chap. 12).

Identifying Taxa for Conservation Management

Selecting taxa on which to focus conservation management activities is fundamental to developing and implementing an overall species conservation program. Species conservation has traditionally focused on a limited number of often well-studied vertebrates, plants, and showy arthropods (e.g., butterflies). Many of these species are referred to as charismatic, reflecting a societal and science bias toward certain taxa (Cassidy and Grue 2001). RLK species can also include a diverse assemblage of taxa less familiar and less charismatic to society and conservation science in general. Limitations of the charismatic species approach to RLK species conservation are discussed further in chapter 6.

Conservation Goals and Objectives

Chapter 2 outlined some goals and objectives typical of conservation planning. Careful consideration and clear statements of goals and objectives are critical because they will determine how and what information is collected, analyzed, and used in decision making.

As noted in previous sections, the initial choice of species and the criteria used to designate them are important scientific and programmatic considerations for setting goals and objectives. Once species are designated, selecting the science-based metrics to gauge species viability and population persistence also becomes a critical consideration for setting conservation objectives. Maintaining "viability" typically relies on use of population viability metrics (viz., population demographic and genetic data) as used in classic population viability analyses (Shaffer 1990). Given the potential number of RLK species, the difficulties of obtaining such information, and ultimately the expense of such an undertaking, it may not be realistic to use such viability metrics.

For example, to protect rare, late-seral forest species on federal public lands in the Pacific Northwest, the Northwest Forest Plan (as described in the FEMAT [1993] analysis) set and evaluated viability objectives within a context of species distribution in reserve and matrix land allocations, and effects of other management guidelines. Initial viability assessments centered on "understanding how provision of habitat on federal lands under each [management] option could contribute to population persistence and distribution over a century" (FEMAT 1993, II-99). Provision of habitat included the amount, quantity, and distribution of habitat. Meeting this objective required combining guidelines for conservation of individual target species with guidelines for geographically based conservation (viz., reserved areas) as later implemented in the Survey and Manage program under the Northwest Forest Plan of the Pacific Northwest as discussed in chapter 4.

The geographic scale at which conservation planning decisions are made also has a major effect on meeting goals and objectives for conserving RLK species (Kintsch and Urban 2002; Schwartz 1999). Monitoring and maintaining the persistence of RLK species at a local (e.g., federal Forest Service district, approximately 100,000 to 200,000 acres) planning level (as in many Forest Service and Bureau of Land Management [BLM] sensitive species programs) differs from maintaining persistence at a large regional scale that might include many local planning units. At the local scale, there is more intimate knowledge that contributes to regional planning, but local resource specialists can have a hard time seeing how they fit into the big picture. That is, resource specialists are most familiar with the planned use of their managed land base, as well as species distribution and location of specific habitat types therein; given that familiarity, conservation strategies can be integrated with local management plans. When regional-scale conservation objectives are chosen, a high degree of regional planning and oversight is needed to both collect needed information and develop management plans that can cross multiownership boundaries. Communicating results and decisions of a regional plan to field units or different landowners and coordinating and monitoring the conservation program becomes a complex task that requires strong interagency or multiowner cooperation.

For example, conducting surveys to determine distribution, abundance, and vegetation associations of RLK species across a broad regional area may entail establishing a stratified random or multistage sampling design (Cutler et al. 2001; Philippi et al. 2001), such that the aggregate of local samples provides a desired level of confidence or power in determining presence and association with specific vegetation conditions at the wider scale of an ecoprovince. Individual administrative units, such as management districts of national forests, may play critical roles in gathering such information even though the statistical inference is to be made to an area broader than their individual administrative boundaries. It is critical that participants understand the broader-scale context and value of such efforts that may, otherwise, seem expensive and pointless at the scale of an individual resource management project or field administrative unit. It is statistically inappropriate to draw inferences to finer scales from coarse-scale data (Plotkin and Muller-Landau 2002). There may also be opportunities to establish multiscale sampling designs and modeling schemes that provide estimates of species distribution and abundance at given confidence levels and at two or more geographic scales or levels of spatial resolution (Borcard et al. 2004).

In fact, determining the appropriate scale for setting conservation goals and objectives; for designing inventories, surveys, and studies; and for developing and applying species–habitat relationships models of rare and little-known species can be a vital early step in the process. Remember that "scale" means several things: geographic extent, spatial resolution, and level of detail on maps and in geographic information systems (GIS) and other databases. Appropriate scales should be determined by (1) clearly identifying conservation issues, including environmental conditions and species (e.g., preservation of an old-growth stand, restoration of a watershed, conservation of overall biodiversity, protection of specific RLK species, etc.); (2) evaluating available data on distribution of environmental conditions and species of conservation interest, to determine their geographic extent and spatial resolution; and (3) consulting any pertinent models or experts to provide further understanding of geographic locations, areal extent, and spatial resolution pertinent to the species' population distribution and potential dispersal and movement patterns. The results can be used to guide over what area, in what specific locations, and at what level of resolution to conduct a set of guidelines for species conservation, inventory, survey, and study or to apply a species prediction model.

Combining conservation approaches to achieve management objectives is covered in detail in chapters 8 and 12; chapter 11 provides details on programmatic implementation considerations.

Sampling Considerations

Detecting and sampling rare species entails special statistical considerations (Thompson 2004). It may be possible to estimate the total number of species in a community by using methods of compiling species accumulation curves (Shen et al. 2003) and rarefaction and related analyses (e.g., Haddad et al. 2001; Mao and Colwell 2005). However, although these curves describe total species richness, they may not provide reliable information on individual RLK species.

It may not be critical (or possible) to survey or monitor all RLK species in a community. One reason is that many such species may be very difficult to detect; that is, the probability of their appearing in a survey or inventory sample given that they are actually present may be very low. Once a threat assessment (see chap. 4) helps determine which RLK species to study, one of the more difficult challenges in conserving the species is developing statistically valid sampling schemes to effectively deal with the detectability issues. Such sampling schemes may entail the use of particular statistical distributions to guide the number and dispersion of samples in the field, depending on the expected density and dispersion of the rare species being sampled. Different statistical distributions have different properties, and some are better suited to help ensure that rare species are adequately included in samples at specified levels of confidence.

For example, Venette et al. (2002) suggested use of the hypergeometric, binomial, and beta-binomial distributions for sampling rare invertebrates. Green and Young (1993) suggested that adequate sampling effort to detect rare species could be determined by use of the Poisson and negative binomial distributions, particularly if the organisms are not spatially aggregated. However, with many species, one form of rarity (see table 5.1) is indeed local aggregation with overall total low population numbers. In such cases, other authors have suggested use of various adaptive sampling strategies (papers in Thompson 2004) such as adaptive cluster sampling (Philippi 2005), which begins at known concentration centers or locations of a species and then extends adjacent samples from those points.

Cutler et al. (2001) described sampling RLK species as a problem of representing rare events in space and time. Because of the many unique attributes of RLK species, they stated, "these characteristics provide unique statistical challenges to the design of surveys to provide quantifiable estimates about the species." They prescribed a three-pronged approach to information gathering: (1) conduct coarse-grained distribution surveys to determine the degree of rarity and habitat association of the species and to determine their probability of occurrence and their habitat associations; (2) conduct midscale (e.g., watershed) distribution surveys to model probability of occurrence and ecological association; and (3) gather persistence-related information, at the scale of the population, to determine parameters such as local population density and, as possible, demographic vital rates (survivorship and mortality rates). They described a suite of sampling designs, several of which have been implemented in the Survey and Manage program (see Cutler et al. 2001; Molina et al. 2003; Molina et al. 2006; and chap. 4 for details) and emphasized three primary considerations in design selection: (1) cost effectiveness, (2) ability to quantify the uncertainty in the estimates of population characteristics, and (3) flexibility in making changes to the designs to obtain different types of needed information.

Determining appropriate sampling frames and sample sizes for RLK species means knowing something about two parameters: the probability P that the organism is present at a site, and the probability Q of detecting the organism given that it is present (MacKenzie et al. 2003). That is, if D represents a detection event, then $E(D|\bar{P}) = 0$, $E(D|P) = Q$, $E(\bar{D}|P) = 1 - Q$, and where E = expected value or probability (and the bar represents "not"). Parameter Q is generally referred to as "detectability," and determining this probability is not easy. For example, Royle (2006) noted that estimating the probability of occurrence of a species can be confounded by detectability itself and may vary due to variation in the species abundance (especially when using methods where user experience, learning, and search image are important). Still, estimating detectability is important because false conclusions of species absence may lead to habitat-disturbing activities that would locally reduce or extirpate the organism, adding to overall threat levels. Thus estimates of detectability should be considered when estimating overall abundance, local presence, or viability of RLK organisms (MacKenzie and Kendall 2002; Royle et al. 2005; see also Beavers and Ramsey 1998; Jenouvrier and Boulinier 2006).

For example, in the Pacific Northwest, preliminary results from the Survey and Manage program's regionwide cross-taxonomic surveys of RLK species, using a stratified random sampling grid, have shown that some species are so rare and spotty in distribution that the numbers of encounters have been too low to reliably extrapolate numbers throughout the region (Molina et al. 2006). However, for some of the less rare species, random grid designs at the plan scale have been effective in finding new locations of individuals (Edwards et al. 2004), although a strict analysis of detectability has not been conducted.

A more broad-brush approach to inventory of RLK species may entail use of rapid survey and ecological assessment methods (Sayre et al. 2000), such as Oliver and Beattie (1996) and Jones and Eggleton (2000) used to inventory invertebrates. Such methods typically entail blanketing a study area with several taxonomic specialists who intensively collect as many specimens or sightings of organisms over a short time period. The samples usually include many (but seldom all) RLK species; the approach will often miss organisms that may be present but that are not detectable during the short time period of the survey. The purpose is to sample and determine at least the presence of individual species, not just to build species accumulation curves and estimate total species richness. The advantage of rapid survey methods is the collection of a relatively large, representative set of organisms from a community in a short time; the disadvantages are that some rare species are still missed in the samples and the statistical properties of the samples may be difficult to assess.

Some species that are secretive or that occur in locations difficult to study may be known from just a few locations, and some of these species might be more abundant than previously suspected. Such seems to be the case with the sharp-tailed snake in the U.S. state of Oregon. The limited inventory previously available on this species suggested that it might be rare, and it is listed as vulnerable by the state of Oregon. Surveys by Hoyer et al. (2006) more recently expanded the known range and number of sightings of this species by almost a factor of 10, suggesting that the secretive nature of this species lent to its being unnecessarily listed as vulnerable. However, because a species is secretive does not necessarily mean that it is more abundant than thought and thus not vulnerable.

Habitat Modeling

One tool that may be useful for assessing and conserving RLK species is that of modeling to predict habitat quality and potential presence of organisms. Species-habitat modeling has a long and rich history in ecology and resource management (e.g., Verner et al. 1986; Scott et al. 2002), but there are special considerations pertaining to RLK species.

Many of the same detectability and sampling issues apply to developing predictive habitat models for RLK species. Failing to account for missing an RLK species during surveys carries similar risks as predicting species absence when the species is actually present. The modeling risks include biasing estimates of site occupancy, colonization, and local extinction probabilities (MacKenzie et al. 2003).

It can be extremely difficult to collect enough habitat information when few sites are available for analysis and when rare species are patchily distributed. Nevertheless, some rare species have been modeled successfully. For instance, Dunk et al. (2004) predicted occurrence of five species of rare mollusks in Northern California, using generalized additive models to estimate each species' distributional range and habitat associations. In the Pacific Northwest, federal land management agencies under the Survey and Manage program (see example later in chapter) have used regionwide potential natural vegetation GIS models (Henderson 2001; Lesher 2005) and Bayesian belief network models (Marcot 2006) to predict Survey and Manage species occurrence and to prioritize landscape and stand-scale surveys.

It is essential to clarify the objectives for modeling RLK species and habitats, including the potential application of the models. One such set of objectives, as used in the Survey and Manage species program, may be to assess quality and spatial distribution of habitat by which to prioritize sites for expensive field surveys for the organisms, as in preproject or predisturbance surveys. In the Survey and Manage example, models of selected RLK species have been developed at two scales: (1) an ecoregional scale using broad topographic and climatic parameters in GIS analyses, by which to map suitability of environments across watersheds (Henderson 2001; Lesher 2005); and (2) a fine scale using within-stand vegetation and substrate characteristics in Bayesian belief network (BBN) models by which to calculate likelihoods of habitat quality classes within the watershed at particular sites (Marcot 2006). These two scales of models can be used in a tiered fashion, first predicting suitable areas broadly at the ecoregional scale, and then within those areas further assessing degree of suitable habitats and substrates within vegetation stands or at particular sites. Alternatively, one could also accrete local model outcomes to model broader-area conditions (Rastetter et al. 1992). Tiering the models may reduce errors of inclusion, that is, avoiding the prediction of suitable environments and potential presence of the species in areas clearly outside the species' range of distribution or range of tolerable conditions.

However, RLK species prediction models may be developed to deliberately err more on the side of commission than omission (presuming that some modeling prediction error is always present), particularly if predicting species absence means that proceeding with ground-disturbing activities would decimate the organisms from the site. The balance between types of errors can be defined according to the real and opportunity costs involved. Predicting high habitat quality or presence of the organism when the habitat is actually low in quality or the organism is absent (a type I error) may unnecessarily trigger expensive and activity-delaying surveys and losing access to valuable resources (e.g., timber). On the other hand, incorrect prediction of low habitat quality or absence of the organism when the habitat is actually high in quality or the organism is present (a type II error) may unnecessarily result in local extirpation of the species.

Models can and should be tested to determine their frequency (and thus probabilities) of type I and type II errors. The land and natural resource manager (hereafter, resource manager) would then need to weigh these tradeoffs associated with each type of error (Marcot 1986; Mapstone 1995) and may also consider opportunity costs and likelihoods of legal appeals and of not meeting other resource management objectives. In this way, habitat models for RLK species could be part of a broader decision-modeling framework in which alternative management actions (e.g., to survey or not; to proceed with ground-disturbing activities or not), their costs and their effects on habitat conditions, and the utility values of resulting habitat states or resources produced, can be integrated into decision trees or other tradeoff analyses.

Another use of habitat modeling of RLK species is to expand our ecological understanding of the species. This can be done using traditional multivariate statistical analyses of inventory and survey data to identify major environmental correlates. It should be emphasized that nothing replaces sound empirical field research for expanding ecological understanding. Models play a subservient role in such ventures, to help identify major correlates and causes.

Habitat modeling of RLK species can also be used to identify and prioritize vegetation or other habitat attributes for conservation or restoration. Once an RLK species habitat model is built (and hopefully calibrated and perhaps further validated with field data), sensitivity analysis (e.g., Keitt et al. 1997) can reveal which prediction variables (vegetation or habitat attributes) most account for variation in response variables (habitat quality or RLK species presence). For instance, rank-ordering prediction variables according to their explanatory power (i.e., the degree to which variation in the species response is accounted for by variation in each prediction variable) tells which vegetation or habitat characteristics have the most influence on RLK species habitat quality or species presence. Of course, the reliability of using this approach depends in part on the set of habitat variables included in the model and their correlations. The importance of any variable is always measured in the context of the other variables.

The explanatory power of prediction variables can be gauged in various ways. A popular approach is the use of Akaike's information criterion (AIC). This index denotes which model—that is, which combination of predictor variables—best fits the observed data without overfitting the data with too many parameters (Burnham and Anderson 2002). The modeler will still need to decide if the best-fit model makes ecological sense. Other approaches to determining the explanatory power of prediction variables and the best model to use include information theoretic modeling (Burnham and Anderson 2002), which entails depicting the relation among variables as likelihoods and comparing the relative values of those likelihoods. This approach too can suggest the best suite of variables that have the best explanatory power for a given response.

Another way to gauge explanatory power of prediction variables uses variance reduction in analysis of model sensitivity. Variance reduction is the degree to which incremental changes in the value of a predictor variable accounts for variation in the response variable. High variance reduction of a predictor variable means much of the variation in the response variable is accounted for by that predictor. Variance reduction is used with continuous variables; an analogous "entropy reduction" is used with categorical variables. The resource manager could then use the rank-order list of site variables to prioritize variables for restoration or conservation. This presumes that the validated model truly represents the set of prediction (site) variables that affect the species response variables. If validation is not done with such models, then the models represent little more than the belief structures of the modelers, so that sensitivity analyses cannot be used to truly determine the main influential parameters. It may still be useful to pose testable hypotheses.

For example, sensitivity analysis of a site-scale habitat model of Townsend's big-eared bat (*Corynorhinus townsendii*) revealed the following site attributes listed in decreasing influence (decreasing values of entropy reduction) on prediction of habitat quality for this species: presence of caves or mines, large snags or live trees, forest edges, cliffs, bridges or buildings, and boulder piles (Marcot et al. 2001). The influence of each of these habitat attributes was positive on habitat quality. Most of the influence came from the first four attributes, so in this case, the resource manager could get the greatest return by focusing initial funding and conservation efforts for this species on gating caves and mines, conserving large snags or live trees, retaining forest edges, and not disrupting cliff environments. However, in this particular example, the bat model was developed from knowledge of bat experts, not from empirical data, so the interpretation of the sensitivity test results should be treated as testable working hypotheses, not as definitive knowledge of which site factors more influence presence of this bat species.

One should realize that testing, validating, updating, and testing sensitivity of such models using correlation analysis does not necessarily reveal the true causal variables. The truism is that correlation is not necessarily causation. However, if the results of testing fundamentally disagree with the conceptual models of experienced biologists, then the basic model structure should probably be reexamined.

For any kind of habitat modeling of RLK species, it may be valuable to record field data when species are not detected as well as when they are detected (e.g., Zielinski and Stauffer 1993; Carroll et al. 1999; Kery 2002). With such data, type I and II errors can both be analyzed, as with so-called confusion matrices, which present error rates of false prediction of presence and absence. Too often, however, empirical data on RLK species are only on presence, although modeling species–habitat relations based on use-availability data can also be of value (e.g., Hirzel et al. 2002). Museum and collection records, and most field inventory and survey records, generally provide only presence data.

With no way to gauge how well models predict absence (i.e., statistical power; Steidl et al. 1997; Thomas 1997), the resource manager would be unsure if prediction of low habitat quality or "lack of presence" of an RLK organism in an area truly implies a high confidence that it is absent. Thus there may be uncertainties associated with gauging risk of inadvertent local extirpation. However, the methods of MacKenzie (2006) provide for ways to adjust species responses for the probability of their detection given a level of sampling effort, and this might help reduce some of the uncertainty associated with lack of data on species absence.

For plants and allies (fungi, lichens, bryophytes), lack of detection in well-surveyed vegetation plots may be taken as a reasonable "absence" for modeling purposes, although some cryptic life forms may still be difficult to detect, and people conducting field surveys may require special training. For any taxa, it may be worth establishing and testing field survey protocols for ensuring that both presence and absence can be determined at acceptable levels of confidence. Also, it is important to denote in species databases whether lack of presence means confirmed absence (searched for, according to a survey protocol, and not found) or unknown absence (a survey protocol was not used to determine absence, and the species was not reported present).

Habitat modeling of RLK species could also distinguish between prediction of habitat quality and prediction of actual presence of the organism (Church et al. 2000). With rare species, habitat may seem entirely suitable but the organism may often still be absent because of many other factors affecting rarity (see chap. 3). Such effects are particularly confounding with RLK species. Even with fully suitable habitat, to the best of expert knowledge and as far as can be identified and predicted from field data, only a small percentage of such sites might actually harbor an RLK organism. The resource manager may find that habitat models predicting habitat quality instead of species presence are entirely sufficient for most purposes, such as to help prioritize sites for field surveys.

As an example, Imm et al. (2001) developed models of rare plants based on resource and vegetation characteristics in southeastern U.S. hardwood forests. They found that the models accurately predicted habitat but, not surprisingly, only when the plants were strongly associated with these variables and the scale of modeling coincided with habitat size. They concluded that habitat prediction models can be useful tools for managing rare plants, especially when combined with information from research and monitoring.

As new information is gathered, it can be integrated back into the models to further calibrate, validate, and amend them to be more accurate. Some modeling approaches, such as BBN models and many "data mining" tools (e.g., associations and pattern discovery methods, Bayesian hierarchical modeling, and probabilistic relational modeling), can handle missing data and can integrate new data to update underlying probability tables or functions (Dominici et al. 1997). In some approaches, databases, even those having some missing data, can be used to generate a set of rules that optimally predict some outcome, such as habitat quality, based on a set of environmental or habitat predictor variables. These are "rule induction" algorithms and they include classification and regression trees and fuzzy-logic-based inductive modeling methods (Stockwell et al. 1990; Jeffers 1991; Uhrmacher et al. 1997). Other updating techniques include sequential and empirical Bayes methods (Johnson 1985, 1989; Ver Hoef 1996) and expectation maximization and other methods (Dempster et al. 1977). The value of such approaches is to create a predictive model that can at least be used to generate hypotheses about causal influences of species presence. The risk, though, with rule induction approaches is in overfitting the data so that the resulting model may work well with the (often limited) data at hand but is not general enough to apply anywhere else or under any other conditions.

These very powerful tools extend the toolkit of traditional ("frequentist") statistical analyses and can be used to construct and update habitat models of RLK species. In particular, using methods that account for missing data—a common situation with RLK species data sets—may be quite useful. The initial modeling, in fact, can be done based on quizzing a species expert and building a prediction model based on the expert's best judgment of what constitutes suitable habitat (building such models has long been called knowledge engineering in the artificial intelligence computer programming literature; e.g., Fox 1984). The reliability of models based solely on expert knowledge—"expert systems" in the broadest sense—may be unknown or difficult to judge without subsequently testing it with field data. However, a rigorous peer-review process (or a sequential interview or rigorous expert panel process) can still be used to build the initial expert-based model *if* there are such other peer experts available who have knowledge of the species.

In some cases, the species may be just so rare, so difficult to detect, or so unknown, that any field surveys are not practical. In such cases, expert-based habitat models should be used with caution and as working hypotheses. Very RLK species may be impossible to model individually. In those cases, developing models of species habitat groups (e.g., old-forest canopy-dwelling arthropods, or mesoscopic soil organisms) or species functional groups (e.g., rock-adhering stream bryophytes that serve to filter water) may be a useful approach (chap. 6), although prediction of individual species within such species groups becomes problematic.

Dealing with Uncertainty

This section characterizes the kinds and sources of uncertainty and discusses ways to address uncertainty to help reduce management risk.

Assessing, modeling, and managing any species carries various kinds of uncertainty, or, more properly, sources of error. Sources of error traditionally include both measurement error and random error. Measurement error can be reduced by using statistically rigorous research or survey designs, by attending to adequate training of individuals conducting the surveys and research, and by calibrating measurement tools, be they calipers, densiometers, eyeballs, identification keys, or models. Random errors can stem from inherent variation in species' distributions and environmental conditions affecting species' presence, and can be reduced (but usually not eliminated) with adequate sample sizes and sampling designs.

Random error introduces unexplainable noise in biological studies. Moller and Jennions (2002) conducted a meta-analysis of a variety of 43 biological studies and found that the mean amount of variance (R^2) by which the response variables were explained by the predictor variables was a rather dismal 2.51 to 5.42%. One implication was that the average sample sizes needed to conclude that a particular relationship was absent with a power of 80%, and alpha = 0.05 (two-tailed) was considerably larger than that usually recorded in studies of evolution and ecology. This has dire implications for studies of RLK species where sample sizes are often far lower than needed for meeting widely accepted levels of statistical confidence and power. Of course, if a survey includes all members of some rare biological population, then it constitutes a complete census of the entire statistical population and there are no errors of estimation involved. This is seldom the case, or at least, it can seldom be demonstrated that an entire biological population has been located or that environmental conditions encountered during surveys would be invariant. Thus some errors of estimation or prediction are nearly always involved.

Beyond these traditional types of uncertainty, assessment and conservation of RLK species also carry some other special considerations. Some of these were mentioned earlier, particularly problems of detectability, taxonomy, and field sampling methods and expertise. When species are inconspicuous or rare, research studies and surveys are expensive, detections tend to be few, and data are sparse by which to project population estimates and determine habitat correlates. Estimators of relative abundance of species can account for low detectability of RLK species (e.g., MacKenzie and Kendall 2002), but with rare species it takes a lot of field effort to reliably develop such correction factors. Further, experts who can effectively locate the species in the field tend to be few, and training of survey technicians needs to be done rigorously to ensure reliable outcomes (e.g., Scott and Hallam 2003; Wilkie et al. 2003), particularly with data on species absence.

When modeling RLK species, many of these sources of uncertainty combine to create greater error in prediction of habitat quality and species presence (and absence). There is no one practical method for calculating such *propagation of error* from all these sources of uncertainty. The traditional analytic method of calculating propagation of error entails expansion of the Taylor series to include terms for covariance among affector variables (Meyer 1986). Practically, however, such an approach is nearly impossible with RLK species. Clark (2003) has proposed a hierarchical approach to dealing with propagation of error in estimating population growth rates. In general, various sources of error can be characterized in a variety of ways.

One of the simplest ways to characterize uncertainty of scientific understanding of RLK species is to have species experts denote a general level of confidence, such as depicting a level of confidence on an ordinal scale (e.g., low, moderate, high) representing the overall level of scientific understanding of the species. Such an approach has been used in several regional species–environment relations databases that include information on RLK species, and has been useful to sort species needing further study (Marcot et al. 1997; O'Neil et al. 2001).

A more detailed listing of species information in addition to identifying information gaps and research needs on each species can be useful for guiding studies. For example, a database on research needs was compiled on terrestrial vertebrate species for the Interior Columbia Basin Ecosystem Management Project (Marcot 1997). The database represents significant gaps in scientific knowledge and includes 482 potential research study topics on 232 individual species and 18 groups of species. The main keyword subjects include basic ecology, distribution, inventory and monitoring, environmental disturbance, effects of land management activities, and other topics. The database can be searched a variety of ways, such as to determine which species share specific information gaps, or the full array of research needs on a specified species or group of species. A research needs assessment can also address a broader array of biodiversity attributes related to RLK species and can be directed at prioritizing studies to meet the needs of resource managers (e.g., Smythe et al. 1996).

If at least some field information is available on RLK species, such as from surveys or collection and voucher records, statistical analyses can aid further understanding of the ecology of the species, and initial models can be built to predict habitat quality and perhaps species presence. With few sample points, however, traditional statistical estimation procedures may be fraught with high levels of error. For example, extrapolating a regional population from a small sample of a rare species may be risky if the sampling design fails to block on, or account for, some major environmental or habitat correlates of species presence, such as presence of a key pollinator or dispersal agent. The estimate may be precise (such as having a low variance or low odds ratio) but it may be biased if the sample does not accurately represent the full range of conditions present throughout the region. Statistical methods of bootstrapping, rarefaction analysis, or cumulative error analysis may help reveal such biases but cannot eliminate such errors. Further, nothing short of solid field autecology studies can truly reveal the mechanisms underlying distribution and abundance patterns of RLK species.

Other kinds of uncertainty in assessing and managing for RLK species pertain to metapopulation dynamics (Thomas 1994). If RLK species are patchily distributed with at least some interchange among patches (populations), two factors can have a major (positive) effect on long-term persistence of the overall metapopulation: (1) if populations tend to vary asynchronously between one another, and (2) if organisms can successfully disperse among habitat patches and recolonize empty patches. These two factors likely vary greatly among many taxonomic groups, rendering it impossible to generalize among all RLK species.

Other factors pertaining to population viability—including demographic vital rates, and genetic isolation and simplification of gene pools—also likely vary greatly among taxa and species. It is not useful to predict viability of an RLK species based solely on general life history characteristics of a taxonomic group or even of a closely allied congeneric (i.e., belonging to the same genus) species (Lehmkuhl et al. 2001; Wilcox and Possingham 2002); there is simply too much variation among species (see chap. 6).

Sampling and analysis schemes can be devised to estimate richness of species, including rare species. Rosenzweig et al. (2003) tested six methods of estimating total species richness of butterflies in Canada and the United States, at the broad scale of biogeographical provinces. According to their findings, methods that rely on extrapolating from frequency of sampled species did not work well to estimate the total number of species. Other methods rely on estimating the asymptote of the species-accumulation curve, which is a graph of the number of species in a set of samples by the number of species occurrences in those samples, where the number of species at the asymptote is the estimated total number of species in the community. Asymptote methods served as reliable estimators of total number of species even with relatively small samples (10% of the ecoregions sampled) when samples were well dispersed spatially but not when samples were clustered. In this way, at least the total number of species in a community can be estimated, although such approaches do not reveal which species, particularly the rare ones, are present.

In summary, despite best efforts and use of analytic and evaluation methods to describe and reduce errors of estimation and to improve understanding and prediction, some uncertainty in scientific knowledge and management of RLK species will remain.

Management Implications

By definition, managing for RLK species carries great uncertainty. How might the resource manager use the information on uncertainty as discussed in the previous section?

This is the case with surveys of northern spotted owls (*Strix occidentalis caurina*), a relatively rare species in conifer forests of the Pacific Northwest, where a strict and rather complicated survey protocol (USDA Forest Service 1993) has been established to determine nesting status (box 5.1). The complexity of this example is not unusual, given that spotted owls are scarce everywhere (rarity condition [a] in fig. 5.1), occupy large home ranges, and are not necessarily highly detectable during any given survey outing, either by direct visual observation or by soliciting responses from playing taped calls and songs. Also, should nesting *not* be confirmed after all the survey visits stipulated in the protocol are conducted, only then can one state with a moderate degree of confidence that nesting is indeed not occurring at that site. Similar, complex survey protocols to determine presence and absence have been established for a wide variety of species and taxa under the Survey and Manage program, including for RLK fungi, lichens, bryophytes, and mollusks, and other species of concern within the Northwest Forest Plan, such as marbled murrelets (*Brachyramphus marmoratus*). For example, the marbled murrelet protocol uses detection probabilities to set the number of site visits to provide a high confidence (95%) of observing the birds if they are present.

Box 5.1. Survey protocol standards for determining nesting status of northern spotted owls (*Strix occidentalis caurina*) in the Pacific Northwest (USDA Forest Service 1993).

Shown here is only part of a much longer protocol that provides guidelines for determining pair status, resident single status, unknown (single owl) status, absence (verified unoccupied site), and reproductive status (nesting status and reproductive success).

1. **Nesting is confirmed** if any of the following conditions are observed. Two observations, at least 1 week apart, are required to determine nesting status if the first observation occurs before 1 May. This is necessary because the owls may show signs of initiating nesting early in the season without actually laying eggs, and their behavior could easily be mistaken for nesting behavior. After 1 May, a single observation is sufficient. Nesting is confirmed if, on two visits before 1 May, or one visit after 1 May:
   a. the female is detected (seen) on the nest; or
   b. either member of a pair carries natural or observer-provided prey to the nest; or
   c. a female possesses a brood patch when examined in hand during mid-April to mid-June. Only one observation is required. Dates may vary with the particular areas. Be careful not to confuse the normal small areas of bare skin (apteria) on the abdomen with the much larger brood patch. A fully developed brood patch covers most of the lower abdomen, extending to the base of the wings. Describe the brood patch on the field form, including length, width, color, and texture of the skin, and any evidence of regenerating feathers around the edge (NOTE: while a scientific research permit from the U.S. Fish and Wildlife Service is not required for calling spotted owls, any capture or handling of spotted owls does require such a permit); or
   d. young are detected in the presence of one or both adults. Because young barred owls look like young spotted owls until late in the summer, the presence of young alone is not sufficient to confirm nesting.
2. **Nonnesting (and nonreproduction) is inferred** if any of the following are observed. Two observations are required during the nest survey period, with at least 3 weeks separating these observations to ensure that late nesting attempts are not missed. The second observation should occur after 15 April. Because nesting attempts may fail before surveys are conducted, the nonnesting status includes owls that did not attempt to nest as well as those that have failed. Nonnesting is inferred if:
   a. the female is observed roosting for 60 minutes, particularly early in the season (1 April to 1 May). (Be aware that nesting females with large nestlings often roost outside the nest during warm weather. If in doubt, be sure to schedule one or more visits in mid-June to check for fledglings); or

(*continues*)

Box 5.1. *Continued*

b. the female does not possess a brood patch when examined in-hand between mid-April and mid-June. Only one observation is required; or
c. the pair is not located after a 4-hour search on two separate visits; or
d. you offer prey to one or both members of the pair and they cache the prey, sit with prey for an extended period of time (30 to 60 minutes), or refuse to take additional prey beyond the minimum of two prey items. To be considered a valid nesting survey, an owl must take at least two prey items.

Surveys where the bird(s) leave the area with prey and you are unable to determine the fate of the prey cannot be classified as to nesting status and do not count toward the required two visits. Banded or radio-marked birds may be reluctant to take prey at all; therefore, nesting status should be inferred from other means (e.g., checking for fledglings later in the season).

3. If nesting is not determined before the latest date (by province) listed for nesting status visits, you cannot classify the owls as nonnesting using the criteria listed above. **Nesting is unknown** if:
   a. owls are found after these dates, without young; or
   b. no owls are found after these dates at those sites where owls were present prior to these dates.

A related and more general question is, how far should assumptions of species' presence and habitat associations be relied upon? Another way of asking this is, what are acceptable levels of type I and II errors when modeling or predicting occurrence of an RLK species? As discussed earlier, a type I error is a false prediction of species presence or adverse effect, and a type II error is false prediction of species absence or no effect. Here, the answer is less clear than simply demanding proof of presence or of adverse effects. A resource manager may wish to determine opportunity costs and values of outcomes ("utilities" in decision modeling), as well as possible harm or benefit to the species, under each possible type of error, and then balance them according to their management risk attitude.

To do this, resource managers would first have to characterize their risk attitude. Risk attitude refers to the degree to which one is risk seeking, risk neutral, or risk averse for particular kinds of decisions. For example, a resource manager may be averse to moderate or high probabilities that an RLK species would become imperiled because it may mean that the species would then become petitioned for listing as a threatened or endangered species, which could then entail expensive and time-consuming revisions of natural resource and land management plans. Or the resource manager may be tolerant (seeking) of such risks if social, administrative, or political pressures were high for maintaining or increasing rates of extraction of other natural resources. In some cases, risk may even be judged according to potential for changes in one's career path, potential promotion or demotion, possibilities of lawsuits and legal actions, and so forth. These are legitimate causes for concern and consideration by decision makers and line officers, and they do figure into their risk attitudes, albeit usually tacitly and not as part of explicit decision criteria. In other cases, one's risk attitude may be dictated, to a degree, by policy or law, such as leaning toward risk aversion when dealing with potential harm to a listed threatened or endangered species because national law and resource and land management agency policy and regulations dictate so.

A risk analysis would then help outline alternative or sequential management decisions and probabilities of various outcomes such as response by an RLK species. In risk analysis parlance, a "policy" becomes a set of alternative management decision options along with the probabilities of outcomes and the values or utilities of each possible outcome. Risk attitudes can determine which decision or sets of sequential decisions are "best" in terms of a stated decision objective, such as minimizing harm to an RLK species.

Uncertainty over the occurrence or response of little-known species can be represented as a range of possible outcomes of decisions and can be depicted as a range of probabilities of species presence or as a general trend of populations or habitats. Depending on their risk attitude, resource managers could then look at the spread of possible outcomes and not just the median or mean outcome, or just the extreme "best" or "worst case" outcome. With RLK species, the spread is likely to be far wider than with more common and better-known species.

Outcomes could pertain to possible effects on species persistence but also to secondary effects on ecological processes resulting from changes to RLK species. Such processes may include the ecological roles that RLK species might play as key pollinators, dispersal agents, contributors to ecosystem resistance to exotic species invasion, decomposers and other roles in nutrient cycling, and so on. The aim would be to view RLK species not just as independent entities but also as dependent players in the broader ecological tapestry.

The implications of managing for RLK species are further discussed in subsequent chapters. Chapters 6 and 7 evaluate how various species- and system-level management approaches might suffice for conservation of RLK species. Chapter 8 then discusses effectiveness of the various management approaches; chapters 9 and 10 address social and economic considerations; chapter 11 reviews implementation considerations; and chapter 12 presents a process for determining appropriate and useful approaches for conservation of RLK species.

Conclusion

RLK species comprise a diverse lot. They span a wide range of taxonomic classes, life histories, ecologies, levels of abundance, and distributional patterns. Some are cryptic, secretive, and difficult to find, and thus there is little to no information available on their abundance, distribution, and autecology. Some are truly rare, and others might be more common than their apparent rarity suggests.

One way to help make sense of this confusing array of conditions is to denote the type of rarity and the reason for poor knowledge levels of a given RLK species (Table 5.1). These categories of rare and little known can help clarify why a species might be rare, explain why it might be little known, and describe what could be done to aid in its conservation and to increase our knowledge.

For example, a species that is rare because it is generally scarce everywhere could benefit from dispersed conservation activities, be they multiple reserves or guidelines to conserve or restore the species' key habitat attributes in selected locations dispersed across the species' range. On the other hand, a species that is rare because it has an overall small total population but that might be locally more abundant, or because it consists of a distinctive or peripheral population, could benefit by concentrating reserves or activities for conserving its key habitat attributes at its population centers.

Likewise, a species that is little known because of incomplete taxonomic descriptions could benefit from systematic studies to better determine its taxonomic status. A species that is little known because limited inventories have been conducted could benefit from expanded survey efforts or surveys with statistically sound protocols to provide more reliable information on its presence and absence. A species that is little known because there is poor understanding about its ecology could benefit from surveys and studies designed to provide information on life history and habitat associations.

In many cases of RLK species, knowledge is power. Knowledge from ecological studies and general surveys can greatly reduce uncertainty over whether a species is truly rare, should really garner status as threatened or endangered, or requires special consideration for conservation reserves or management guidelines. Data from well-designed surveys can be used to develop models that predict the habitat and perhaps occurrence of RLK species, for use in land management and for prioritizing species for conservation focus. Examples we have cited suggest that with additional information, conservation of some species may become less stringent than initially suspected and may need to be reaffirmed or increased for other species. The next two chapters explore potential approaches to species conservation and their effectiveness for providing for RLK species.

RLK species should not be discounted in terms of the potential ecological importance they may play in ecosystem functions. Certainly, many poorly known micro- and mesoarthropods play crucial ecological roles in soils and forest canopies. Many such species are uncharistmatic and may seldom be chosen for specific conservation focus. But they may be contributing important functions to help ensure productivity of ecosystems and provision of services to people.

References

Beavers, S. C., and F. L. Ramsey. 1998. Detectability analysis in transect surveys. *Journal of Wildlife Management* 62:948–57.

Borcard, D., P. Legendre, C. Avois-Jacquet, and H. Tuomisto. 2004. Dissecting the spatial structure of ecological data at multiple scales. *Ecology* 85:1826–32.

Burnham, K. P., and D. Anderson. 2002. Model selection and multimodel inference. New York: Springer Verlag.

Butler, R. S., and R. L. Mayden. 2003. Cryptic biodiversity. *Endangered Species Bulletin* 28:24–26.

Carroll, C., W. J. Zielinski, and R. F. Noss. 1999. Using presence–absence data to build and test spatial habitat models for the fisher in the Klamath Region, U.S.A. *Conservation Biology* 13:1344–59.

Cassidy, K. M., and C. E. Grue. 2001. The role of private and public lands in conservation of at-risk vertebrates in Washington state. *Wildlife Society Bulletin* 28:1060–76.

Church, R., R. Gerrard, A. Hollander, and D. Stoms. 2000. Understanding the tradeoffs between site quality and species presence in reserve site selection. *Forest Science* 46:157–67.

Clark, J. S. 2003. Uncertainty and variability in demography and population growth: A hierarchical approach. *Ecology* 84:1370–81.

Cutler, R., T. C. Edwards, Jr., J. Alegria, and D. McKenzie. 2001. A sample design framework for Survey and Manage species under the Northwest Forest Plan. *Proceedings of the Section on Statistics and Environment, Joint Statistical Meetings, August 5–9, 2001*. Atlanta: American Statistical Association. 8 pp. CD-ROM.

Dempster, A., N. Laird, and D. Rubin. 1977. Maximum likelihood from incomplete data via the EM algorithm. *Journal of the Royal Statistical Society* 39:1–38.

Dominici, F., G. Parmigiani, R. Wolpert, and K. Reckhow. 1997. Combining information from related regressions. *Journal of Agricultural, Biological, and Environmental Statistics* 2:313–32.

Dunk, J. R., W. J. Zielinski, and H. K. Preisler. 2004. Predicting the occurrence of rare mollusks in northern California forests. *Ecological Applications* 14:713–29.

Edwards, T. C., D. R. Cutler, L. Geiser, J. Alegria, and D. McKenzie. 2004. Assessing rarity of species with low detectability: Lichens in Pacific Northwest forests. *Ecological Applications* 14:414–24.

FEMAT (Forest Ecosystem Management Assessment Team). 1993. Forest ecosystem management: An ecological, economic, and social assessment. Washington, DC: U.S. Government Printing Office 1993-793-071. Available at: Regional Ecosystem Office, P.O. Box 3623, Portland, OR 97208.

Fox, J. 1984. A short account of knowledge engineering. *Knowledge Engineering Review* 1:4–14.

Green, R. H., and R. C. Young. 1993. Sampling to detect rare species. *Ecological Applications* 3:351–56.

Haddad, N. M., D. Tilman, J. Haarstad, M. E. Ritchie, and J. M. H. Knops. 2001. Contrasting effects of plant richness and composition on insect communities: A field experiment. *American Naturalist* 158:17–35.

Henderson, J. A. 2001. The PNV model-a gradient model for predicting environmental variables and potential natural vegetation across a landscape. Unpublished report. Mountlake Terrace, WA: USDA Forest Service, Mt. Baker–Snoqualmie National Forest.

Hirzel, A. H., J. Hausser, D. Chessel, and N. Perrin. 2002. Ecological-niche factor analysis: How to compute habitat-suitability maps without absence data? *Ecology* 83:2027–36.

Hoyer, R. F., R. P. O'Donnell, and R. T. Mason. 2006. Current distribution and status of sharp-tailed snakes. *Northwestern Naturalist* 97:195–202.

Imm, D. W., H. E. Shealy, Jr., K. W. McLeod, and B. Collins. 2001. Rare plants of southeastern hardwood forests and the role of predictive modeling. *Natural Areas Journal* 21:36–49.

International Union for Conservation of Natural Resources (IUCN). 2006. IUCN Red List of Threatened Species. http://www.redlist.org/search/details.php?species=39987.

Jeffers, J. N. R. 1991. Rule induction methods in forestry research. *AI Applications* 5:37–44.

Jenouvrier, S., and T. Boulinier. 2006. Estimation of local extinction rates when species detectability covaries with extinction probability: Is it a problem? *Oikos* 113:132–38.

Johnson, D. H. 1985. Improved estimates from sample surveys with empirical Bayes methods. *Proceedings of the American Statistical Association*, 395–400.

———. 1989. An empirical Bayes approach to analyzing recurring animal surveys. *Ecology* 70:945–52.

Jones, D. T., and P. Eggleton. 2000. Sampling termite assemblages in tropical forests: Testing a rapid biodiversity assessment protocol. *Journal of Applied Ecology* 37:191–203.

Keitt, T. H., D. L. Urban, and B. T. Milne. 1997. Detecting critical scales in fragmented landscapes. *Conservation Ecology* 1:4. http://www.ecologyandsociety.org/vol1/iss1/art4.

Kery, M. 2002. Inferring the absence of a species: A case study of snakes. *Journal of Wildlife Management* 66:330–38.

Kintsch, J. A., and D. L. Urban. 2002. Focal species, community representation, and physical proxies as conservation strategies: A case study in the Amphibolite Mountains, North Carolina, U.S.A. *Conservation Biology* 16:936–47.

Lehmkuhl, J. F., B. G. Marcot, and T. Quinn. 2001. Characterizing species at risk. Pp. 474–500 in *Wildlife–Habitat Relationships in Oregon and Washington*, ed. D. H. Johnson and T. A. O'Neil. Corvallis: Oregon State University Press.

Leonard, W. P., and M. A. Leonard. 1998. Occurrence of the sharptail snake (*Contia tenuis*) at Trout Lake, Klickitat County, Washington. *Northwestern Naturalist* 79:75–76.

Lesher, R. D. 2005. An environmental gradient model predicts the spatial distribution of potential habitat for *Hypogymnia duplicata* in the Cascade Mountains of northwestern Washington. PhD diss., University of Washington, Seattle.

MacKenzie, D. I. 2006. Modeling the probability of resource use: The effect of, and dealing with, detecting a species imperfectly. *Journal of Wildlife Management* 70:367–74.

MacKenzie, D. I., and W. L. Kendall. 2002. How should detection probability be incorporated into estimates of relative abundance? *Ecology* 83:2387–93.

MacKenzie, D. I., J. D. Nichols, J. E. Hines, M. G. Knutson, and A. B. Franklin. 2003. Estimating site occupancy, colonization, and local extinction when a species is detected imperfectly. *Ecology* 84:2200–7.

Mao, C. X., and R. K. Colwell. 2005. Estimation of species richness: Mixture models, the role of rare species, and inferential challenges. *Ecology* 86:1143–53.

Mapstone, B. D. 1995. Scalable decision rules for environmental impact studies: Effect size, type I, and type II errors. *Ecological Applications* 5:401–10.

Marcot, B. G. 1986. Summary: Biometric approaches to modeling—the manager's viewpoint. Pp. 203–4 in *Wildlife 2000: Modeling habitat relationships of terrestrial vertebrates*, ed. J. Verner, M. L. Morrison, and C. J. Ralph. Madison: University of Wisconsin Press.

———. 1997. Research information needs on terrestrial vertebrate species of the interior Columbia River Basin and northern portions of the Klamath and Great basins. Research Note PNW-RN-522. Portland, OR: USDA Forest Service. Abstract and database available online: http://www.fs.fed.us/pnw/marcot.html.

———. 2006. Characterizing species at risk, I: Modeling rare species under the Northwest Forest Plan. *Ecology and Society* 11 (2): 10. http://www.ecologyandsociety.org/vol11/iss2/art10.

Marcot, B. G., M. A. Castellano, J. A. Christy, L. K. Croft, J. F. Lehmkuhl, R. H. Naney, K. Nelson, et al. 1997. Terrestrial ecology assessment. Pp. 1497–1713 in *An assessment of ecosystem components in the interior Columbia Basin and portions of the Klamath and Great Basins*, vol. 3, ed. T. M. Quigley and S. J. Arbelbide. USDA Forest Service General Technical Report PNW-GTR-405. Portland, OR: U.S. Department of Agriculture, Forest Service, Pacific Northwest Research Station.

Marcot, B. G., R. S. Holthausen, M. G. Raphael, M. M. Rowland, and M. J. Wisdom. 2001. Using Bayesian belief networks to evaluate fish and wildlife population viability under land management alternatives from an environmental impact statement. *Forest Ecology and Management* 153:29–42.

Meyer, S. L. 1986. *Data analysis for scientists and engineers*. Evanston, IL: Peer Management Consultants, Ltd.

Molina, R., D. McKenzie, R. Lesher, J. Ford, J. Alegria, and R. Cutler. 2003. Strategic survey framework for the Northwest Forest Plan Survey and Manage Program. General Technical Report PNW-GTR-573. Portland, OR: U.S. Department of Agriculture, Forest Service, Pacific Northwest Research Station.

Molina, R., B. Marcot, and R. Lesher. 2006. Protecting rare, old-growth-forest-associated species under the Survey and Manage Program guidelines of the Northwest Forest Plan. *Conservation Biology* 20:306–18.

Moller, A. P., and M. D. Jennions. 2002. How much variance can be explained by ecologists and evolutionary biologists? *Oecologia* 132:492–500.

Oliver, I., and A. J. Beattie. 1996. Designing a cost-effective invertebrate survey: A test of methods for rapid assessment of biodiversity. *Ecological Applications* 6:594–607.

O'Neil, T. A., D. H. Johnson, C. Barrett, M. Trevithick, K. A. Bettinger, C. Kiilsgaard, M. Vander Heyden, et al. 2001. Matrixes for wildlife–habitat relationships in Oregon and Washington. CD-ROM in *Wildlife–habitat relationships in Oregon and Washington*, ed. D. H. Johnson and T. A. O'Neil. Corvallis: Oregon State University Press.

Philippi, T. 2005. Adaptive cluster sampling for estimation of abundances within local populations of low-abundance plants. *Ecology* 86:1091–1100.

Philippi, T., B. Collins, S. Guisti, and P. M. Dixon. 2001. A multistage approach to population monitoring for rare plant populations. *Natural Areas Journal* 21:111–16.

Plotkin, J. B., and H. C. Muller-Landau. 2002. Sampling the species composition of a landscape. *Ecology* 83:3344–56.

Raphael, M. G., M. J. Wisdom, M. M. Rowland, R. S. Holthausen, B. C. Wales, B. G. Marcot, and T. D. Rich. 2001. Status and trends of habitats of terrestrial vertebrates in relation to land management in the interior Columbia River basin. *Forest Ecology and Management* 153:63–87.

Rastetter, E. B., A. W. King, B. J. Cosby, G. M. Hornberger, R. V. O'Neill, and J. E. Hobbie. 1992. Aggregating fine-scale ecological knowledge to model coarser-scale attributes of ecosystems. *Ecological Applications* 2:55–70.

Rosenzweig, M. L., W. R. Turner, J. G. Cox, and T. H. Ricketts. 2003. Estimating diver-

sity in unsampled habitats of a biogeographical province. *Conservation Biology* 17:864–74.
Royle, J. A. 2006. Site occupancy models with heterogeneous detection probabilities. *Biometrics* 62:97–102.
Royle, J. A., J. D. Nichols, and M. Kerry. 2005. Modelling occurrence and abundance of species when detection is imperfect. *Oikos* 110:353–59.
Sayre, R., E. Roca, G. Sedaghatkish, B. Young, S. Keel, R. Roca, and S. Sheppard. 2000. Nature in focus: rapid ecological assessment. Washington, DC: Island Press.
Schwartz, M. W. 1999. Choosing the appropriate scale of reserves for conservation. *Annual Review of Ecology and Systematics* 30:83–108.
Scott, J. M., P. J. Heglund, M. L. Morrison, J. B. Haufler, M. G. Raphael, W. Wall, and F. B. Samson. 2002. *Predicting species occurrences: Issues of scale and accuracy*. Washington, DC: Island Press.
Scott, W. A., and C. J. Hallam. 2003. Assessing species misidentification rates through quality assurance of vegetation monitoring. *Plant Ecology* 165:101–15.
Shaffer, M. L. 1990. Population viability analysis. *Conservation Biology* 4:39–40.
Shen, T., A. Chao, and C. Lin. 2003. Predicting the number of new species in further taxonomic sampling. *Ecology* 84:798–804.
Smythe, K. D., J. C. Bernabo, T. B. Carter, and P. R. Jutro. 1996. Focusing biodiversity research on the needs of decision makers. *Environmental Management* 20:865–72.
Steidl, R. J., J. P. Haynes, and E. Schauber. 1997. Statistical power analysis in wildlife research. *Journal of Wildlife Management* 61:270–79.
Stockwell, D. R. B., S. M. Davey, J. R. Davis, and I. R. Noble. 1990. Using induction of decision trees to predict greater glider density. *AI Applications in Natural Resource Management* 4:33–43.
Thomas, C. D. 1994. Extinction, colonization, and metapopulations: Environmental tracking by rare species. *Conservation Biology* 8:373–78.
Thomas, L. 1997. Retrospective power analysis. *Conservation Biology* 11:276–80.
Thompson, W. L. 2004. *Sampling rare or elusive species: Concepts, designs, and techniques for estimating population parameters*. Washington, DC: Island Press.
Uhrmacher, A. M., F. E. Cellier, and R. J. Frye. 1997. Applying fuzzy-based inductive reasoning to analyze qualitatively the dynamic behavior of an ecological system. *AI Applications* 11:1–10.
USDA Forest Service. 1993. *Protocol for surveying for spotted owls in proposed management activity areas and habitat conservation areas*. March 12, 1991, revised February 1993. San Francisco, CA: U.S. Department of Agriculture, Forest Service, Pacific Southwest Region.
Venette, R. C., R. D. Moon, and W. D. Hutchinson. 2002. Strategies and statistics of sampling for rare individuals. *Annual Review of Entomology* 47:175–205.
Ver Hoef, J. M. 1996. Parametric empirical Bayes methods for ecological applications. *Ecological Applications* 6:1047–55.
Verner, J., M. L. Morrison, and C. J. Ralph. 1986. *Wildlife 2000: Modeling habitat relationships of terrestrial vertebrates*. Madison: University of Wisconsin Press.
Vilella, F. J., and E. R. Garcia. 1995. Post-hurricane management of the Puerto Rican parrot. Pp. 618–21 in *Integrating people and wildlife for a sustainable future*, ed. J. A. Bissonette and P. R. Krausman. Bethesda, MD: Wildlife Society.
Warren, M. L., Jr., B. M. Burr, S. J. Walsh, H. L. Bart, Jr., R. C. Cashner, D. A. Etnier, B.

J. Freeman, et al. 2000. Diversity, distribution, and conservation status of the native freshwater fishes of the southeastern United States. *Fisheries* 25:7–29.

Wilcox, C., and H. Possingham. 2002. Do life history traits affect the accuracy of diffusion approximations for mean time to extinction? *Ecological Applications* 12:1163–79.

Wilkie, L., G. Cassis, and M. Gray. 2003. A quality control protocol for terrestrial invertebrate biodiversity assessment. *Biodiversity and Conservation* 12:121–46.

Wilson, M. H., C. B. Kepler, N. F. R. Snyder, S. R. Derrickson, F. J. Dein, J. W. Wiley, J. M. Wunderle, Jr., A. E. Lugo, D. L. Graham, and W. D. Toone. 1994. Puerto Rican parrots and potential limitations of the metapopulation approach to species conservation. *Conservation Biology* 8:114–23.

Zielinski, W. J., and H. B. Stauffer. 1993. Monitoring *Martes* populations in California: Survey design and power analysis. *Ecological Applications* 6:1254–67.

6

Species-Level Strategies for Conserving Rare or Little-Known Species

Bruce G. Marcot and Curtis H. Flather

Conservation science is concerned, in part, with anticipating how natural or human-caused disturbance affects the pattern of commonness and rarity among the biota of a given area (Lubchenco et al. 1991; also see chap. 1). In this and the next chapter, we review various species- and system-level approaches proposed or applied to species conservation planning and assessment. We review the literature on approaches to conservation and management of individual species, groups of species, communities, and ecosystems, as published in peer-reviewed ecological outlets or in reports by land management agencies and nongovernmental organizations. We summarize conservation goals and objectives, discuss the scientific basis, and identify strengths and weaknesses of each approach for addressing conservation needs of rare or little-known (RLK) species of plants and animals. This review is intended to provide land and natural resource managers (hereafter, just "resource managers") access to this diverse literature and the basic information needed to select those approaches that best suit their conservation objectives and ecological context. A companion chapter (see chap. 8) outlines a process to evaluate how well each approach meets conservation objectives.

Categories of Approaches

In our review of the literature and drawing upon our knowledge and experience, we developed a classification system of approaches to conservation.

We classify the approaches in two broad categories. *Species approaches* result in conservation actions focused on providing for individual species or groups of species with common needs or common ecological characteristics. *System approaches* result in conservation actions focused on providing for community or ecosystem composition, structure, or function. These categories are artificial, in that species and system approaches are not necessarily mutually exclusive or independent. However, this dichotomy is useful for systematically presenting the numerous approaches appearing in the literature.

Other classifications of conservation approaches exist in the literature. One familiar to many resource managers is the dichotomy of coarse- and fine-filter management (Noss 1987). Although this classification has high recognition, it is also characterized by a high degree of ambiguity. For example, Overton et al. (2006) used these concepts to characterize landscape (coarse-filter) and microhabitat (fine-filter) use by pigeons (*Columba fasciata*) in western Oregon, whereas other authors use the concepts to refer to geographic scales of land-use planning and resource management. Schwartz (1999) defined a fine-filter approach as one where conservation efforts are focused on genetic, population, or species levels, and coarse-filter approaches are aimed at the community, ecosystem, or landscape levels. Under this simple distinction, coarse-filter approaches may correspond to what we refer to here as system approaches (e.g., Reyers et al. 2001; Armstrong et al. 2003), and fine-filter approaches may correspond to species approaches.

An underlying assumption in common among at least some of the species-level approaches is that meeting the needs of one or more species would serve, to a degree, to provide for other species and for broader ecological communities or systems. The operative phrase is *to a degree,* which begs for a risk analysis of threats to the species (i.e., a threats assessment; see chap. 4) and a risk management framework (see chap. 9) by which the likelihood of successfully meeting these assumptions is gauged.

As we have defined them, species approaches focus on meeting the needs of individual species or groups of species and include those focused on managing for the viability of individual species, surrogate species, or groups of species, plus geographically based approaches (table 6.1). For each of these approaches we describe the theory, concepts, and scientific basis for the approach and provide examples, where available, of their use in technical assessments, management or planning, and monitoring.

Table 6.1. *Summary of species approaches showing descriptions and main assumptions*

Name of Approach	Description	Main Assumptions
	VIABILITY APPROACH	
Conservation of individual species based on concepts of population viability	Likelihood of persistence of a population over a specified time period and geographic area	• The ecology, dynamics, demography, and/or genetics of a species is known well enough to estimate persistence probabilities • Viability analyses are realistic enough to incorporate and correctly represent the major factors influencing population size over time • Population isolation and fragmentation and lowering of genetic diversity have deleterious effects that can be modeled and predicted • Additional ecological considerations, including ecology and requirements of obligate symbionts, are known and accounted for
	SURROGATE SPECIES	
Umbrella species	A single species that represents the requirements of a portion or all of a species assemblage or community	• Single species somehow represent the requirements of other species or its biotic community • Successful conservation of an umbrella species confers protection of its species associates and its ecological community
Focal species	Target species that are identified for the purpose of guiding management of environments, habitats, and landscape elements in a tractable way	• Target species represent the response to stressors by other species • There is a greater degree of such representation if the focal species has stringent requirements of resource use and dispersal capability but that use the broadest area • Patterns of resource use, ecological processes, and aspects of system status associated with focal species are closely correlated with those of other species that the focal species is intended to somehow represent
Guild surrogates	One member or a subset of members serves as a surrogate for other members of the guild	• The status of one member or a subset of a guild is closely correlated with status of other members of the guild • Changes in environmental conditions affecting one or a subset of a guild would affect the other members in the same way or at least that population responses would be similar and significantly positively correlated

(*continues*)

Table 6.1. *Continued*

Name of Approach	Description	Main Assumptions
Habitat assemblage surrogates	One of a group of species that share common macrohabitats	• One or a subset of species of a species assemblage represents the full assemblage • All members of an assemblage respond the same to availability of macrohabitats and, more specifically, the response of one member is closely and positively correlated with responses of other members • Macrohabitat is the major factor affecting presence, distribution, and trend of associated species
Management indicator species	A variety of categories: federally listed species; species with special habitat needs; species that are hunted, fished, or trapped; species of special or social interest; and species for which population changes might indicate effects on or status of other species	• A set of species chosen as management indicator species somehow represent the array of management issues, conditions, objectives, and conservation concerns • Population status and trend of ecological indicator species represents others; that is, there is significant positive correlation in at least population trend (and possibly also size, distribution, and persistence likelihoods) between the indicator species and other species
Biodiversity indicator species	One or more species selected to indicate areas of high biodiversity or high biological productivity	• Indicator taxa are easier to locate and/or identify than indicated taxa • The number of species of indicator taxa serve as an index of the number of species of the indicated taxa • Locations of indicator taxa are highly correlated with locations of indicated taxa
Flagship species	Species that carry high public interest and garner social or cultural concern for their well-being	• Individual species can be identified in a given ecosystem and promoted in highly visible conservation programs that would entice positive interest and support by the general public
	MULTIPLE SPECIES	
Entire guilds	Species that share some common attribute of resource use or ecological role defining the guild	• All species of a guild would respond in like manner to presence or changes in environmental conditions
Entire habitat assemblages	Species that share common usage of some macrohabitat	• If the macrohabitat is provided, the requirements of the entire habitat group of species will be met

Name of Approach	Description	Main Assumptions
		• Species needs are mostly defined by the macrohabitat(s) they occupy
	GEOGRAPHICALLY BASED APPROACHES	
Locations of target species at risk	Known site locations of species at risk, which can include federally listed, heritage ranked, regionally sensitive, and other species	• Managing for known locations of target species likely contributes to ensuring persistence of the species and viability of its populations • Known locations are where the organism finds suitable resources and habitat conferring local persistence • Lack of "protection" of such sites confers a high likelihood of adverse anthropogenic stress on the organism and its environment, leading to local extirpation • Such organisms have low resilience or resistance, and high sensitivity, to disturbance
Species hot spots or concentrations of biodiversity	Global areas where a high number of endemic or other species are found in a relatively small geographic area	• Species' distributions are structured and nonrandom • Geographic patterns of species richness covary among at least some taxa • Hot spots represent ecological communities and ecosystems and their persistence
Reserves or protected areas	Areas delineated to represent, complement, or otherwise efficiently include species and biodiversity elements	• Protecting such areas from specific human activities or disturbances will provide for species or system persistence

Conservation of Individual Species Based on Concepts of Population Viability

This approach entails developing a conservation strategy for a given geographic area that incorporates quantitative or qualitative population viability analyses of the affected species. Such conservation strategies provide management guidelines based on an understanding of the life history, habitat needs, and factors that threaten the species. This usually includes recommendations for maintaining or restoring habitats necessary for various life functions such as breeding, foraging, roosting, dispersal, and migration. It may also include guidelines for managing conditions that would otherwise place a species at risk, such as discouraging the species' predators or competitors, or mitigating human disturbance factors (Holthausen et al. 1994). Such guidelines can take the form of designing reserves for particular species.

Methods for quantifying population viability of target species have become common in the conservation biology literature (see review by Beissinger and McCullough 2002). Population viability is generally defined as the likelihood of population persistence over a specified time period, over a specific geographic area, and over a specified population size. Population viability is thus seen as a probabilistic event, and viability outcomes are cast as odds or likelihoods. Quantitative population viability models can be useful in comparing relative population growth rates among populations (Beissinger and Westphal 1998) and can provide insights into how environmental factors induce fluctuations in rates of reproduction, mortality, and growth (Burgman et al. 1993).

A rather massive and relatively recent literature exists on metapopulation dynamics, modeling, demography, and viability evaluation of most taxonomic groups. This literature is, by definition, species-specific and based on autecological data or assumptions. Some examples include viability assessments of vascular plants (Satterthwaite et al. 2002), mollusks (Taylor et al. 2003), insects (Ranius 2000), fish (Ratner et al. 1997), reptiles (Doak et al. 1994), birds (Haig et al. 1993), and mammals (Armbruster and Lande 1993).

A main assumption of the population viability approach is that enough is known about the ecology, dispersal and colonization dynamics, demography, and/or genetics of a species to reasonably predict future sizes of populations and distributions of organisms, from which probabilities of persistence (or, conversely, extinction) can be estimated. Another important assumption is that viability analyses are realistic enough to incorporate and correctly represent the major factors influencing population size over time. Such factors include population isolation, fragmentation, and genetic diversity.

In practice, conducting a formal population viability analysis (PVA) entails knowing a lot about autecology, including demography and population genetics. For example, viability of plant and animal populations might be affected by obligate, or strongly facultative, mutualistic relationships (e.g., pollination, mychorrizal associations), dispersal agents and mechanisms, hybridization and out-crossing with nontarget species, episodic reproduction, and many other factors. Thus PVAs perform best with data-rich species.

Formal PVAs have come under some recent criticism in the conservation biology literature as the precision and performance of PVA models have been tested or compared. Like any modeling approach, PVAs may express the bias of the modeler (Brook 2000) and specific model structure (Pascual et al. 1997). Lindenmayer et al. (2000) tested the efficacy of the commercially available VORTEX PVA program to predict viability attributes of three marsupials in Australia and concluded that the model predicted reasonably well only when a rather unrealistic amount of detail was known about the spatial distribution and dynamics of habitat patches. They concluded that conservation biologists should proceed with caution when using PVA models to predict population responses in fragmented systems, even when the species is well known and has a relatively simple life history. We suggest that such caution would include disclosing key uncertainties regarding the species' habitat associations and population persistence estimates.

Earlier "rules" regarding what constitutes viable populations have also been more recently viewed as too simplistic to rely on for real-world conservation programs. One such early rule was the "50–500 rule" whereby populations were deemed to have short-term viability if they consisted of at least 50 individuals, and long-term with at least 500. These numbers have appeared in some earlier land planning and habitat management documents. However, further scientific work suggested that they should apply to effective population sizes (of breeding individuals only, correcting overall population size for various factors of distribution, variability in population size, dispersal, sex ratio, and other things). The 50–500 rule was actually initially derived from laboratory-bred *Drosophila* (fruit flies) in controlled conditions and was based on minimal numbers of individuals needed to avoid genetic introgression or inbreeding depression. The "rule" is generally no longer used in real-world PVAs that typically entail estimating persistence or extinction likelihoods from population projections tailored to the life history of each species.

In a similar way, early contentions that there exist minimum viable population (MVP) sizes (e.g., Reed et al. 1986) that can be discretely calculated for a given species have also given way to more realistic projections of persistence likelihoods. The 50–500 rule was an early attempt at defining a blanket MVP size. The concept of MVP originally arose in the context of maintaining genetic diversity within a population and avoiding potentially deleterious effects of inbreeding depression. That is, an MVP is assumedly a size at or above which the population is secure from loss of genetic variation, and below which it is doomed to eventual extinction because of simplification of the gene pool caused by inbreeding depression, genetic drift, and related conditions.

Correctly calculating effective population size is one of the steps in estimating the potential loss of genetic variation as resulting from inbreeding depression or genetic drift in small, isolated populations, and thus in estimating the minimum size of a population that still maintains its genetic variability over time. There is no standardized method for calculating effective population size in all cases, and many variations exist (Waples 2002). Essentially, the main problem is that there are no fixed minimum population sizes below which a given species or population is doomed to extinction and above which it is assured of persistence. Although it still occasionally appears (e.g., Wielgus 2002), MVP analysis has given way in the scientific literature to more species-specific probabilistic approaches where the focus is on quantifying the degree of viability (likelihoods of persistence) given specified conditions, time periods, and geographic extents.

How well do PVAs perform in predicting population persistence? Clinchy et al. (2002) concluded, by using simulations, that recolonized habitat patches are often occupied ephemerally and thus that overall population persistence will be overestimated if static or declining patterns of patch occupancy are mistakenly attributed to dynamically stable metapopulations. However, in a test of Australian treecreepers, McCarthy et al. (2000) reported that PVA models underestimated occupancy of habitat patches.

Schiegg et al. (2005) tested the predictive accuracy of a stochastic, spatially explicit, individual-based population model of red-cockaded woodpeckers (*Picoides borealis*) by comparing simulation results with field data on population dynamics. They reported that their population models performed well and were reliable but could be improved by further including dynamics of dispersal and colonization. Johnson (2005) tested a patch occupancy model of metapopulation response of a tropical beetle species to synchrony of disturbance (flooding), and concluded that a simpler logistic regression approach provided adequate predictability over a more complicated Monte Carlo modeling approach.

McCarthy et al. (2000) used validation results to refine their initial models, an iterative procedure and information source not usually available in PVAs. McCarthy et al. (2001) also suggested using such an iterative process in modeling viability. In a similar vein, Foley (2000) recognized the uncertainty in PVA modeling and suggested using a sequential, empirical Bayesian approach in modeling population viability as new biological data are gathered on target species over time.

Coulson et al. (2001) pointed out that PVA modeling cannot predict catastrophes, and this failure creates great imprecision in viability projections. However, Brook et al.'s (2002) rebuttal to such criticism was that PVAs, despite their limitations, are still the most useful and complete methods for basing management on species demography.

Wahlberg et al. (1996) presented a successful prediction of the patchy distribution of an endangered butterfly by using a spatially realistic metapopulation model, which included first-order effects of patch area and isolation on local extinction and colonization within habitat patches. They concluded that such a model has utility for study and conservation of species in highly fragmented landscapes. Brook et al. (2000) reported that PVA predictions were "surprisingly accurate" when five PVA programs were tested on animal data sets to predict quasi-extinction probabilities and population sizes (quasi extinction occurs when a population size falls below some specified nonzero level [Ginzburg et al. 1982]). Their meta-analysis suggested that when PVAs are conducted using detailed demographic data and commercially available software, they can provide accurate and unbiased short-term projections of population viability.

In summary, results of validating formal PVA approaches to date suggest that predictions of population trend and persistence likelihoods (especially above some predefined quasi-extinction level) may be more accurate than predictions of time to extinction. The more detailed are the empirical data on biology and environmental conditions, the more accurate or precise and reliable are the viability projections. In the absence of such data, the accuracy and precision of PVA predictions may be, at best, uncertain, but PVAs could still be used to compare management strategies (e.g., Raphael and Holthausen 2002).

When data are lacking, as will be the case for little-known species, a "softer" population viability assessment approach may be used that relies more on expert judgment and qualitative ranking of potential stressor effects than it does on empirical data and modeling of population demographics, genetics, and other factors. Several decision-modeling approaches are applicable to data-poor situations, including use of Bayesian belief networks to incorporate expert judgment for evaluating population viability (e.g., Raphael et al. 2001). Levels of confidence can be expressed in expert-based viability assessments, and various sources of uncertainty can be described using a variety of methods. However, few if any such expert-based assessments of population viability have been formally tested with field validation research and experiments.

A twist to the formal PVA approach is in predicting level of viability or vulnerability of a species based on its general life history characteristics (Bernt-Erik and Bakke 2000). Some authors suggest that general life history characteristics can tell a lot about viability of a species and its populations (Verheyen et al. 2003). For example, the now-classic approach by Rabinowitz (1981) depicts rarity and potential viability risk of species according to their degree of habitat specificity, geographic extent, and local concentration (also see chap. 3). Some authors have integrated some life history characteristics into methods for ranking species vulnerability (e.g., Mace and Lande 1991). Other authors have used life history traits to predict invasiveness of plants (Higgins and Richardson 1999), birds (Cassey 2002), and other taxa. Davies et al. (2004) found that natural abundance and degree of specialization acted synergistically with beetle species so that rare and specialized beetles were especially vulnerable to extinction. They concluded that some combinations of traits increased vulnerability.

However, some authors have found that life history traits by themselves may be poor predictors of viability or risk levels of species (e.g., modeling validation analyses by Lehmkuhl et al. 2001; and simulation modeling by Wilcox and Possingham 2002) and therefore should be used with great caution. Several approaches have been proposed that explicitly incorporate uncertainty into ranking species vulnerability levels (e.g., Regan et al. 2000).

Conservation of Surrogate Species

This subcategory of species approaches is organized around the concept of surrogate species, where knowledge and conservation of one or a few species are presumed to provide for the needs of other species. This is a different focus than species indicating some aspect of ecosystem status, which is discussed in chapter 7 as a systems approach.

The value of using surrogate species is to reduce complexity of analysis, to provide a better-organized framework for management direction, to make monitoring tasks realistic, and to provide a way to draw inference to other species that cannot feasibly be monitored directly (Caro and O'Doherty 1999). The concept presumes that the surrogate species is one that is well known and easily sampled and provides a direct correlation with the species for which it is serving as surrogate. Surrogates are often relatively abundant species (Caro and O'Doherty 1999), but Warman et al. (2004) found some success in using rare Canadian species listed as threatened or endangered as surrogates to conserve terrestrial vertebrates.

Categories of surrogate species proposed in the literature include umbrella species, focal species, guild surrogates, management indicator species, biodiversity indicator species, and flagship species. Strictly speaking, the assumption that one species can be a surrogate to and represent others is by definition false based on the concept of the niche (Hutchinson 1978). Basic ecological theory says that each species is unique with respect to the environmental conditions occupied, its ecological role, and its resource-utilization functions. For this reason, no two species are entirely interchangeable or mutually replaceable (Caro et al. 2005). The surrogate species notion, however, is usually applied less stringently, looking at patterns of co-occurrence within some land area rather than niche overlap and, as such, the various categories of surrogate species explored here have varying degrees of utility, efficacy, and validity.

Umbrella Species

"Umbrella species" is a term that refers to a single species, typically a wide-ranging and widely distributed species, that represents the requirements of a portion or all of a species assemblage or community. It is the use of species with extensive spatial distributions that sets this approach apart from other surrogate species such that its area of occupancy on the landscape physically encompasses the occurrence of other, less widely distributed sympatric species (Caro and O'Doherty 1999). Suter et al. (2002) suggested that an umbrella species should have habitat requirements that are similar to those of other species, but that its range extent should be broader than other species it is intended to cover. The distinguishing, and often untested, assumption of the umbrella species approach is that satisfying conditions for the persistence of the umbrella species confers persistence to other co-occurring species of its biotic community.

Umbrella species have been used in a variety of conservation assessments and management plans. Launer and Murphy (1994) used a threatened butterfly as an umbrella species for conservation of a threatened grassland ecosystem. In the U.S. Pacific Northwest, the northern spotted owl (*Strix occidentalis caurina*) has been used as an umbrella species to further conservation of a wide array of old-growth species and communities. Invertebrates have been proposed both as umbrellas and as the benefactors of the umbrella approach. New (2004) suggested using velvet worms (phylum Onychophora) as potential umbrella species for other soil macroarthropods, while Rubinoff (2001) examined how well a vertebrate insectivore, California gnatcatcher (*Polioptila californica*), functioned as an umbrella for arthropods.

Several authors have explicitly tested the umbrella species concept. Berger (1997) tested the efficacy of using black rhinos (*Diceros bicornis*) as an umbrella species for conservation of other desert ungulates in Namibia. The study found that the space used by black rhinos alone would be unlikely to assure the existence of other ungulate populations since rhinos did not use all ungulate habitats during all seasons and rainfall levels. Similarly, Rowland et al. (2006) compared the distribution of their umbrella species, greater sage-grouse (*Centrocercus urophasianus*), to 39 other species also occurring as sagebrush obligates and found that their umbrella overlapped well for some species, such as the pygmy rabbit (*Brachylagus idahoensis*) and poorly for others, such as the lark sparrow (*Chondestes grammacus*). The poor overlap was again attributed to basic differences in habitat associations and overall geographic ranges.

Some studies have specifically looked at the value of rare species as umbrellas. In a test of the "umbrella effect" attributed to endemic and threatened bird species in South Africa and Lesotho, Bonn et al. (2002) found that a reserve network designed for these umbrella species may be insufficient to preserve overall species diversity.

The absence of strong evidence supporting the umbrella species approach across this admittedly small subset of studies is not atypical. In reviews of the umbrella species concept, Andelman and Fagan (2000), along with Roberge and Angelstam (2004), concluded that umbrella species too seldom overlap with other species of interest and rarely work to protect all species in the area of conservation concern.

Evidence of success in using the umbrella species concept is not completely absent in the literature. Bani et al. (2006) found that birds selected according to their sensitivity to habitat patch isolation, patch size, edge effect, and habitat structure were effective as umbrellas for other co-occurring birds. Similarly, Rondinini and Boitani (2006) found that amphibians and mammals (1654 species) together served as effective umbrellas for most species in the other taxa, and that developing reserves based on suitable habitat areas rather than overall distribution ranges improved the utility of one taxon serving as an umbrella for the other. Finally, Poiani et al. (2001) tested the umbrella species approach to conservation of natural grassland habitats in Minnesota using greater prairie chicken (*Tympanuchus cupido pinnatus*) as the umbrella species. They found that the umbrella species approach functioned well when used in conjunction with a conservation approach that also considered conservation of the largest native habitat patches. The combination of umbrella and habitat size criteria provided for greater conservation benefits than did either approach alone. Favorable evaluations of the umbrella species approach appear to be related to studies that have defined several (as opposed to a single) umbrella species, or have used the umbrella concept in combination with other conservation approaches.

Focal Species

"Focal species" refers to a set of target species that are identified for the purpose of guiding management of environments, habitats, and landscape elements in a tractable way. The Committee of Scientists (1999) defined a focal species as a species whose measurement provides substantial information beyond its own status. The key aspect of the focal species concept, at least as it relates to species conservation approaches, is that the distribution and abundance of focal species provide inference on the status of other species. Clearly the focal and umbrella species approaches are similar. What sets them apart is that focal species are defined using a variety of ecological attributes and are not constrained by the strong focus of umbrella species selection on species with large area requirements.

The seminal work on defining focal species has been attributed to Lambeck (1997). He defined focal species as those that are resource, area, or dispersal limited, and that respond to similar stressors. Moreover, focal species are a strict subset of the total species pool present in a landscape, and they possess the most stringent requirements. For example, focal species can be those whose distribution is restricted by specific disturbance events such as fire, who require the largest habitat patch sizes or landscape areas, and whose dispersal requirements are most limiting. In this way, the set of focal species should be those whose requirements collectively encompass the needs of all other species.

In addition to the common assumption for all surrogate approaches (i.e., reflecting the ecology of other species), the focal species approach assumes that a greater degree of such representation can be attained if the focal species has stringent requirements of resource use and dispersal capability. This characteristic assumption has received relatively little empirical testing among a variety of ecosystems.

There are many examples where focal species have been used to guide biodiversity conservation on public or private land. In Australia, Watson et al. (2001) used Lambeck's (1997) focal species approach to identify species that were sensitive to woodland fragmentation stressors in southeastern Australia. They identified a set of focal species based on those with the most restrictive requirements for minimum habitat patch size, habitat structural complexity, and habitat connectivity across the landscape. They concluded that the focal species approach was effective for efficiently developing landscape planning guidelines for woodland birds—noting that 95% of the resident woodland bird species in the region should be accommodated. However, they did not explicitly test this assertion with actual field trials and follow-on research.

Other authors have used the focal species approach to derive landscape designs. Snaith and Beazley (2002) used moose (*Alces alces*) as a focal species to derive a reserve design in Nova Scotia, Canada. Bani et al. (2002) proposed using the focal species approach to plan woodland ecological networks in Italy based on their central assumption that focal species encompass the structural and functional needs of entire ecological communities. They developed maps of habitat core areas and connections for their set of focal species. However, they did not explicitly test their main assumption.

Several authors have tested or reviewed the assumptions of the focal species approach. Rothley (2002) found that detailed descriptions of focal species and their conservation interests were needed to resolve conflicts among objectives. Lindenmayer et al. (2002) suggested that the underlying theoretical basis of the focal species approach is problematic, citing the limited utility of surrogate-species approaches in general. They also found that lack of data guiding selection of focal species sets is often a problem for practical implementation. They suggested using a mix of strategies in any given landscape to spread the risk of failure of any one approach. They cautioned that resource managers should be aware that the focal species approach might not result in the conservation of all biota in a landscape. In this spirit, Noss et al. (2002) used three complementary approaches—protecting special elements, representing environmental variation, and securing habitat for focal species—to identifying highest-priority "megasites" for conservation in the Greater Yellowstone Ecosystem.

Although the focal species concept has a foundation in the ecological literature, the varied criteria that can be used to define them has led to a diverse set of applications and considerable equivocation on the expected conservation benefits. For this reason it should not be surprising that some have cautioned against relying too heavily on the focal species approach to biodiversity conservation (Campbell et al. 2002). Confidence in this approach will remain reserved until more extensive testing has occurred.

Guild Surrogates

The guild surrogate approach entails one member or a subset of members serving as a surrogate for other members of the guild. It differs from an umbrella or focal species in that the surrogacy effect is restricted to guild members. Strictly speaking, the term "guild," as originated by Root (1967), refers to a set of species having a common diet and foraging mode. However, many other authors have appropriated and greatly extended the term to mean a set of species that share virtually anything in common, including habitat association, behavior, trophic orientation, diet, or resource selection patterns. We use this broader definition in our discussion of the guild surrogate species approach.

The main assumption of the guild approach is that the response of one member (or a small subset of members) to environmental conditions or changes is closely correlated with the response of all guild members. A corollary is that changes in environmental conditions affecting one or a subset of a guild would affect the other members in the same way. Consequently, population responses are expected to be similar in direction if not in magnitude—that is, significantly positively correlated.

In general, wildlife–habitat relationships databases can be used to generate lists of species sharing common attributes (Patton 1992). Any such list can be thought of as a guild in this broader sense of the term. Overall, the literature suggests that (1) correctly placing species into guilds that are assumed to contain species with similar ecological characteristics may necessitate detailed autecological, including behavioral, information; (2) guilds can be structured by a wide variety of mechanisms, including interspecific resource competition (mechanisms of resource and habitat overlap and partitioning) and common response to general habitat structure, but no one mechanism operates invariantly across all guilds; and (3) often, and even if guilds are defined based on detailed autecological information, guilds tend to consist of species that have disparate responses to changes in environmental and habitat conditions. Therefore, identifying guilds may require much ecological information, likely not available for RLK species.

In a study that sought to test the guild surrogate approach among bird communities, Block et al. (1987) concluded that one species of a guild does not represent the habitat specificity and population patterns of other members. Mannan et al. (1984) and Morrison et al. (2006) noted that individual species of a guild respond differently to changing environmental conditions, whereas the guild as a whole shows little or no variation in its presence, total species richness, or abundance. Morrison et al. (2006) concluded that grouping species into guilds may be useful for depicting species with similar functions or trophic relations, but the guild approach to management may not be useful for specifically predicting responses of individual species to environmental conditions and changes.

Based on these empirical tests, using guilds to predict common responses among all component species to environmental changes and management activities may not have a strong scientific foundation. Some have suggested that the lessons learned from the guild surrogate failures point to treating the entire guild as the response unit for management and monitoring—such an approach is discussed under the multiple species approaches.

Management Indicator Species

"Management indicator species" (MIS) is a broad term that has been used by the U.S. Department of Agriculture (USDA) Forest Service in its planning regulations for national forest and national grassland management. "MIS" refers to any species or set of species of management or conservation concern, and can include federally listed species; species with special habitat needs; species that are hunted, fished, or trapped; species of special interest; and species for which population changes are presumed to reflect the status of other species (R. Holthausen, pers. comm.). MIS have been used in U.S. national forest planning since the 1980s. For example, northern flying squirrels (*Glaucomys sabrinus*) have been used as a management indicator species of temperate rain forest (Smith et al. 2005). Milledge et al. (1991) suggested use of large owls and gliders as MIS in ash forests of Victoria, Australia.

For U.S. federal land management, habitat objectives are typically established for MIS, planning alternatives are evaluated based on their effects on MIS, and population trends of MIS are monitored and relations to habitat changes determined. It is difficult to derive a simple set of assumptions about the use of MIS, other than the following: a set of species chosen as MIS somehow represent the array of management issues, conditions, objectives, and conservation concerns. When used strictly to reflect the status of other species, a major assumption of the management indicator species approach is that the population status and trend of one species can represent others; that is, there is significant positive correlation in at least population trend (and possibly also size, distribution, and persistence likelihoods) between the indicator species and other species.

Niemi et al. (1997) examined the distribution of birds proposed as "management indicators" or "sensitive" species in a national forest in Wisconsin. They found that few bird species were consistently associated with habitats for which they were deemed to be indicators, and few sensitive species were positively or negatively associated with other species. Also, they found that most MIS were either too rare or too difficult to practically monitor.

Biodiversity Indicator Species

Biodiversity indicator species are single species or taxonomic groups used to identify areas of high species diversity or of high biological productivity (Caro and O'Doherty 1999). We extend the definition here to include relatively common species that are used to identify locations of rare species.

Kintsch and Urban (2002) tested the value of more common indicator plant species as predictors of the occurrence of rare plant species in the Amphibolite Mountains of North Carolina. (They termed these indicators focal species, but we believe they are more properly referred to as biodiversity indicators.) The indicators proved successful in predicting occurrences of the rare species, with 62 to 100% of the rare species locations predicted by the occurrence of the indicator species. However, there were also overprediction rates as high as 275%.

Fleishman et al. (2000, 2001) tested the value of a set of criteria that could be used to select biodiversity indicators. (They termed these species umbrellas, but again they seem better described as biodiversity indicators.) In one test Fleishman et al. (2000) used these criteria to identify two butterfly species as indicators of other butterflies. They found that protection of canyons where the two species occurred would protect sites of 97% of all other butterfly species. In a second test (Fleishman et al. 2001) they used the same criteria to select biodiversity indicators for both plant species and butterfly species in California and Ohio. Again they found that protection of sites of the indicators would protect a large proportion of sites for most other species within the same taxonomic group, and that protecting sites with the umbrella species was more efficient (few sites needed) than protecting random sites. They also tested the effectiveness of the indicators for cross-taxonomic protection. Here they found that the indicators were no more effective than random sites for cross-taxonomic protection.

Butchart et al. (2006) and Quayle and Ramsay (2006) suggested using trends in conservation status of species as indicators of biodiversity conditions. This approach would use selected species with particular trends in distribution or abundance as indexes to overall biodiversity conditions but would not necessarily extend reliable inference to individual species, particularly rare or little-known species. Oertli et al. (2005) tested the overlap in distribution of bees, grasshoppers, and wasps in the Swiss Alps and determined that none of the test taxa reflected species richness or community similarity of the other taxa well enough to be used as biodiversity indicator species.

Flagship Species

Flagship species are those that carry high public interest and garner social or cultural concern for their well-being. They are rallying points for conservation of broader ecological communities and systems (Walpole and Leader-Williams 2002). The main assumption of the flagship species approach is that individual species can be identified in a given ecosystem and promoted in highly visible conservation programs that would entice positive interest and support by the general public. Flagships are often claimed to have surrogacy benefits, but their distinction is in being able to elicit strong public support for conservation (Caro and O'Doherty 1999).

Examples of flagship species include the northern spotted owl, grizzly bear (*Ursus arctos horribilis*), bald eagle (*Haliaeetus leucocephalus*), and tiger (*Panthera tigris*), and in southern Africa the "Big Five"—African elephant (*Loxodonta africana*), leopard (*P. pardus*), lion (*P. leo*), African buffalo (*Syncerus caffer*), and rhino (*Diceros bicornis* or *Ceratotherium simum*) are touted as such in many African ecotourism and conservation activities. In India, entire national conservation programs have been founded on flagship species, including Project Tiger and more recently Project Elephant, the latter for the Asian elephant (*Elephas maximus*). Clark et al. (2003) proposed using Baird's tapir (*Tapirus bairdii*) as a flagship species in Costa Rica in an interdisciplinary approach to resolve questions about conservation policies entailing reasonableness, political practicality, and moral justification.

In summary, flagship species are used to rally conservation support. They are useful for spurring social concern and public interest for conservation, usually of large-bodied, charismatic species (usually birds or mammals) and their habitats. They might also be useful for educating the public about broader concerns for ecosystem health. For example, in a general news article on primate conservation, the Canadian Broadcasting Corporation reported that "Primates act as a flagship species that indicate the health of their surrounding ecosystem" (CBC 2002). To indicate or include conservation of RLK species, flagship species may be useful to initially garner some general protection of habitats or geographic locations.

However, relying on flagships to fully provide for all RLK species may produce some of the same problems as experienced with other surrogate approaches: lack of specificity of resources, environmental conditions, and habitats selected by the RLK species; lack of assurance that sites used by RLK species will be protected; and lack of assurance that natural disturbances and human activities will not harm RLK species. Flagships may generally not serve as reliable indicators of other species, particularly RLK species. Further, many of the species identified in the literature as umbrella species, focal species, or management indicator species might be better thought of as flagship species. That is, the motivation for choosing them has large elements of public concern rather than ecological representation or sensitivity to environmental change.

Conservation of Multiple Species

In this variant of the species approach, the aim is to ensure the viability or continuance of multiple species as an assemblage or selected subset of the ecological community. The two main categories pertinent to this are the use of entire guilds and the use of entire habitat groups.

Entire Guilds

As defined in the "Guild Surrogate" section, a guild is a set of species that share some common attribute of resource use or ecological role. In the entire guild approach, the focus is not on a species selected to represent the rest of the guild, but rather on all species of the guild. The guild as a whole becomes the response that is managed and monitored to determine the success or failure of conservation management. The main assumption of the entire guild approach is that conservation of all guild members can be judged by monitoring the total number of individuals making up the guild. One of the strengths of this approach is derived from pooling the counts from all guild member species, thus increasing the resource manager's power to detect trends in the guild as a whole (Verner 1984).

Entire guilds have been used for assessing habitat disturbance effects and, in some cases, guiding management. For example, federal land resource managers have used guilds of avian cavity excavators and Neotropical migrants as a means of focusing forest and grassland management for all species of these guilds. Knopf et al. (1988) studied the effects of seasonal cattle grazing on the guild structure of a riparian avifauna. Maurer et al. (1981) studied effects of logging on the guild structure of a forest bird community in West Virginia. Rewa and Michael (1984) used habitat evaluation procedures (HEPs) to assess habitat values of wildlife guilds. O'Connell et al. (2000) used 16 bird guilds based on their behavioral and physiological response as indicators of ecological conditions in the central Appalachians.

However, few studies have tested the basic assumption that the status of all members of a guild could be used to infer the conservation status of guild members, and that conditions providing for or affecting the guild as a whole would equally influence all members of the guild. Steffan-Dewenter et al. (2002) observed that the correlation of various bee pollinator guilds with landscape attributes varied depending on the scale of the analysis. They reported that disruption of local landscape attributes affected solitary wild bees more than social bees.

Lopez et al. (2002) reported that guilds of wetland plant species responded differentially to changes in landscape habitat attributes. In their detailed study of ungulates in East and Southern Africa, Fritz et al. (2002) identified a series of herbivore guilds based on their common trophic relations and responses to environmental variations. They discovered that the guilds respond differently to competitive interaction with large herbivores and to soil and climatic factors.

In a Brazilian Atlantic rainforest, Aleixo (1999) found that forest bird guilds delineated on the basis of foraging behavior were similar in species richness and diversity but strongly differed in species composition between primary and selectively logged forest, and that within guilds, understory and terrestrial insectivore birds were most sensitive to habitat changes. Bishop and Myers (2005) determined that there are tight associations among bird species with common primary habitat, area sensitivity, migratory status, and nest placement, and that species so grouped could serve usefully to identify areas of high bird species richness as candidate areas for conservation.

The lesson, like the individual-species indicator approach to the use of guilds (discussed earlier), is that species within a guild typically respond differently to environmental change, particularly anthropogenic stressors. Thus, even if a guild as a whole remains unchanged or shows some group responses, one cannot assert generally that this response pertains to each component species of the guild.

Entire Habitat Assemblages

In the entire habitat assemblage approach, species are identified that belong to a habitat group; that is, they share common usage of some macrohabitat, such as a general vegetation type and vegetation structural or successional stage. The approach focuses on providing for these macrohabitat types under the assumption that if the macrohabitat is provided, the requirements of the entire habitat group of species will be met. Implicit in this approach is the assumption that species needs are mostly defined by the macrohabitat(s) they occupy. It is distinguished from the entire guild approach in that this approach focuses on monitoring macrohabitat status rather than monitoring species.

A major effort to identify and evaluate entire habitat groups has been the "source habitats" analysis by Wisdom et al. (2000) for the interior Columbia River basin. Source habitats of terrestrial vertebrate species, as analyzed in this study, are primarily vegetation conditions that assumedly provide for the requirements of the organisms. In other analyses, Fauth et al. (2000) modeled landscape patterns to identify source habitats of Neotropical migrants in the U.S. Midwest, and Dyer et al. (2001) explored definitions of functional groups (see later discussion) of plants based on their source habitats and growing conditions.

The validity of the basic underlying assumptions to the entire habitat group approach has not been explicitly tested in the field and reported in the literature. Many approaches to depicting and researching wildlife–habitat relationships suggest, however, that habitat is defined for each species individually and with multiple parameters at various scales of space and time, including at macro- as well as microhabitat levels.

However, for habitats that are discrete and highly limited geographically, such as caves, ocean-floor thermal vents, vernal pools, and thermal hot springs, the tenet underlying this approach may be entirely valid. That is, conserving such extreme and isolated habitats may conserve its full biota, including RLK species.

Geographically Based Approaches

A common strategy used to conserve species, regardless of their degree of commonness or rarity, is to establish management areas (e.g., parks, reserves, refuges) where the overarching objective is the protection or restoration of natural conditions and biological diversity (Beazley et al. 2005). Numerous approaches now exist to assist conservation scientists in the identification and prioritization of areas that should be the focus of resource management. A characteristic common to these approaches is that they are geographically explicit—that is, all lead ultimately to the delineation of geographic areas on the landscape that warrant conservation focus. Furthermore, all are based on the key assumption that by managing the delineated areas appropriately, the species, their biophysical environments, and the ecological system's characteristic processes will be preserved (Flather et al. 1997).

Three broad classes of geographically based approaches may be described: (1) approaches based on management for locations of target species at risk, (2) approaches that focus on the identification of so-called hot spots or centers of concentration of rare species, and (3) approaches based on reserves or protected areas established for biodiversity conservation. The difference among the categories is often subtle, and membership in each category is not necessarily mutually exclusive. Often, geographically based approaches are applied at a coarse scale with the intent of identifying broad geographic areas that compel conservationists to take a finer-scale examination of the biodiversity pattern within these areas (Shriner et al. 2006).

Locations of Target Species at Risk

Management for locations of target species at risk refers to providing guidelines for protecting sites known to be occupied by individuals of a species deemed to be at risk. This is the basic concept behind the Survey and Manage species program of the Forest Service and the U.S. Department of the Interior Bureau of Land Management in the Pacific Northwest (Molina et al. 2006). In that program, locations of target species have been the basis of local protection and "known site surveys" in the Survey and Manage species program of the interagency Northwest Forest Plan (see chap. 4). This approach differs from the PVA approach in that the PVA approach entails modeling population demography or genetics, whereas managing for locations of target species at risk does not and is based primarily on known occupancy patterns.

The main tenet of this approach is that managing for known locations of target species contributes to ensuring persistence of the species. That is, it is presumed that known locations are where the organism finds suitable resources and habitat conferring local persistence. This is an important assumption because management guidelines for sustaining the species will focus on sustaining habitats where the species is known to occur. In general, the validity of this assumption is largely untested and unknown for many of the rare and little-known taxa.

This presents two potential problems. First, if known locations represent suboptimal habitat, managing to sustain these habitats may do little to sustain the species. Second, even if locations where the species currently occur do, in fact, represent optimal habitat, simply protecting these locations may not sustain the habitat—an understanding of the processes behind habitat formation is required (Everett and Lehmkuhl 1999; Ferrier 2002). Sustaining habitats then becomes linked with restoring various natural disturbance regimes (see chap. 7).

For RLK species of fungi, cryptogams, vascular plants, and invertebrates, this approach may rely on site location information stored in State Heritage Program databases and other agency location data. The approach thus relies on the completeness, accuracy, and recency of such location data; on being able to identify land-disturbing management activities to be avoided at each location; and on an appropriate protection buffer around each location (Aubry and Lewis 2003). Unfortunately, species databases are often incomplete for many taxa, extremely sparse for many regions, and characterized by survey biases toward easily accessible sites (Ferrier 2002). Although managing for locations of target species at risk can be a critical component of planning, such data are lacking for many if not most RLK species.

Species "Hot Spots" or Concentrations of Biodiversity

Distributions of plant and animal species are inherently heterogeneous. One manifestation of this heterogeneity is that some areas (whether examined at a global, continental, regional, or local scale) support more species than other areas. This pattern is observed whether one is counting (or estimating) all species that occur within some locale, or is counting a subset of species that share some attribute of conservation interest (e.g., rare species, threatened species, or endemic species). Given spatial structure in species' distributions, resource managers are faced with deciding which geographic areas should be the focus for biodiversity conservation.

The term "hot spot" was coined by Myers (1989) to denote those global areas where a high number of endemic species were found in a relatively small geographic area. Myers argued that greater conservation benefits could be realized if efforts were focused in those areas where endemics were concentrated—that is, the hot spots. Since Myers' specific usage with endemism, the term "hot spot" has become generalized to refer to an area, or set of areas, that ranks high on any number of ecosystem attributes, including species richness (Scott et al. 1993), threatened or endangered species (Flather et al. 1998), imperiled species (Chaplin et al. 2000), or indicators of ecosystem condition (Hof et al. 1999).

The concepts underlying hot spot analysis have been implemented in the United States under an approach generally referred to as gap analysis (Burley 1988). Gap analysis uses cartographic techniques to identify underrepresented elements of biodiversity in an existing network of protected areas—that is, hot spots that lie outside existing protected areas. As described by Scott et al. (1993), in gap analysis vegetation maps and animal distributions are used to determine if vegetation types are inadequately represented within protected areas, or if centers of species richness fall outside such areas currently managed to conserve biodiversity. Those areas of vegetation types or high species richness that fall outside the existing conservation network thus define the geographic areas that could be considered for future conservation efforts.

The appeal of mapping the occurrence pattern of plants and animals as a way of setting conservation priorities is derived largely from the simplicity of the approach. This simplicity, however, belies an important constraint associated with this approach—distributional data for most taxa are incomplete or unavailable (Ferrier 2002). One means of overcoming this data constraint is to assume that the diversity pattern of relatively well studied taxa indicate the pattern among other taxonomic groups (Reid 1998). Although there is some evidence that the geographic pattern of species richness covary among some taxa, this pattern is certainly not generally true. Empirical evidence to date cautions conservation planners against using hot spots for a few taxa to indicate where the overall diversity hot spots would occur if we had taxonomically comprehensive and spatially extensive inventories (Flather et al. 1997).

Another consideration in mapping hot spots or centers of concentration of any set of species is errors in species occurrence. This is particularly important with some RLK species that may be difficult to detect and thus may appear absent in seemingly suitable habitat (see chap. 5 for further discussion of detectability). Since such errors apply to each species, the overlaying of individual species occurrence maps can cause these errors to compound (Dean et al. 1997), leading to suboptimal conservation decisions

(Conroy and Noon 1996). Also, areas of high species richness might represent ecotones where edges of species' ranges meet, rather than more optimal conditions toward the center of species' ranges and ecosystems (Araújo and Williams 2001). Thus focusing conservation on those hot spots might overlook conservation of optimal habitat conditions for some species.

Apart from the problems associated with the hot spot approach caused by limited distributional data and locational errors, there is also concern that hot spot analysis focuses too much on ecological pattern at the expense of the ecological processes that have generated the pattern. This could be a problem insofar as conservation of a specific pattern would fail to provide for the future dynamics of the processes, such as future wildfires, that would produce mosaics of vegetation age classes outside of current locations of such mosaics. As currently applied, hot spot analysis implicitly assumes that geographic delineation of areas based on patterns of species occurrence will also "capture" the important processes that have caused, and will presumably maintain, that pattern. However, theoretical and empirical support for this assumption is weak (McNeely 1994).

A final limitation of the hot spot approach is that it often focuses on counts of species rather than species composition. Because areas that rank high on species counts may share many species, the typical hot spot analysis does not permit the teasing out of the relative contribution of each area to the preservation of some overall species pool. Consequently, it does not necessarily follow that conservation priority should always be given to the most species-rich areas. A solution to this problem is to consider measures that take into account how completely the species pool within a geographic area is represented within a reserve system (Vane-Wright et al. 1991). Over the last couple of decades, a number of methods have been developed that attempt to efficiently (greatest conservation benefit for the least cost) identify priority conservation areas (Margules and Pressey 2000). These approaches are the subject of the next section.

Reserves or Protected Areas Established for Biodiversity Conservation

One of the next great challenges for conservation science is the design and implementation of comprehensive and ecologically adequate reserve networks (Ferrier 2002). Although reserves have been used to address systemwide conservation needs, they are often created for one or a few species that are commonly RLK or in peril. We have considered reserves in this species context and discuss reserves as a species-level strategy rather than as one of the system-level strategies that are reviewed in chapter 7. Worldwide, the proportion of land that is strictly managed for the conservation of biological diversity is small—about 3% of the terrestrial land base (McNeely 1994).

Because conservation reserve areas are rare (at least in areal extent) and human impacts on natural ecosystems are increasing, there is a growing realization that the choice should be optimal when there are alternatives on where to locate reserve lands (Flather et al. 2002). A major assumption of the reserve or protected area approach is that protecting such areas from specific human activities or disturbances will provide for species or system persistence over time. In this approach, it is important to clarify specifically what "protection" refers to; that is, what elements or aspects of ecological communities and ecosystems are to be protected, and what conditions or (usually anthropogenic) stressors they are to be protected from. Most of the literature on protected area approaches, however, fails to specify or clarify these basic criteria.

Much of the recent literature on optimal reserve designs has focused on species assemblages. As such, the metric of interest concerns a collection of species and often attempts to maximize the number of different species that are conserved within some fixed or minimal area or set of areas (e.g., Camm et al. 1996). However, the reserve design problem also has important aspects that are focused on the conservation of individual species. The establishment of wildlife refuges in the United States has often been tied directly to conservation of a single species. Moreover, reserve design criteria for a single target species of conservation concern, such as the black-footed ferret (*Mustela nigripes*) (Bevers et al. 1997), have been used to address resource management conflicts on lands managed for multiple uses. Reserve designs for single species often consider how population viability varies as a function of habitat size and layout (Hof and Flather 1996), and often account for metapopulation dynamics (Holloway et al. 2003).

Early reserve design principles directed at the conservation of species assemblages were derived largely from the equilibrium theory of island biogeography (MacArthur and Wilson 1967), where the number of species expected to be supported on an island or habitat patch is predicted by the interplay between colonization and extinction rates. Key design principles ostensibly derived from this theory can be summarized as follows: (1) a single large reserve is better than several small ones of equal total area (the so-called SLOSS [single large or several small] debate); (2) reserve shapes that are compact are better than those that are convoluted or elongated; (3) multiple reserves that are close together are better than multiple reserves that are far apart; and (4) reserve units that are connected via corridors are better than reserve units that are isolated (Wilson and Willis 1975, 529). However, the SLOSS debate may have revealed that none of these principles followed strictly from the theory of island biogeography, and that most are empirically untrue in many of the archipelagoes and reserve systems for which there are adequate data, although these design rules live on (J. Quinn, pers. comm.).

The applicability of these reserve design recommendations based on island-biogeographic principles has been questioned for a number of reasons, including unrealistic simplifying assumptions and the inability to track species identity. Island-biogeographic design recommendations assume that habitat is identical from place to place—an assumption that will never be met in practice (Soulé and Simberloff 1986). Habitats vary spatially and this heterogeneity almost certainly accounts for the fact that many empirical studies have found that species counts from several dispersed sites are at least as large as counts from a single contiguous site of equal total area (Simberloff 1998). Island biogeography theory predicts the number of species that could be supported in habitat of a certain size. It does not predict the identity of those species. For this reason, it is not possible to predict the total number of species in a complex of multiple habitat patches because the number of species shared among units is indeterminate. More contemporary reserve selection algorithms that explicitly account for spatial variation in habitat and species composition address these two shortcomings (Higgs and Usher 1980).

The reserve selection problem can be simply stated as a question: How do we best locate reserve units on the landscape such that they contain the greatest number of biodiversity elements as possible (Pressey et al. 1993)? Three concepts have emerged as key in addressing that problem: representativeness, complementarity, and efficiency (for reviews see Margules and Pressey 2000; Sarkar et al. 2006). Under the concept of representativeness, conservation is focused on ensuring that some target set of species are adequately addressed in the conservation plan. The concept of complementarity is closely related to representativeness and measures the gain in new species that receive coverage as individual reserve units are added to an existing or candidate reserve network. The notion of efficiency is important because it recognizes that resources available to conservation efforts are limited, and therefore reserve designs need to maximize representation for the least cost. These concepts have emerged as important reserve design principles since scoring procedures that rank conservation priority based on simple counts of biodiversity elements (e.g., species hot spots) may rank areas with high counts as having high conservation priority even if many of the species are shared among the reserve units making up the network.

Although the literature is characterized by what may seem to be an overwhelming variety of reserve selection algorithms, the approaches do fall into two broad classes—those that define reserve networks based on iterative or stepwise algorithms, and those that are based on optimization techniques from operations research. Both approaches generally attempt to define the set of reserve units that will meet the biodiversity targets (e.g., occurrence of all species, rare species, or endemic species at least n times in the reserve network) for the minimum amount of total reserve area or reserve acquisition cost. They differ fundamentally in that iterative procedures are often referred to as heuristic since they can only approximate a maximally efficient design (Underhill 1994); operations research approaches do prescribe a truly optimal design.

The literature is dominated by heuristic reserve design algorithms (for review see Williams 1998) and their frequent use is related to the fact that they are intuitive and simple, they seem to be applicable to a broad and realistic set of conservation problems, and they seem to provide reasonably good solutions when compared to optimization approaches (Pressey et al. 1997). Attempts to use optimization methods sometimes fail or take too long to solve reserve selection problems, although this limitation may be overcome by creative formulation or software advances (see Rodrigues and Gaston 2002).

Since we are primarily concerned with RLK species, a logical question is how well reserve selection algorithms perform if rarity, rather than simple richness, is the biodiversity element of interest. A few investigators have compared the relative efficiency of simple richness-based algorithms with rarity-based algorithms. They found that rarity-based algorithms tended to be more efficient at finding the minimum reserve area necessary to represent all species at least once (Kershaw et al. 1994). However, this finding is hardly general (Csuti et al. 1997) and appears to be contingent on the size of the species pool, scale (i.e., size of individual reserve selection units), and degree of species endemism (Rodrigues and Gaston 2001). Consequently, differences in efficiency between richness- and rarity-based algorithms are not generalizable and depend on the distribution of species abundances, the size of reserve units considered, and the pattern of species occurrence on the landscape (Williams 1998).

Data limitations have long been recognized as an important constraint associated with geographically based conservation efforts (Botsford et al. 2003). Like hot spot analysis approaches, there have been attempts to quantify the degree to which reserve designs based on well-studied taxa also meet the conservation needs of poorly known groups (Flather et al. 1997; Reid 1998). For example, Launer and Murphy (1994) found that of all sites where a rare butterfly occurred, in an attempt to conserve genotypic representation, greater than 98% of native spring-flowering forbs also received protection. Similarly, Csuti et al. (1997) indicated that reserve designs resulting in complete representation of one major taxon may adequately represent occurrence patterns of other unrelated taxa. Although these findings do offer some hope that biodiversity benefits from reserves that are designed based on well-known taxa will transfer to little-known and unmeasured taxa, opposing evidence does exist in the literature. For example, in deciduous woodlands of western Norway, Sætersdal et al. (1993) found that among the 32 reserve units with complete representation of plants, only 5 units were shared among the set of 12 units with complete representation of birds. Such contradictory findings in the literature point to an important research need—namely, to identify those ecological circumstances, if any, when it is tenable to use occurrence patterns of one or a few taxa to represent the pattern for other taxa.

Another important limitation with reserve selection algorithms is that the results are contingent upon the pattern of species occurrence across the region where the reserve design is being planned. Since most algorithms consider only presence/absence data, it is difficult to determine whether the reserve will conserve the species pool as land use activities change the landscape context within which the reserve is embedded (Soulé and Simberloff 1986).

Potential approaches to address this concern include consideration of land suitability and anthropogenic threats (Margules and Pressey 2000) or explicit measures of persistence probability (Williams and Araújo 2000; Gaston et al. 2002). However, estimates of land suitability and persistence are often based on current conditions or recent historical data, and it is not clear that reserve designs will be robust (i.e., continue to conserve species) to future land use changes in surrounding nonreserve lands. Incorporation of such concerns will require more mechanistic models of species persistence which acknowledge that reserve units are not independent of each other, nor are they independent of the landscape matrix within which they reside.

Conclusion

A plethora of species-level approaches to conservation has been advanced, dealing with population viability, species surrogacy, multiple species groups, and geographic targets. Most of these approaches were developed to help streamline difficult, and at times intractable, tasks of simultaneously managing for many species, although the viability approaches usually pertain to single species.

The approaches are not clearly independent. Thus any attempt at categorizing these approaches is not clean because of vague and ambiguous definitions. For example, the literature has many examples of where authors have mixed and overlapped the definitions and use of many of the species approaches we reviewed here. We suggest that whatever names are used, the purpose of, and assumptions behind any given approach be clearly articulated. In this way, the pertinence and applicability to conservation of RLK species can be more accurately assessed.

Few approaches have had their fundamental assumptions rigorously tested. This is particularly problematic when extending inference to conservation of RLK species. Objectives, assumptions, implementation, and actual performance of any given approach for RLK species conservation should not be conflated but should be evaluated individually. For example, the objective of some surrogate species approach, such as the use of indicator species, could be to focus management on a species that is not rare and is well known, to indicate the status and conservation of some RLK species. The assumption would be that if the indicator species is provided for, then the RLK species will be provided for. The implementation of the approach might be feasible and economical. However, the veracity and actual performance of the approach might be unknown or suspect; that is, whether providing for the target indicator species truly serves to provide for the RLK species. Research and monitoring can help determine the degree of veracity and performance, which should not be presumed simply from the stated objective and assumption. Likewise, the implications of uncertainty (i.e., errors and lack of basic information on RLK species) need to be admitted and their influence on performance of any given conservation approach evaluated.

In general, RLK species have characteristics that make demonstrably successful implementation of any of these approaches challenging. To a certain degree, perhaps the best we can reasonably expect is to use research and monitoring to determine which approaches, alone or in combination, could best meet conservation objectives for some management situations or geographic areas or species groups, and then extend inference of those results elsewhere. It may not be reasonable to conduct enough study on RLK species to fully determine their true rarity and make them well known.

The next chapter continues a review of system-level approaches and will draw fuller conclusions on application of both species- and system-level approaches.

Acknowledgments

We thank Carolyn Sieg for suggestions, discussions, and contributions. We thank Richard Holthausen, Randy Molina, Dede Olson, and Martin Raphael for helpful discussions. We appreciate the useful reviews of the manuscript by Chris Iverson, Nancy Molina, and Jim Quinn.

References

Aleixo, A. 1999. Effects of selective logging on a bird community in the Brazilian Atlantic forest. *Condor* 101:537–48.

Andelman, S. J., and W. F. Fagan. 2000. Umbrellas and flagships: Efficient conservation surrogates, or expensive mistakes? *Proceedings of the National Academy of Sciences (USA)* 97:5954–59.

Araújo, M. B., and P. H. Williams. 2001. The bias of complementarity hotspots toward marginal populations. *Conservation Biology* 15:1710–20.

Armbruster, P., and R. Lande. 1993. A population viability analysis for African elephant (*Loxodonta africana*): How big should reserves be? *Conservation Biology* 7:602–10.

Armstrong, G. W., W. L. Adamowicz, J. A. Beck, S. G. Cumming, and F. K. A. Schmiegelow. 2003. Coarse filter ecosystem management in a nonequilibrating forest. *Forest Science* 49:209–23.

Aubry, K. B., and J. C. Lewis. 2003. Extirpation and reintroduction of fishers (*Martes pennanti*) in Oregon: Implications for their conservation in the Pacific states. *Biological Conservation* 114:79–90.

Bani, L., M. Baietto, L. Bottoni, and R. Massa. 2002. The use of focal species in designing a habitat network for a lowland area of Lombardy, Italy. *Conservation Biology* 16:826–31.

Bani, L., D. Massimino, L. Bottoni, and R. Massa. 2006. A multiscale method for selecting indicator species and priority conservation areas: A case study for broadleaved forests in Lombardy, Italy. *Conservation Biology* 20:512–26.

Beazley, K., L. Smandych, T. Snaith, F. MacKinnon, P. Austen-Smith, and P. Duinker. 2005. Biodiversity considerations in conservation system planning: Map-based approach for Nova Scotia, Canada. *Ecological Applications* 15:2192–2208.

Beissinger, S. R., and M. I. Westphal. 1998. On the use of demographic models of population viability in endangered species management. *Journal of Wildlife Management* 62:821–41.

Beissinger, S. R., and D. R. McCullough, eds. 2002. *Population viability analysis.* Chicago: University of Chicago Press.

Berger, J. 1997. Population constraints associated with the use of black rhinos as an umbrella species for desert herbivores. *Conservation Biology* 11:69–78.

Bernt-Erik, S., and O. Bakke. 2000. Avian life history variation and contribution of demographic traits to the population growth rate. *Ecology* 81:642–53.

Bevers, M., J. Hof, D. W. Uresk, and G. L. Schenbeck. 1997. Spatial optimization of prairie dog colonies for black-footed ferret recovery. *Operations Research* 45:495–507.

Bishop, J. A., and W. L. Myers. 2005. Associations between avian functional guild response and regional landscape properties for conservation planning. *Ecological Indicators* 5:33–48.

Block, W. M., L. A. Brennan, and R. J. Gutierrez. 1987. Evaluation of guild-indicator species for use in resource management. *Environmental Management* 11:265–69.

Bonn, A., A. S. L. Rodrigues, and K. J. Gaston. 2002. Threatened and endemic species: Are they good indicators of patterns of biodiversity on a national scale? *Ecology Letters* 5:733–41.

Botsford, L. W., F. Micheli, and A. Hastings. 2003. Principles for the design of marine reserves. *Ecological Applications* 13:S25–S31.

Brook, B. W. 2000. Pessimistic and optimistic bias in population viability analysis. *Conservation Biology* 14:564–66.

Brook, B. W., J. J. O'Grady, A. P. Chapman, M. A. Burgman, H. R. Akçakaya, and R. Frankham. 2000. Predictive accuracy of population viability analysis in conservation biology. *Nature* 404:385–87.

Brook, B. W., M. A. Burgman, H. R. Akçakaya, J. J. O'Grady, and R. Frankham. 2002. Critiques of PVA ask the wrong questions: throwing the heuristic baby out with the numerical bath water. *Conservation Biology* 16:262–63.

Burgman, M. A., S. Ferson, and H. R. Akçakaya. 1993. *Risk assessment in conservation biology*. London: Chapman and Hall.

Burley, F. W. 1988. Monitoring biological diversity for setting priorities in conservation. Pp. 227–30 in *Biodiversity*, ed. E. O. Wilson. Washington, DC: National Academy Press.

Butchart, S. H., H. R. Akçakaya, E. Kennedy, and C. Hilton-Taylor. 2006. Biodiversity indicators based on trends in conservation status: Strengths of the IUCN Red List Index. *Conservation Biology* 20:579–81.

Camm, J. D., S. Polasky, A. Solow, and B. Csuti. 1996. A note on optimal algorithms for reserve site selection. *Biological Conservation* 80:83–97.
Campbell, S. P., J. A. Clark, L. H. Crampton, A. D. Guerry, L. T. Hatch, P. R. Hosseini, J. J. Lawler, and R. J. O'Connor. 2002. An assessment of monitoring efforts in endangered species recovery plans. *Ecological Applications* 12:674–81.
Caro, T. M., and G. O'Doherty. 1999. On the use of surrogate species in conservation biology. *Conservation Biology* 13:805–14.
Caro, T., J. M. Eadie, and A. Sih. 2005. Use of substitute species in conservation biology. *Conservation Biology* 19:1821–26.
Cassey, P. 2002. Life history and ecology influences establishment success of introduced land birds. *Biological Journal of the Linnean Society* 76:465–80.
Canadian Broadcasting Corporation (CBC). 2002. Great apes among species threatened by extinction: Red list. CBC News Online Staff. CBC News on-line, Tuesday, 8 October 2002. Toronto: Canadian Broadcasting Centre. http://www.cbc.ca/storyview/CBC/2002/10/08/endangered021008.
Chaplin, S. J., R. A. Gerrard, H. M. Watson, L. L. Master, and S. R. Flack. 2000. The geography of imperilment. Pp. 159–99 in *Precious heritage: The status of biodiversity in the United States*, ed. B.A. Stein, L. S. Kutner, and J. S. Adams. New York: Oxford University Press.
Clark, T. W., M. Wishnie, and G. Gorman. 2003. An interdisciplinary approach to natural resources conservation: A flagship species example from Costa Rica. *Journal of Sustainable Forestry* 16:161–90.
Clinchy, M., D. T. Haydon, and A. T. Smith. 2002. Pattern does not equal process: What does patch occupancy really tell us about metapopulation dynamics? *American Naturalist* 159:351–62.
Committee of Scientists. 1999. Sustaining the people's lands: Recommendations for stewardship of the national forests and grasslands into the next century. *Committee of Scientists Report*, March 15, 1999. Washington, DC: U.S. Department of Agriculture, 193 pp.
Conroy, M. J., and B. R. Noon. 1996. Mapping of species richness for conservation of biological diversity: Conceptual and methodological issues. *Ecological Applications* 6:763–73.
Coulson, T., G. M. Mace, E. Hudson, and H. Possingham. 2001. The use and abuse of population viability analysis. *Trends in Ecology and Evolution* 16:219–21.
Csuti, B., S. Polasky, P. H. Williams, R. L. Pressey, J. D. Camm, M. Kershaw, A. R. Kiester, et al. 1997. A comparison of reserve selection algorithms using data on terrestrial vertebrates in Oregon. *Biological Conservation* 80:83–97.
Davies, K. F., C. R. Margules, and J. F. Lawrence. 2004. A synergistic effect puts rare, specialized species at greater risk of extinction. *Ecology* 85:265–71.
Dean, D. J., K. R. Wilson, and C. H. Flather. 1997. Spatial error analysis of species richness for a gap analysis map. *Photogrammetric Engineering and Remote Sensing* 63:1211–17.
Doak, D., P. Kareiva, and B. Klepteka. 1994. Modeling population viability for the desert tortoise in the western Mojave desert. *Ecological Applications* 4:446–60.
Dyer, A. R., D. E. Goldberg, R. Turkington, and C. Sayre. 2001. Effects of growing conditions and source habitat on plant traits and functional group definition. *Functional Ecology* 15:85–95.
Everett, R. L., and J. F. Lehmkuhl. 1999. Restoring biodiversity on public lands

through disturbance and patch management irrespective of land-use allocation. Pp. 87–105 in *Practical Approaches to the Conservation of Biological Diversity*, ed. R. K. Baydack, H. Campa III, and J. B. Haufler. Washington, DC: Island Press.

Fauth, P. T., E. J. Gustafson, and K. N. Rabenold. 2000. Using landscape metrics to model source habitat for Neotropical migrants in the midwestern U.S. *Landscape Ecology* 15:621–31.

Ferrier, S. 2002. Mapping spatial pattern in biodiversity for regional conservation planning: where do we go from here? *Systematic Biology* 51:331–63.

Flather, C. H., K. R. Wilson, D. J. Dean, and W. C. McComb. 1997. Identifying gaps in conservation networks: Of indicators and uncertainty in geographic-based analyses. *Ecological Applications* 7:531–42.

Flather, C. H., M. S. Knowles, and I. A. Kendall. 1998. Threatened and endangered species geography: Characteristics of hot spots in the conterminous United States. *BioScience* 48:365–76.

Flather, C. H., M. Bevers, and J. Hof. 2002. Prescribing habitat layouts: Analysis of optimal placement for landscape planning. Pp. 428–53 in *Applying landscape ecology in biological conservation*, ed. K. J. Gutzwiller, New York: Springer-Verlag.

Fleishman, E., D. D. Murphy, and P. F. Brussard. 2000. A new method for selection of umbrella species for conservation planning. *Ecological Applications* 10:569–79.

Fleishman, E., R. B. Blair, and D. D. Murphy. 2001. Empirical validation of a method for umbrella species selection. *Ecological Applications* 11:1489–1501.

Foley, P. 2000. Problems in extinction model selection and parameter estimation. *Environmental Management* 26:55–73.

Fritz, H., P. Duncan, I. J. Gordon, and A. W. Illius. 2002. Megaherbivores influence trophic guild structure in African ungulate communities. *Oecologia* 131:620–25.

Gaston, K. J., R. L. Pressey, and C. R. Margules. 2002. Persistence and vulnerability: Retaining biodiversity in the landscape and in protected areas. *Journal of Biosciences* 27 (4 Suppl 2): 361–84.

Ginzburg, L. R., L. B. Slobodkin, K. Johnson, and A. G. Bindman. 1982. Quasi-extinction probabilities as a measure of impact on population growth. *Risk Analysis* 2:171–81.

Haig, S. M., J. R. Belthoff, and D. H. Allen. 1993. Population viability analysis for a small population of red-cockaded woodpeckers and an evaluation of enhancement strategies. *Conservation Biology* 7:289–301.

Higgins, S. I., and D. M. Richardson. 1999. Predicting plant migration rates in a changing world: The role of long-distance dispersal. *American Naturalist* 153:464–75.

Higgs, A. J., and M. B. Usher. 1980. Should nature reserves be large or small? *Nature* 285:568–69.

Hof, J., and C. H. Flather. 1996. Accounting for connectivity and spatial correlation in the optimal placement of wildlife habitat. *Ecological Modelling* 88:143–55.

Hof, J., C. Flather, T. Baltic, and S. Davies. 1999. Projections of forest and rangeland condition indicators for a national assessment. *Environmental Management* 24:383–98.

Holloway, G. J., G. H. Griffiths, and P. Richardson. 2003. Conservation strategy maps: A tool to facilitate biodiversity action planning illustrated using the heath fritillary butterfly. *Journal of Applied Ecology* 40:413–21.

Holthausen, R. S., M. J. Wisdom, J. Pierce, D. K. Edwards, and M. M. Rowland. 1994.

Using expert opinion to evaluate a habitat effectiveness model for elk in western Oregon and Washington. Research Paper PNW-RP-479. Portland, OR: US Department of Agriculture, Forest Service, Pacific Northwest Research Station.

Hutchinson, G. E. 1978. *An introduction to population ecology*. New York: Yale University Press.

Johnson, D. M. 2005. Metapopulation models: An empirical test of model assumptions and evaluation methods. *Ecology* 86:3088–98.

Kershaw, M., P. H. Williams, and G. C. Mace. 1994. Conservation of Afrotropical antelopes: consequences and efficiency of using different site selection methods and diversity criteria. *Biological Conservation* 3:354–72.

Kintsch, J. A., and D. L. Urban. 2002. Focal species, community representation, and physical proxies as conservation strategies: A case study in the Amphibolite Mountains, North Carolina, U.S.A. *Conservation Biology* 16:936–47.

Knopf, F. L., J. A. Sedgwick, and R. W. Cannon. 1988. Guild structure of a riparian avifauna relative to seasonal cattle grazing. *Journal of Wildlife Management* 52:280–90.

Lambeck, R. J. 1997. Focal species: A multispecies umbrella for nature conservation. *Conservation Biology* 11:849–56.

Launer, A. E., and D. D. Murphy. 1994. Umbrella species and the conservation of habitat fragments: A case of a threatened butterfly and a vanishing grassland ecosystem. *Biological Conservation* 69:145–53.

Lehmkuhl, J. F., B. G. Marcot, and T. Quinn. 2001. Characterizing species at risk. Pp. 474–500 in *Wildlife–habitat relationships in Oregon and Washington*, ed. D. H. Johnson and T. A. O'Neil. Corvallis: Oregon State University Press.

Lindenmayer, D. B., R. C. Lacy, and M. L. Pope. 2000. Testing a simulation model for population viability analysis. *Ecological Applications* 10:580–97.

Lindenmayer, D. B., A. D. Manning, P. L. Smith, H. P. Possingham, J. Fischer, I. Oliver, and M. A. McCarthy. 2002. The focal-species approach and landscape restoration: A critique. *Conservation Biology* 16:338–45.

Lopez, R. D., C. B. Davis, and M. S. Fenessy. 2002. Ecological relationships between landscape change and plant guilds in depressional wetlands. *Landscape Ecology* 17:43–56.

Lubchenco, J., A. M. Olson, L. B. Brubaker, S. R. Carpenter, M. M. Holland, S. P. Hubbell, S. A. Levin, et al. 1991. The sustainable biosphere initiative: An ecological research agenda. *Ecology* 72:318–25.

MacArthur, R. H., and E. O. Wilson. 1967. *The theory of island biogeography*. Princeton: Princeton University Press.

Mace, G. M., and R. Lande. 1991. Assessing extinction threats: Toward a reevaluation of IUCN threatened species categories. *Conservation Biology* 5:148–57.

Mannan, R. W., M. L. Morrison, and E. C. Meslow. 1984. Comment: The use of guilds in forest bird management. *Wildlife Society Bulletin* 12:426–30.

Margules, C. R., and R. L. Pressey. 2000. Systematic conservation planning. *Nature* 405:243–53.

Maurer, B. A., L. B. McArthur, and R. C. Whitmore. 1981. Effects of logging on guild structure of a forest bird community in West Virginia. *American Birds* 35:11–13.

McCarthy, M. A., D. B. Lindenmayer, and H. P. Possingham. 2000. Testing spatial PVA models of Australian treecreepers (Aves: Climacteridae) in fragmented forest. *Ecological Applications* 10:1722–31.

McCarthy, M. A., H. P. Possingham, J. R. Day, and A. J. Tyre. 2001. Testing the accuracy of population viability analysis. *Conservation Biology* 15:1030–38.

McNeely, J. A. 1994. Lessons from the past: Forests and biodiversity. *Biodiversity and Conservation* 3:390–405.

Milledge, D., C. Palmer, and J. Nelson. 1991. "Barometers of change": The distribution of large owls and gliders in mountain ash forests of the Victorian Central Highlands and their potential as management indicators. Pp. 53–66 in *Conservation of Australia's forest fauna,* ed. D. Lunney. New South Wales, Australia: Royal Zoological Society of New South Wales.

Molina, R., B. G. Marcot, and R. Lesher. 2006. Protecting rare, old-growth forest associated species under the survey and manage guidelines of the Northwest Forest Plan. *Conservation Biology* 20:306–18.

Morrison, M. L., B. G. Marcot, and R. W. Mannan. 2006. *Wildlife–habitat relationships: Concepts and applications.* 3rd ed. Washington, DC: Island Press.

Myers, N. 1989. Threatened biotas: "Hotspots" in tropical forests. *Environmentalist* 8:1–20.

New, T. R. 2004. Velvetworms: Charismatic invertebrates for conservation. *Wings* 27:12–15.

Niemi, G. J., J. M. Hanowski, A. R. Lima, T. Nicholls, and N. Weiland. 1997. A critical analysis on the use of indicator species in management. *Journal of Wildlife Management* 61:1240–52.

Noss, R. 1987. From plant communities to landscapes in conservation inventories: A look at the Nature Conservancy (USA). *Biological Conservation* 41:11–37.

Noss, R. F., C. Carroll, K. Vance-Borland, and G. Wuerthner. 2002. A multicriteria assessment of the irreplaceability and vulnerability of sites in the Greater Yellowstone Ecosystem. *Conservation Biology* 16:895–908.

O'Connell, T. J., L. E. Jackson, and R. P. Brooks. 2000. Bird guilds as indicators of ecological condition in the central Appalachians. *Ecological Applications* 10:1706–21.

Oertli, S., A. Müller, D. Steiner, A. Breitenstein, and S. Dorn. 2005. Cross-taxon congruence of species diversity and community similarity among three insect taxa in a mosaic landscape. *Biological Conservation* 126:195–205.

Overton, C. T., R. A. Schmitz, and M. L. Casazza. 2006. Linking landscape characteristics to mineral site use by band-tailed pigeons in western Oregon: Coarse-filter conservation with fine-filter tuning. *Natural Areas Journal* 26:38–46.

Pascual, M. A., P. Kareiva, and R. Hilborn. 1997. The influence of model structure on conclusions about the viability and harvesting of Serengeti wildebeest. *Conservation Biology* 11:966–76.

Patton, D. R. 1992. *Wildlife–habitat relationships in forested ecosystems.* Portland, OR: Timber Press.

Poiani, K. A., M. D. Merrill, and K. A. Chapman. 2001. Identifying conservation-priority areas in a fragmented Minnesota landscape based on the umbrella species concept and selection of large patches of natural vegetation. *Conservation Biology* 15:513–22.

Pressey, R. L., C. J. Humphries, C. R. Margules, R. I. Vane-Wright, and P. H. Williams. 1993. Beyond opportunism: Key principles for systematic reserve selection. *Trends in Ecology and Evolution* 8:124–28.

Pressey, R. L., H. P. Possingham, and J. R. Day. 1997. Effectiveness of alternative heuristic algorithms for identifying indicative minimum requirement for conservation reserves. *Biological Conservation* 80:207–19.

Quayle, J. F., and L. R. Ramsay. 2006. Biodiversity indicators based on trends in conservation status: advancing the science. *Conservation Biology* 20:582–83.

Rabinowitz, D. 1981. Seven forms of rarity. Pp. 205–17 in *The biological aspects of rare plant conservation,* ed. H. Synge. Chichester: Wiley.

Ranius, T. 2000. Minimum viable metapopulation size of a beetle, *Osmoderma eremita,* living in tree hollows. *Animal Conservation* 3:37–43.

Raphael, M. G., M. J. Wisdom, M. M. Rowland, R. S. Holthausen, B. C. Wales, B. G. Marcot, and T. D. Rich. 2001. Status and trends of habitats of terrestrial vertebrates in relation to land management in the interior Columbia River basin. *Forest Ecology and Management* 153:63–87.

Raphael, M. G., and R. S. Holthausen. 2002. Using a spatially explicit population model to analyze effects of habitat management on northern spotted owls. Pp. 701–712 in *Predicting species occurrences: issues of scale and accuracy,* ed. J. M. Scott, P. J. Heglund, M. L. Morrison, J. B. Haufler, M. G. Raphael, W. A. Wall, and F. B. Samson. Washington, DC: Island Press.

Ratner, S., R. Lande, and B. B. Roper. 1997. Population viability analysis of spring chinook salmon in the South Umpqua River, Oregon. *Conservation Biology* 11:879–89.

Reed, J. M., P. D. Doerr, and J. R. Walters. 1986. Determining minimum population sizes for birds and mammals. *Wildlife Society Bulletin* 14:255–61.

Regan, H. M., M. Colyvan, and M. A. Burgman. 2000. A proposal for fuzzy International Union for the Conservation of Nature (IUCN) categories and criteria. *Biological Conservation* 92:101–108.

Reid, W. V. 1998. Biodiversity hotspots. *Trends in Ecology and Evolution* 13:275–80.

Rewa, C. A., and E. D. Michael. 1984. Use of habitat evaluation procedures (HEP) in assessing guild habitat value. *Transactions of the Northeast Section of the Wildlife Society* 41:122–29.

Reyers, B., D. H. K. Fairbanks, A. S. van Jaarsveld, and M. Thompson. 2001. Priority areas for conservation of South African vegetation: A coarse filter approach. *Diversity and Distributions* 7:79–95.

Roberge, J. M., and P. Angelstam. 2004. Usefulness of the umbrella species concept as a conservation tool. *Conservation Biology* 18:76–85.

Rodrigues, A. S. L., and K. J. Gaston. 2001. How large do reserve networks need to be? *Ecology Letters* 4:602–609.

———. 2002. Optimisation in reserve selection procedures: Why not? *Biological Conservation* 107:123–29.

Rondinini, C., and L. Boitani. 2006. Differences in the umbrella effects of African amphibians and mammals based on two estimators of the area of occupancy. *Conservation Biology* 20:170–79.

Root, R. B. 1967. The niche exploitation pattern of the blue-gray gnatcatcher. *Ecological Monographs* 37:317–50.

Rothley, K. D. 2002. Dynamically-based criteria for the identification of optimal bioreserve networks. *Environmental Modeling and Assessment* 7:123–28.

Rowland, M. M., M. J. Wisdom, L. H. Suring, and C. W. Meinke. 2006. Greater sage-grouse as an umbrella species for sagebrush-associated vertebrates. *Biological Conservation* 129:323–35.

Rubinoff, D. 2001. Evaluating the California gnatcatcher as an umbrella species for conservation of southern California coastal sage scrub. *Conservation Biology* 15:1374–83.

Sætersdal, M., J. M. Line, and H. J. B. Birks. 1993. How to maximize biological diversity in nature reserve selection: Vascular plants and breeding birds in deciduous woodlands, western Norway. *Biological Conservation* 66:131–38.

Sarkar, S., R. L. Pressey, D. P. Faith, C. R. Margules, T. Fuller, D. M. Stoms, A. Moffett, et al. 2006. Biodiversity conservation planning tools: Present status and challenges for the future. *Annual Review of Environment and Resources* 31:123–59.

Satterthwaite, W. H., E. S. Menges, and P. F. Quintana-Ascencio. 2002. Assessing scrub buckwheat population viability in relation to fire using multiple modeling techniques. *Ecological Applications* 12:1672–87.

Schiegg, K., J. R. Walters, and J. A. Priddy. 2005. Testing a spatially explicit, individual-based model of red-cockaded woodpecker population dynamics. *Ecological Applications* 15:1495–1503.

Schwartz, M. W. 1999. Choosing the appropriate scale of reserves for conservation. *Annual Review of Ecology and Systematics* 30:83–108.

Scott, J. M., F. Davis, B. Csuti, R. Noss, B. Butterfield, C. Groves, H. Anderson, et al. 1993. Gap analysis: A geographic approach to protection of biological diversity. *Wildlife Monographs* 123:1–41.

Shriner, S. A., K. R. Wilson, and C. H. Flather. 2006. Reserve networks based on richness hotspots and representation vary with scale. *Ecological Applications* 16:1660–73.

Simberloff, D. 1998. Flagships, umbrellas, and keystones: Is single-species management passe in the landscape era? *Biological Conservation* 83:247–57.

Smith, W. P., S. M. Gende, and J. V. Nichols. 2005. The northern flying squirrel as an indicator species of temperate rain forest: Test of any hypothesis. *Ecological Applications* 15:689–700.

Snaith, T. V., and K. F. Beazley. 2002. Moose (*Alces alces americana* [Gray Linnaeus Clinton] Peterson) as a focal species for reserve design in Nova Scotia, Canada. *Natural Areas Journal* 22:235–40.

Soulé, M. E., and D. Simberloff. 1986. What do genetics and ecology tell us about the design of nature reserves? *Biological Conservation* 35:19–40.

Steffan-Dewenter, I., U. Münzenberg, C. Bürger, C. Thies, and T. Tscharntke. 2002. Scale-dependent effects of landscape context on three pollinator guilds. *Ecology* 83:1421–32.

Suter, W., R. F. Graf, and R. Hess. 2002. Capercaillie (*Tetrao urogallus*) and avian biodiversity: Testing the umbrella-species concept. *Conservation Biology* 16:778–88.

Taylor, R. J., T. Regan, H. Regan, M. Burgman, and K. Bonham. 2003. Impacts of plantation development, harvesting schedules and rotation lengths on the rare snail *Tasmaphena lamproides* in northwest Tasmania: A population viability analysis. *Forest Ecology and Management* 175:455–66.

Underhill, L. G. 1994. Optimal and suboptimal reserve selection algorithms. *Biological Conservation* 70:85–87.

Vane-Wright, R. I., C. J. Humphries, and P. H. Williams. 1991. What to protect? Systematics and the agony of choice. *Biological Conservation* 55:235–54.

Verheyen, K., O. Honnay, G. Motzkin, M. Hermy, and D. R. Foster. 2003. Response of forest plant species to land-use change: A life-history trait-based approach. *Journal of Ecology* 91:563–77.

Verner, J. 1984. The guild concept applied to management of bird populations. *Environmental Management* 8:1–14.

Wahlberg, N., A. Moilanen, and I. Hanski. 1996. Predicting the occurrence of endangered species in fragmented landscapes. *Science* 273:1536–38.
Walpole, M. J., and N. Leader-Williams. 2002. Tourism and flagship species in conservation. *Biodiversity and Conservation* 11:543–47.
Waples, R. S. 2002. Definition and estimation of effective population size in the conservation of endangered species. Pp. 147–68 in *Population viability analysis,* ed. S. R. Beissinger and D. R. McCullough. Chicago: University of Chicago Press.
Warman, L. D., D. M. Forsyth, A. R. E. Sinclair, K. Freemark, H. D. Moore, T. W. Barrett, R. L. Pressey, and D. White. 2004. Species distributions, surrogacy, and important conservation regions in Canada. *Ecology Letters* 7:374–79.
Watson, J., D. Freudenberger, and D. Paull. 2001. An assessment of the focal-species approach for conserving birds in variegated landscapes in southeastern Australia. *Conservation Biology* 15:1364–73.
Wielgus, R. B. 2002. Minimum viable population and reserve sizes for naturally regulated grizzly bears in British Columbia. *Biological Conservation* 106:381–88.
Wilcox, C., and H. Possingham. 2002. Do life history traits affect the accuracy of diffusion approximations for mean time to extinction? *Ecological Applications* 12:1163–79.
Williams, P. H. 1998. Key sites for conservation: area-selection methods for biodiversity. Pp. 211–49 in *Conservation in a changing world,* ed. G. M. Mace, A. Balmford, and J. R. Ginsberg. Cambridge: Cambridge University Press.
Williams, P. H., and M. B. Araújo. 2000. Using probability of persistence to identify important areas for biodiversity conservation. *Proceedings of the Royal Society of London, Series B* 267:1959–66.
Wilson, E. O., and E. O. Willis. 1975. Applied biogeography. Pp. 522–34 in *Ecology and evolution of communities,* ed. M. L. Cody and J. M. Diamond. Cambridge, MA: Belknap Press.
Wisdom, M. J., R. S. Holthausen, B. C. Wales, C. D. Hargis, V. A. Saab, D. C. Lee, W. J. Hann, et al. 2000. *Source habitats for terrestrial vertebrates of focus in the Interior Columbia Basin: broad-scale trends and management implications.* 3 vols. General Technical Report PNW-GTR-485. Portland, OR: U.S. Department of Agriculture, Forest Service, Pacific Northwest Research Station.

7

System-Level Strategies for Conserving Rare or Little-Known Species

Bruce G. Marcot and Carolyn Hull Sieg

In this chapter we review the literature on system-level strategies for conserving rare or little-known (RLK) species, continuing from the species-level approaches addressed in the previous chapter. We define system-level approaches as those that result in conservation actions focused on providing for community or ecosystem composition, structure, or function. See table 7.1 for a description and main assumptions of the system approaches we review.

As used generally in planning and management, system-level approaches to conservation do not necessarily or specifically pertain to individual RLK species or groups of species per se. Instead, they tend to emphasize the maintenance of ecosystem structure, function, and integrity as at-risk entities in their own right or as surrogates for individual species. We discuss the scientific merit and strengths and weaknesses of each approach for conservation of RLK species.

The organization in this section is based on criteria used in designing the approach. That is, if structural attributes of an ecosystem were used in developing an approach, we would classify the approach as "structural," even though the approach also influences the functioning of the system. Likewise, if the approach is based mostly on disturbance regimes or functional aspects of a system, we would tend to classify it as a "functional" approach.

Table 7.1. *Summary of system approaches showing descriptions and main assumptions*

Name of Approach	Description	Main Assumptions
	MAINTAINING SYSTEM STRUCTURE AND COMPOSITION	
	Maintaining System Structure Composition	
Range of natural variability (RNV)	Managing within a range of historic vegetation and environmental conditions	• Range of historic conditions represents conditions under which native species persisted, and under which modern humans had little anthropogenic or technological impact • Such conditions would provide for future persistence of species and systems
Key habitat conditions	Maintain a mix of conditions including types and successional stages of plant communities	• Appropriate habitat conditions can be identified by focusing on selected species, including species playing key ecological roles • A mix of habitat conditions will support the diversity of associated species
	Species that play critical ecological roles	
Keystone species	A species that regulates local species diversity in lower trophic levels	• Their effect on species diversity in lower trophic levels is disproportionate to their abundance • They perform roles not performed by other species or processes • Only a relatively few species within an ecosystem have such effects
Ecosystem engineers	Species whose activities modulate physical habitat	• Their effect on physical habitats and therefore ecosystem structure and function is disproportionate to their abundance • They perform roles not performed by other species or processes • Only a relatively small subset of species in an ecosystem serve these influences
	MAINTAINING SYSTEM PROCESSES AND FUNCTIONS	
	Maintaining disturbance regimes	
Fire	Restoration of historic timing, severity, seasonality, and size of fires	• Restoration of the fire regime will result in structural characteristics upon which many species depend and will ultimately result in a mosaic of structural classes that occurred historically
Herbivory	Restoration of historic patterns, intensity, and types of herbivory	• Restoration of historic herbivory patterns is important in systems that evolved with grazing and in those systems that were largely devoid of large herbivores

Name of Approach	Description	Main Assumptions
	Maintaining Other Ecosystem Processes	
Key ecological functions of organisms	Patterns of the major ecological roles played by species in their ecosystems	• Providing habitat for natural or desired numbers of species with specific key ecological function categories would ensure functionality of the system • Greater levels of functional redundancy impart greater stability and resilience of the system and ecological communities to external environmental perturbations • All species within a community perform some key ecological functions that can influence the environment of, or resources used by, other species
Food webs	Flows of substance and energy among feeding links and across trophic levels	• Providing for food chains within ecological communities can maintain the stability and natural dynamics of the community

Maintaining System Structure and Composition

This set of approaches pertains to maintaining the structure or composition of species assemblages or ecological communities, or maintaining the dynamics of ecosystems.

Range of Natural Variability

This approach focuses on maintaining the mix of ecosystems across the landscape within the historic range of natural variability (RNV) (Landres et al. 1999). An assumption of this approach is that native species of a region adapted to and occurred within some historic range of variability that can be estimated and potentially replicated by land management. This approach also assumes that maintaining the mix of communities within this natural range will provide for the current needs of associated species (Morgan et al. 1994; Agee 2003).

This approach is based on an understanding of "natural" structure, composition and disturbance regimes that existed when ecosystems were relatively unaffected by people. Such reconstructions of historic conditions depend greatly on the geographic area, time period, and specific conditions considered. Selecting a starting time, that is, deciding how far to look back, is an important consideration in historical reconstructions and in deciding how to depict the natural range of variability. Morgan et al. (1994) recommended that reconstructions should consider only the time period during which the current vegetation was in relative equilibrium with the current climate and other biotic factors. Based on this recommendation, Hann et al. (1997) used the last 2000 years for the Interior Columbia Basin Ecosystem Management Project.

Of course, this method presumes that equilibrium could be, and had been, reached. Longer reconstructions are useful for appreciating the role of climate change on ecosystems and disturbance regimes (Swetnam et al. 1999) and for considering evolution of adaptive phenotypes and behaviors of species. Millar and Woolfenden (1999) argued that a more appropriate analog for western North America might be the Medieval Warm Period (AD 900–1350). Gill and McCarthy (1998) used longevity of plant species that reproduce only after fire to help define the upper limit of fire return intervals, to help in fire planning; they suggested that this approach could be useful in cases with little presettlement information on fire return intervals. Unfortunately, such reconstructions—representing *pre*historic (i.e., prior to written history) conditions—typically suffer from a "fading record" problem characterized by decreasing reliability with increasing time before the present (Swetnam et al. 1999).

The common assumption in this approach is that recent historic conditions represent conditions under which native species evolved. This assumption may very well be false, given the very long span of time (or variable spans of time among different taxa) over which evolution proceeds, as well as the high level of even recent (Holocene) variation in climate and vegetation conditions. Thus it may be more correct to refer to the historic range of variation as representing conditions over which native species may have *persisted* rather than evolved.

An analysis of the RNV is often used to describe desired (future) conditions. However if this connotes static, fixed conditions of vegetation structure, composition, and distribution, and of other environmental factors, to be consistent with the concept of variation, including natural (and anthropogenic) disturbance regimes, the term may better be revised as "desired (future) dynamics" (Hansson 2003). Focusing on dynamic elements of an ecosystem, such as variation in fire regimes, climate, and hydrology, helps describe a range of possible future vegetation and environmental conditions. In this approach, the land and natural resource manager could then decide the degree to which the fit between historic and future variations would be acceptable, given management goals and objectives.

In one version of the RNV approach, historical reconstruction techniques have been used to examine a number of ecosystem attributes. In forested ecosystems, these include forest composition and structural attributes such as the relative distribution and patch size of forest types and even the tree and age class distribution within and among stands. We discuss here approaches focusing on structure and composition of ecosystems separately from those focusing on restoring disturbance processes (discussed later). But, in reality, most efforts to restore historic structural attributes rely on an understanding of disturbance regimes that created or maintained this structure.

Incorporating an understanding of the dynamic nature of vegetation is an important attribute of strategies to manage within some historic range of variability (Hunter 1991). Some RNV approaches begin with an objective of providing the diversity of structural elements in variable configurations and quantities, with the ultimate objective of maintaining the dynamic patterns and processes that are integral to ecological integrity (Aplet and Keeton 1998; also discussed later in this chapter in the context of maintenance of disturbance regimes).

Sources of information for reconstructing historical structural and community attributes of ecosystems include information from biological archives such as packrat middens (Betancourt et al. 1990), opal phytoliths (Kerns and Moore 1997), and plant macrofossils and pollen deposited in soils, lakes, and bogs (Jacobson and Grimm 1986). Other sources of information used in reconstructing structural and community attributes of ecosystems include documentary archives, such as photographs and diaries of early explorers, naturalists, scientists, and settlers. The usefulness of these sources of information is limited by uneven spatial and temporal distribution, and as a result might not represent all, and in some cases even modal, earlier conditions. In addition, these sources may reveal relative abundances of major species groups but cannot be used to address questions about the patchiness of vegetation types, the area or age class distributions of forest types, or occurrence and prevalence of plant species that are poorly represented in midden or pollen samples. These limitations are particularly troubling when dealing with rare species because specific mention of even their presence in historical documents is uncommon.

In spite of the difficulties in reconstructing historical structure and composition of native landscapes, this approach is widely applied in conservation efforts. Strategies aimed at ultimately restoring the natural distribution of ecosystems and seral stages must characterize how these patterns have changed. Central to such approaches is first assessing, at least regionally (and not just within the planning area), how coverage by major vegetation types has been altered by humans. Such analyses are useful in identifying vegetation types that are rare throughout their range and not just locally rare in the planning unit. This information can be used to prioritize conservation efforts on the more globally rare types.

Large-scale regional assessments may fail to address landscape features that occupy only a small area but support biodiversity elements, including RLK species. Simply preserving uncommon habitats such as wetlands, fens, bogs, springs, riverine systems, and other unique habitats is an important step in meeting species diversity goals. For example, protection of springs and sphagnum bogs was deemed necessary for maintaining some rare insects in eastern Oregon and Washington forests (LaBonte et al. 2001). Maintenance of habitat features such as wetlands, riverine systems, sandhills, and caves, plus prairie dog (*Cynomys* spp.) colonies and concentrations of Richardson's ground squirrels (*Spermophilus richardsonii*) was needed to support a number of threatened and endangered species of the Great Plains (Sieg et al. 1999).

A number of strategies using an understanding of RNV are aimed at one specific vegetation type. For example, Prober (1996) quantified floristic composition and soil types in remnant *Eucalyptus* woodlands in Australia. These data were then used to design reserves that would include sites representative of the natural variation and that would compensate for high losses of woodlands on soils suitable for agriculture. In another example, the U.S. National Park Service used a combination of repeat photographs and tree ring evidence to document recent tree invasion of a montane meadow in New Mexico (Swetnam et al. 1999). This information was then used to develop a strategy to restore the grassland character to this region. Reconstruction of historic stand densities, as well as tree size and arrangement, and fuel loadings are key components of strategies designed to restore southwestern U.S. ponderosa pine (*Pinus ponderosa*) forests to conditions more typical of late 1800s reference conditions (e.g., Fulé et al. 2002).

The U.S. Department of Agriculture (USDA) Forest Service has used various versions of RNV-based strategies in developing land and natural resource management plans. For example, management guidelines for an area in southern Idaho were designed to maintain habitat types or fire groupings of stands within a historic range of conditions. Other examples of applications where RNV principles have been used focus on identifying and restoring within-stand structural components. For example, a comparison of historic to modern photographs taken from the same photo points can be used in developing range management and restoration plans (Skovlin and Thomas 1995).

RNV approaches can also be used to aid understanding of historic vertebrate and invertebrate species composition. Such information may be used in conservation strategies in a number of ways. First, this information can identify species that have been extirpated from the planning area, and consideration can be given to their reintroduction. This information can also help identify species that are now more or less common relative to presettlement periods. The presence of specific wildlife species can also provide valuable information about habitat conditions at the time. For example, lists of birds and estimates of their abundances compiled by naturalists in the late 1800s and early 1900s along the Missouri River provided insights into the composition and extent of riparian woodlands at that time (Rumble et al. 1998). Observations on the presence of ducks, fish, and frogs throughout the year and in drought years give some sense for whether historic lakes supported persistent populations of aquatic animals and species (Severson and Sieg 2006). This type of information is potentially useful in setting more informed restoration goals.

Regardless of the approach, efforts designed to restore the structure and composition of native landscapes are fraught with implementation problems. Areal coverage of some ecosystem types such as tallgrass and rough fescue (*Festuca hallii*) prairies and longleaf pine (*Pinus palustris*) forests have been reduced to such a high degree that restoration of even a portion of the historic range of variation seems unlikely (Klopatek et al. 1979). Species extinctions, geographic fragmentation, continued human expansion, exotic species introductions, and changes in atmospheric gases are among the other impediments to restoring historic structural and compositional diversity to native landscapes (McPherson 1997), as are social and economic costs of resource (re)allocation, and possibly climate change.

Effects of all of these impediments on understanding historic distributions of RLK species and their restoration or conservation are largely unknown for most such species.

Diversity of Habitat Conditions

This set of approaches focuses on how species use diversity of habitats and conditions, but not from the perspective of historic or natural variation. We describe two approaches: managing for key habitat conditions, and managing for species that play critical ecological roles.

Key Habitat Conditions

An approach based on key habitat conditions is an example of a "coarse filter" strategy (see chap. 4), where the goal is to maintain a mix of habitat conditions at an appropriate management scale (Kaufman et al. 1994). However, this approach is not specifically tied to an understanding of historic range of variation. The assumption is that restoring or maintaining a mix of seral stages will provide the range of habitat conditions to support the diversity of associated species (e.g., Oliver 1992; Hof and Raphael 1993). In some regions, dense human populations and fragmentation of remaining habitats prevent management designed around natural disturbance regimes (Litvaitis 2003). Examples using this approach include Haufler et al.'s (2002) "adequate ecological representation" approach, whereby the management goal is to maintain a minimum of 10% of the maximum of the range of historical conditions of each community type. To aid in the recovery of Pitcher's thistle (*Cirsium pitcheri*), a threatened species in the United States and Canada, early-successional dune habitats are maintained (McEachern et al. 1994).

Species That Play Critical Ecological Roles

These approaches focus on restoring species that have large effects on community structure or ecosystem function that is disproportionate to their abundance or biomass and perform roles not performed by other species (Power et al. 1996). A number of schemes have been proposed for classifying species that play critical ecological roles. We discuss two examples: *key-*

stone species, which regulate local species diversity and competition, and *ecosystem engineers*, or species that modulate physical habitat.

The keystone species approach relies on the management for specific species that play major ecological roles in their ecosystems that influence many other species. Keystone species are species that regulate local species diversity in lower trophic levels, and whose removal results in significant shifts (increases or decreases) in the presence, distribution, or abundance of other species (Bond 1994). The term "keystone" was initially suggested by Paine (1966, 1974) in experiments in which mussels released from starfish predation took over the intertidal environment, and in doing so, reduced or excluded other species. Paine thus defined a keystone as a species (like the starfish) that kept other species in the system by keeping mussel populations in check.

The main assumption of the keystone approach is that some individual species disproportionately affect the distribution and abundance of either resources for other species or other species directly. The effect is directly causal from the keystone species' behaviors and is not just correlational. Further, the removal or decline in keystone species results in significant (and presumably undesired) changes in those resources or other species, and only a relatively few species within an ecosystem have such effects.

Since Paine's (1969) experiments, numerous species in many environments have been proposed as keystone species, including the coyote (*Canis latrans*) as a keystone terrestrial vertebrate predator (Henke and Bryant 1999), salamanders as keystone aquatic invertebrate predators (Wilbur 1997), and prairie dogs (*Cynomys* spp.) as keystone providers of ground burrows for other animals (Van Putten and Miller 1999). Even rare species can play keystone roles in some systems. Lyons and Schwartz (2001) reported that experimental removal of rare plant species, as compared with the same removal of dominants, led to greater invasion by an exotic grass species. However, this is not to say that all rare species play equally influential roles on community structure. Also, many little-known species (not necessarily rare) play keystone roles, particularly soil invertebrates that determine the structure and abundance of soil microbial populations (Moldenke et al. 1994).

Ecosystem engineers are species whose activities modulate physical habitats and thus have salient influence over community or ecosystem structure, composition, or function. Jones et al. (1996) defined ecosystem engineers as "organisms that directly or indirectly control the availability of resources to other organisms by causing physical state changes in biotic or abiotic materials." Butler (1995) described such animals as "geomorphic agents" and coined the study of animal ecosystem engineers as "zoogeomorphology." A main assumption of the ecosystem engineer approach is that, by their behaviors, particular species modulate physical habitats and thus have major influence over resources available to other species. Another assumption is that only a relatively small subset of species in an ecosystem serves these influences.

The American beaver (*Castor canadensis*) is a classic ecosystem engineer whose water-impoundment activities serve to increase plant and animal species richness within landscapes (Wright et al. 2002). Soil invertebrates are considered engineers of the soil environment, determining the structure and abundance of soil microbial populations and thereby influencing soil productivity (Moldenke et al. 1994). Smallwood et al. (1998) showed that the key ecological function of burrowing by animals is crucial to soil formation and is a critical link between geologic regolith material and organic soil life.

In summary, conservation approaches based on species that play critical ecological roles entail managing for species that have large effects on community structure or ecosystem function that are disproportionate to their abundance and perform roles not performed by other species or processes. Examples include keystone species that regulate species diversity in lower trophic levels or ecosystem engineers that modulate physical habitats. However, these functional-based conservation approaches tend to pertain to maintaining some aspects of ecosystem function rather than the occurrence and persistence of RLK species per se, unless those RLK species are themselves the species that play critical ecological roles.

Maintaining System Processes and Functions

These approaches are more process oriented; that is, they seek to restore and maintain the dynamics of ecosystem processes. We discuss two broad approaches: maintaining disturbance regimes, and maintaining other ecosystem processes such as food webs. Since it is not possible to discuss all terrestrial disturbances required to maintain ecosystem structure and function, we limit our discussion to two disturbance types for which we found conservation approaches: fire and herbivory. Similarly, we provide two examples of approaches designed to provide for other ecosystem processes: one related to providing key ecological functions and a food web strategy.

Maintaining Disturbance Regimes

Restoration of historic disturbances is increasingly being recognized as an important aspect of maintaining ecosystem structure and function. In disturbance-prone areas, environmental variability itself can be important to the life histories of many organisms (Aplet and Keeton 1998). Conservation strategies for rare species are increasingly recognizing the need to restore appropriate disturbance regimes.

Recognizing the importance of disturbance in maintaining heterogeneity of ecosystems has prompted increased efforts to assess, to the degree possible, how various disturbances in the past, such as fire and grazing, interacted with climatic fluctuations in maintaining community patterns. Characterizing and restoring historic disturbance regimes can be one facet of the RNV approach, discussed earlier. The main difference between these two approaches is that the RNV approach compares current or expected *conditions* to the range of historic or natural conditions, whereas maintenance of disturbance regimes focuses on the dynamics of *systems*, whether or not specific resulting conditions match some other historic or natural conditions.

Attributes of disturbance regimes that influence community structure, composition, and function include the type, severity, frequency, and seasonality of the disturbance and the size and shape of resulting habitat patches (Pickett and White 1985). Often, disturbances interact and their joint effects might be totally or partially additive (Frelich 2002). Altering disturbance regimes can have cascading effects on ecosystems—not just on disturbance-dependent species but also on species that occur in areas that are rarely disturbed. Studies are beginning to elucidate how disturbance processes at site, watershed, and even drainage basin scales can interact to maintain biodiversity (Everett and Lehmkuhl 1999), and how exotic species can influence plant community trajectories following disturbances (Hobbs and Huenneke 1992).

Studies of vegetation patterns, stand composition and age structure, and gap analysis and tree ring analyses of fire and insect events have been used to characterize historic disturbance regimes in forests (e.g., Dean et al. 1997). Charcoal layers in lakes and soils (Gavin et al. 2003) provide insights into the occurrence of past fires in some systems. In grassland ecosystems, tree ring studies from adjacent forests (Sieg 1997) and information from documentary archives can provide insights into past fire and grazing disturbance regimes (Severson and Sieg 2006). Long-term climatic reconstructions from tree rings often provide an understanding of how historic disturbances change with varying climates (Swetnam and Betancourt 1998).

As already discussed under restoring structure and composition based on RNV approaches, reconstructing past disturbance patterns is often difficult. Some ecosystems have crossed thresholds such that returning them to a previous state, or reinstating historic disturbance regimes, can be extremely challenging or impossible. In some cases, past management, introduction of exotic species, rarity of native species, and continuously expanding human populations make reintroduction of historic disturbance regimes quite difficult.

In some situations, it might make greater sense to focus on restoring disturbance regimes and letting specific conditions result as they may; in other situations, the focus may be on altering the conditions and then reintroducing natural disturbance dynamics. Stephenson (1999) characterized the tension between these two approaches as "structural restoration" versus "process restoration." Both may be useful in different circumstances. For example, reintroducing historic fire (a natural disturbance regime) into grasslands of the inland U.S. West that are currently heavily dominated by spotted knapweed (*Centaurea maculosa*) may result only in more knapweed and not serve to restore historic conditions per se. In this case, restoration activities might first focus on alerting the structure of the plant community (that is, removing the knapweed) and then reintroducing the historic disturbance regime. Stephenson (1999) cited an opposite example, where reintroduction of a fire regime without a preceding structural (mechanical) restoration might serve to restore historic conditions in groves of giant sequoia trees (*Sequoiadendron giganteum*), a globally rare but locally common species in the Sierra Nevada, California. The degree to which such structural versus process (disturbance regime) restoration approaches would be useful for conservation of RLK species would likely need to be determined on a case- and context-specific basis.

Much of the information used in maintenance of disturbance regimes pertains to the dynamics of systems and not to the presence and condition of RLK species. Thus the degree to which restoring historic disturbances to a system serves to conserve or restore any given RLK species may be largely unknown.

Fire

Many conservation plans for fire-prone ecosystems call for restoration of historic fire regimes. Restoration of fire regimes is confounded not only by an incomplete understanding of past fires but also by the difficulties of returning fires to areas that are greatly altered in many other ways.

Even for well-studied ecosystems such as the tallgrass prairie of the central U.S. Great Plains, the limited area of remaining grasslands confines the scale at which fire treatments can be implemented and studied. In other systems, such as the pitch pine (*Pinus rigida*)–scrub oak (*Quercus ilicifolia*) barrens in the northeastern United States, restoring large-scale crown fires is hampered by the proximity of human settlements as well as by the small amount of barrens remaining (Jordan et al. 2003). For much of the ponderosa pine type in the western United States, there is a growing recognition that return of frequent, low-intensity surface fires is difficult without first mechanically removing trees and perhaps fuels (Fulé et al. 1997).

Some strategies that are based on understanding historic fire regimes attempt to emulate the structure that results from past fires without burning (Buddle et al. 2006). For example, Cissel et al.'s (1999) landscape plan prescribes timber harvesting in the Pacific Northwest to emulate past burning regimes. Similar strategies have also been attempted on other biomes.

A number of conservation plans incorporate burning guidelines that are modified to account for the needs of rare species. Gill and McCarthy (1998) recognized that variability in fire intervals in nature is inevitable and desirable in prescribed burning plans designed to conserve biodiversity, and that understanding the life histories and habitat requirements (and associated fire regimes) of rare species can be used to propose appropriate fire intervals. Lamont et al. (1991) estimated that a mean fire interval of 20 to 30 years would maintain a rare plant species (*Banksia cuneata*) in western Australia. Brooker and Brooker (2002) modeled the population of the splendid fairy-wren (*Malurus s. splendens*), and concluded that at least 20 years between fires allowed birds to establish territories (3 to 14 years). In another example, managing endangered Kirtland's warblers (*Dendroica kirtlandii*) in the Great Lakes states has entailed periodic logging and burning of jack pine (*Pinus banksiana*) stands to maintain early successional pine forests favored by the bird (Marshall et al. 1998). In many such examples, restoring fire patchiness and variability in burn intensity may also be important.

Herbivory

Historic patterns of terrestrial herbivory by vertebrate species have been altered in many, if not most, native ecosystems in North America. Alterations include changes in native herbivore composition and abundance as well as introductions of nonnative grazers. As a result, in combination with a declining habitat base and efforts to sustain economic livestock operations, the timing, duration, and intensity of herbivory by terrestrial vertebrate species in many systems is likely different from presettlement patterns. Plants (perhaps including some RLK species) in many forested and nonforested ecosystems evolved with some type of herbivory and developed a tolerance and even a dependency on herbivory (Collins et al. 1998), but the impacts of herbivory can be greatly altered by herbivore density as well as forage availability (Laurenroth et al. 1994). The introduction of bison (*Bison bison*) to the Konza Prairie Research Natural Area is an example of an approach that attempted to restore historic grazing patterns based on a perceived range of natural variability (Knapp et al. 1999). In an attempt to introduce heterogeneity in grazing levels, the USDA Forest Service (2001) Northern Great Plains plan proposed to alter livestock grazing management strategies to provide a range of levels of residual cover, based on the recognition that some rare species are dependent on heavily grazed sites and others require higher levels of residual cover for nesting cover.

Maintaining Other Ecosystem Processes

"Other ecosystem processes" includes a range of approaches to considering other ecosystem processes and ecological functions. We provide two examples here: one relating to key ecological functions (KEFs) of organisms and the other relating to maintenance of food webs.

Key Ecological Functions of Organisms

KEFs are the major ecological roles played by species in their ecosystems (Marcot and Vander Heyden 2001; Marcot et al. 2006). Ecological functions of organisms support the trophic structure of ecosystems (energy flows, food webs, predator–prey relations, and nutrient cycling) as well as interspecific symbioses and other interactions that serve to structure ecological communities (Chapin et al. 1996).

Multiple species serving a similar function define the functional redundancy of the system. Theory suggests that ecosystems with greater levels of functional redundancy may be more resilient or resistant to adverse changes in their structure, composition, or function than are ecosystems with lower functional redundancy levels (Rosenfeld 2002). KEFs of species can ultimately affect ecosystem productivity, diversity, resource sustainability, and levels of ecosystem services.

One major assumption of the key ecological functions approach is that providing habitat for natural or desired numbers of species with specific KEF categories would ensure functionality of the system. Another assumption is that greater levels of functional redundancy impart greater stability and resilience of the system and ecological communities to external environmental perturbations. Yet another assumption is that all species within a community perform some key ecological functions that can influence the environment of, or resources used by, other species.

In the KEF approach, categories of KEFs, along with habitats used, are listed for individual species in a species–environment (or wildlife–habitat) relations database. The KEF approach was used in the terrestrial ecology assessment of the Interior Columbia Basin Ecosystem Management Project (ICBEMP) of the Forest Service and the U.S. Department of the Interior (USDI) Bureau of Land Management (Marcot et al. 1997). Several categories and patterns of functions and species were suggested as possible foci for ecosystem management under ICBEMP, such as imperiled KEFs (functions that may vanish from a system when performed by one or a few species that are themselves at risk), functional specialist species (species that perform only one or very few functions), critical functional link species (species that are the only ones in a system that perform specific KEFs), and functional diversity and richness of ecological communities (the number and variety of KEFs in a system). Further, an ecosystem that has retained all its native KEF categories can be defined as "fully functional." Resource managers could then determine the collective array of macrohabitats and habitat elements required by individual species (including RLK species, where their functions are known or suspected) or by a functional species group, pertaining to any of these patterns, and then craft management plans or prescriptions accordingly to provide for such habitats and thus the species and functions they support.

The KEF approach and its key assumptions are largely untested and, as currently implemented, rely on categorical and qualitative data on species' ecological roles. As presented, it is complementary to species- and habitat-specific management and thus does not presume to ensure any specific degree of protection or provision for all RLK species. Only in a general way, by fostering conservation of all ecological functions, does this approach provide for some (unspecified) degree of protection for RLK species.

Food Webs

Studies of food webs are similar to those of energy flow insofar as they relate flows of substance and energy among feeding links and across trophic levels. Food web studies have been a mainstream of ecological research for many decades. The assumption of the food web approach is that providing for food chains within ecological communities can maintain the stability and natural dynamics of the community.

Much has also been written on management of food webs (e.g., papers in Kitchell 1992). For example, Berlow et al. (1999) provided a set of indices by which food webs can be evaluated for use in management, including indices of predator functional response and the strength and importance of species interactions (keystone roles). However, there are few examples of land management plans, particularly for RLK species, that are based on food webs. The management strategy for northern goshawks in southwestern U.S. forests is a notable exception. These guidelines were structured around a food web strategy whereby the habitat needs of the primary prey species of the northern goshawk (*Accipiter gentilis*) were used to develop management guidelines (Reynolds et al. 2006).

One common theme in the ecological literature on food webs is that top predators in a food chain are usually uncommon and often highly sensitive to perturbations affecting the trophic structure of their ecosystem and their prey base (e.g., Lodé et al. 2001). Insofar as some RLK animal species are top predators, this principle may be used as a general guideline for their conservation, that is, by ensuring that their prey base and trophic web are maintained. Sergio et al. (2006) confirmed that top predators, used as flagship or umbrella species (see chap. 6), can justifiably be used as a basis for selecting reserves for overall biodiversity conservation.

Conclusion

The sets of system approaches reviewed here include those that focus on maintaining system structure and composition, as well as approaches that focus on maintaining ecosystem processes and functions. Collectively, such ecosystem approaches may provide for landscape conditions and ecosystem processes that match some natural, historic template, or otherwise normative and desired state, within which some RLK species may find suitable resources and habitats. These approaches account for the dynamics of entire systems in ways that species-specific approaches cannot. However, these approaches may entail managing for disturbance regimes that might harm conditions required by RLK species at specific locations. Also, instituting disturbance management guidelines or managing to maintain species functional groups may necessarily be based on imperfect knowledge of natural conditions and may suffer from problems of practicality of application.

Most system-level approaches to conservation strive to replicate some natural or baseline condition, ecosystem process, or dynamic disturbance regime. RLK species may play roles in mediating or determining such conditions, processes, and regimes, but system-level approaches may not necessarily ensure the conservation of specific RLK species. For example, an ecosystem might be performing within some natural range of variability of some parameter or disturbance regime, and it might contain a diversity of habitat conditions, and it might consist of desired ecosystem processes, but none of this ensures that particular RLK species would be conserved. That is, managing for RLK species is not necessarily the same as managing for biodiversity or for broader system-level performance criteria, conditions, and dynamics.

That said, it is likely true that system-level conditions and dynamics determine the quality of habitats and environments for conservation of RLK species. Thus the best approach to ensuring RLK species conservation may lie in some combination of species- and system-level approaches. The following chapter provides an evaluation of the effectiveness of the various species and system approaches for RLK conservation, as well as other management objectives, and chapter 12 provides an overall process for selecting among approaches.

Acknowledgments

We thank Curt Flather for helpful discussions, reviews, and contributions to this chapter. We also appreciate discussions and guidance from Richard Holthausen, Randy Molina, Dede Olson, and Martin Raphael. We thank Chris Iverson, Nancy Molina, and Jim Quinn for helpful reviews of the manuscript.

References

Agee, J. K. 2003. Historical range of variability in eastern Cascades forests, Washington, USA. *Landscape Ecology* 18:725–40.

Aplet, G. H., and W. S. Keeton. 1998. Application of historical range of variability concepts to biodiversity conservation. Pp. 71–86 in *Practical approaches to the conservation of biological diversity*, ed. R. K. Baydack, H. Campa III, and J. B. Haufler. Washington DC: Island Press.

Berlow, E. L., S. A. Navarrete, C. J. Briggs, M. E. Power, and B. A. Menge. 1999. Quantifying variation in the strengths of species interactions. *Ecology* 80:2206–24.

Betancourt, J. L., T. R. Van Devender, and P. S. Martin. 1990. *Packrat middens: the last 40,000 years of biotic change*. Tucson: University of Arizona Press.

Bond, W. J. 1994. Keystone species. Pp. 237–53 in *Biodiversity and ecosystem function*, ed. E. D. Schulze and H. A. Mooney. Berlin: Springer-Verlag.

Brooker, L., and M. Brooker. 2002. Dispersal and population dynamics of the blue-breasted fairy-wren, *Malurus pulcherrimus*, in fragmented habitat in the Western Australian wheatbelt. *Wildlife Research* 29:225–33.

Buddle, C. M., D. W. Langor, G. R. Pohl, and J. R. Spence. 2006. Arthropod responses to harvesting and wildfire: Implications for emulation of natural disturbance in forest management. *Biological Conservation* 128:346–57.

Butler, D. R. 1995. *Zoogeomorphology: Animals as geomorphic agents*. Cambridge: Cambridge University Press.

Chapin, F. S., III, H. L. Reynolds, C. M. D. Antonio, and V. M. Eckhart. 1996. The functional role of species in terrestrial ecosystems. Pp. 403–28 in: B. H. Walker and W. L. Steffen, eds. Global change and terrestrial ecosystems. Cambridge: Cambridge University Press.

Cissel, J. H., F. J. Swanson, and P. J. Weisberg. 1999. Landscape management using historical fire regimes: Blue River, Oregon. *Ecological Applications* 9:1217–31.

Collins, S. L., A. K. Knapp, J. M. Briggs, J. M. Blair, and E. M. Steinauer. 1998. Modulation of diversity by grazing and mowing in native tallgrass prairie. *Science* 280:745–47.

Dean, D. J., K. R. Wilson, and C. H. Flather. 1997. Spatial error analysis of species richness for a gap analysis map. *Photogrammetric Engineering and Remote Sensing* 63:1211–17.

Everett, R. L., and J. F. Lehmkuhl. 1999. Restoring biodiversity on public lands through disturbance and patch management irrespective of land-use allocation. Pp. 87–105 in *Practical approaches to the conservation of biological diversity*, ed. R. K. Baydack, H. Campa III, and J. B. Haufler. Washington, DC: Island Press.

Frelich, L. F. 2002. *Forest dynamics and disturbance regimes: Studies from temperate evergreen-deciduous forests*. Cambridge: Cambridge University Press.

Fulé, P. Z., W. W. Covington, and M. M. Moore. 1997. Determining reference conditions for ecosystem management of southwestern ponderosa pine forests. Ecological Applications 7:895–908.

Fulé, P. Z., W. W. Covington, H. B. Smith, J. D. Springer, T. A. Heinlein, K. D. Huisinga, and M. M. Moore. 2002. Comparing ecological restoration alternatives: Grand Canyon, Arizona. *Forest Ecology and Management* 170:19–41.

Gavin, D. G., L. B. Brubaker, and K. P. Lertzman. 2003. Holocene fire history of a coastal temperate rain forest based on soil charcoal radiocarbon dates. *Ecology* 84:186–201.

Gill, A. M., and M. A. McCarthy. 1998. Intervals between prescribed fires in Australia: What intrinsic variation should apply? *Biological Conservation* 85:161–69.

Hann, W. J., J. L. Jones, M. G. Karl, P. F. Hessburg, R. E. Keane, D. G. Long, J. P. Menakis, C. et al. 1997. Landscape dynamics of the Basin. Pp. 336–1055 in *An assessment of ecosystem components in the Interior Columbia Basin including portions of the Klamath and Great Basin*. General Technical Report PNW-GTR-405. Vol. 2, ed. T. M. Quigley and S. J. Arbelbide. Portland, OR: U.S. Department of Agriculture, Forest Service, Pacific Northwest Research Station.

Hansson, L. 2003. Why ecology fails at application: Should we consider variability more than regularity? *Oikos* 100:624–27.

Haufler, J. B., R. K. Baydack, H. Campa III, B. J. Kernohan, C. Miller, L. J. O'Neil, and L. Waits. 2002. *Performance measures for ecosystem management and ecological sustainability*. Wildlife Society Technical Review 02-1. Bethesda, MD: Wildlife Society.

Henke, S. E., and F. C. Bryant. 1999. Effects of coyote removal on the faunal community in western Texas. *Journal of Wildlife Management* 63:1066–81.

Hobbs, R., and L. Huenneke. 1992. Disturbance, diversity, and invasion: Implications for conservation. *Conservation Biology* 6:324–37.

Hof, J. G., and M. G. Raphael. 1993. Optimizing timber age class distributions to meet multispecies wildlife population objectives. *Canadian Journal of Forest Research* 23:828–34.

Hunter, M. L. 1991. Coping with ignorance: The coarse-filter strategy for maintaining biodiversity. Pp. 266–81 in *Balancing on the brink of extinction: The Endangered Species Act and lessons for the future*, ed. D. A. Kohm. Washington, DC: Island Press.

Jacobson, G. L., and E. C. Grimm. 1986. A numerical analysis of Holocene forest and prairie vegetation in central Minnesota. *Ecology* 67:958–66.

Jones, C. G., J. H. Lawton, and M. Shachak. 1996. Organisms as ecosystem engineers. Pp. 130–47 in *Ecosystem management: Selected readings*, ed. F. B. Samson and F. L. Knopf. New York: Springer.

Jordan, M. J., W. A. Patterson, III, and A. G. Windisch. 2003. Conceptual ecological models for the Long Island pitch pine barrens: implications for managing rare plant communities. *Forest Ecology and management* 185(1-2): 151–68.
Kaufman, M. R., R. T. Graham, D. A. Boyce Jr., W. H. Moir, L. Perry, R. T. Reynolds, R. L. Bassett, et al. 1994. *An ecological basis for ecosystem management.* General Technical Report RM-246. Fort Collins, CO: U.S. Department of Agriculture, Forest Service, Rocky Mountain Research Station.
Kerns, B. K., and M. M. Moore. 1997. Use of soil characteristics and opal phytoliths to examine vegetation stability in a ponderosa pine–bunchgrass community. *Bulletin of the Ecological Society of America* 78:73.
Kitchell, J. F., ed. 1992. *Food web management.* New York: Springer-Verlag.
Klopatek, J. M., R. J. Olson, C. J. Emerson, and J. L. Joness. 1979. *Land-use conflicts with natural vegetation in the United States.* Publication No. 1333. Oak Ridge, TN: Oak Ridge National Laboratory, Environmental Sciences Division.
Knapp, A. K., J. M. Blair, J. M. Briggs, S. L. Collins, D. C. Hartnett, L. C. Johnson, and E. G. Towne. 1999. The keystone role of bison in North American tallgrass prairie. *BioScience* 49:39–50.
LaBonte, J. R., D. W. Scott, J. D. McIver, and J. L. Hayes. 2001. Threatened, endangered, and sensitive insects in eastern Oregon and Washington forests and adjacent lands. *Northwest Science* 75:185–98.
Lamont, B. B., S. W. Connell, S. M. Bergi. 1991. Seed bank and population dynamics of *Banksia cuneata*: The role of time, fire and moisture. *Botanical Gazette* 152:114–22.
Landres, P. B., P. Morgan, and F. J. Swanson. 1999. Overview of natural variability concepts in managing ecological systems. *Ecological Applications* 9:1179–88.
Laurenroth, W. K., D. G. Milchunas, J. L. Dodd, R. H. Hart, R. K. Heitschmidt, and L. R. Rittenhouse. 1994. Effects of grazing on ecosystems of the Great Plains. Pp. 69–100 in *Ecological implications of livestock grazing*, ed. M. Vavra, W. A. Laycock, and R. D. Pieper. Denver, CO: Society for Range Management.
Litvaitis, J. A. 2003. Are pre-Columbian conditions relevant baselines for managed forests in the northeastern United States? *Forest Ecology and Management* 185:113–26.
Lodé, T., J. Cormier, and D. Le Jacques. 2001. Decline in endangered species as a indication of anthropic pressures: The case of European mink *Mustela lutreola* western population. *Environmental Management* 28:727–35.
Lyons, K. G., and M. W. Schwartz. 2001. Rare species loss alters ecosystem function: Invasion resistance. *Ecology Letters* 4: 358–65.
Marcot, B. G., M. A. Castellano, J. A. Christy, L. K. Croft, J. F. Lehmkuhl, R. H. Naney, K. Nelson, et al. 1997. Terrestrial ecology assessment. Pp. 1497–1713. in *An assessment of ecosystem components in the interior Columbia Basin and portions of the Klamath and Great Basins.* Vol. 3, ed. T. M. Quigley and S. J. Arbelbide. General Technical Report PNW-GTR-405. Portland, OR: U.S. Department of Agriculture, Forest Service, Pacific Northwest Research Station.
Marcot, B. G., and M. Vander Heyden. 2001. Key ecological functions of wildlife species. Pp. 168–86 in *Wildlife–habitat relationships in Oregon and Washington*, ed. D. H. Johnson and T. A. O'Neil. Corvallis: Oregon State University Press.
Marcot, B. G., T. A. O'Neil, J. B. Nyberg, J. A. MacKinnon, P. J. Paquet, and D. Johnson. 2006. Analyzing key ecological functions as one facet of transboundary

subbasin assessment. Pp. 37–50 in *Watersheds across boundaries: Science, sustainability, security*, ed. C. W. Slaughter and N. Berg. Proceedings of the Ninth Biennial Watershed Management Council Conference, November 3–7, 2002, Stevenson, Washington. Center for Water Resources Report No. 107. Riverside: University of California.

Marshall, E., R. Haight, and F. R. Homans. 1998. Incorporating environmental uncertainty into species management decisions: Kirtland's warbler habitat management as a case study. *Conservation Biology* 12:975–85.

McEachern, A. K., M. L. Bowles, and N. B. Pavlovic. 1994. A metapopulation approach to Pitcher's thistle (*Cirsium pitcheri*) recovery in southern Lake Michigan dunes. Pp. 194–218 in *Restoration of endangered species: Conceptual issues, planning and implementation*, ed. M. L. Bowles and C. J. Whelan. Cambridge: Cambridge University Press.

McPherson, G. R. 1997. *Ecology and management of North American savannas.* Tucson: University of Arizona Press.

Millar, C. I., and W. B. Woolfenden. 1999. The role of climate change in interpreting historical variability. *Ecological Applications* 9:1207–16.

Moldenke, A. R., N. Baumeister, E. Estrada-Venegas, and J. Wernz. 1994. Linkages between soil biodiversity and above-ground plant performance. *Transactions of the 15th World Congress on Soil Science* 4a:186–204.

Morgan, P. G., H. Aplet, J. B. Haufler, H. C. Humphries, M. M. Moore, and W. D. Wilson. 1994. Historical range of variability: A useful tool for evaluating ecosystem change. *Journal of Sustainable Forestry* 2:87–111.

Oliver, C. D. 1992. A landscape approach: Achieving and maintaining biodiversity and economic productivity. *Journal of Forestry* 90:20–25.

Paine, R. T. 1966. Food web complexity and species diversity. *American Naturalist* 100:65–75.

———. 1969. The *Pisaster–Tegula* interaction: Prey patches, predator food preference, and intertidal community structure. *Ecology* 50:950–61.

———. 1974. Intertidal community structure. *Oecologia* 15:93–120.

Pickett, S. T. A., and P. S. White. 1985. *The ecology of natural disturbance and patch dynamics*. Orlando, FL: Academic Press.

Power, M. E., D. Tilman, J. A. Estes, B. A. Milne, W. J. Bond, L. S. Mills, G. Daily, J. C. Castilla, J. Lubchenco, and R. T. Paine. 1996. Challenges in the quest for keystones. *Bioscience* 46:609–20.

Prober, S. M. 1996. Conservation of the Grassy White Box woodlands: Rangewide floristic variation and implications for reserve design. *Australian Journal of Botany* 44:57–77.

Reynolds, R. T., R. T. Graham, and D. A. Boyce Jr. 2006. An ecosystem-based conservation strategy for the Northern Goshawk. *Studies in Avian Biology* 31:299–311.

Rosenfeld, J. S. 2002. Functional redundancy in ecology and conservation. *Oikos* 98:156–62.

Rumble, M. A., C. H. Sieg, D. W. Uresk, and J. Javersak. 1998. *Native woodlands and birds of South Dakota: Past and present*. Research Paper RMRS-RP-8. U.S. Department of Agriculture, Forest Service, Rocky Mountain Research Station.

Sergio, F., I. Newton, L. Marchesi, and P. Pedrini. 2006. Ecologically justified charisma: preservation of top predators delivers biodiversity conservation. *Journal of Applied Ecology* 43:1049–55.

Severson, K. E., and C. H. Sieg. 2006. *Pre-1880 historical ecology of eastern North Dakota*. Fargo, ND: Institute for Regional Studies.

Sieg, C. H. 1997. The role of fire in conserving the biological diversity on native rangelands of the Northern Great Plains. Pp. 31–38 in *Proceedings of the symposium on conserving diversity of the northern Great Plains*, ed. D. W. Uresk and G. L. Schenbeck. General Technical Report RM-298. Fort Collins, CO: U.S. Department of Agriculture, Forest Service, Rocky Mountain Research Station.

Sieg, C. H., C. H. Flather, and S. McCanny. 1999. Recent biodiversity patterns in the Great Plains: Implications for restoration and management. *Great Plains Research* 9:277–313.

Skovlin, J. M., and J. W. Thomas. 1995. *Interpreting long-term trends in Blue Mountain ecosystems from repeat photography*. Research Paper PNW-GTR-315. Portland, OR: U.S. Department of Agriculture, Forest Service.

Smallwood, K. S., M. L. Morrison, and J. Beyea. 1998. Animal burrowing attributes affecting hazardous waste management. *Environmental Management* 22:831–47.

Stephenson, N. L. 1999. Reference conditions for giant sequoia forest restoration: Structure, process, and precision. *Ecological Applications* 9:1253–65.

Swetnam, T. W., and J. L. Betancourt. 1998. Mesoscale disturbance and ecological response to decadal climatic variability in the American southwest. *Journal of Climate* 11:3128–47.

Swetnam, T. W., C. D. Allen, and J. L. Betancourt. 1999. Applied historical ecology: Using the past to manage for the future. *Ecological Applications* 9:1189–1206.

USDA (U.S. Department of Agriculture), Forest Service. 2001. *Land and resource management plan for the Nebraska National Forest and associated units*. Denver: U.S. Department of Agriculture, Forest Service, Rocky Mountain Region.

Van Putten, M., and S. D. Miller. 1999. Prairie dogs: The case for listing. *Wildlife Society Bulletin* 27:1113–20.

Wilbur, H. M. 1997. Experimental ecology of food webs: Complex systems in temporary ponds. *Ecology* 78:2279–2302.

Wright, J. P., C. G. Jones, and A. S. Flecker. 2002. An ecosystem engineer, the beaver, increases species richness at the landscape scale. *Oecologia* 132:96–101.

8

Effectiveness of Alternative Management Strategies in Meeting Conservation Objectives

Richard S. Holthausen and Carolyn Hull Sieg

This chapter evaluates how well various management strategies meet a variety of conservation objectives, summarizes their effectiveness in meeting objectives for rare or little-known (RLK) species, and proposes ways to combine strategies to meet overall conservation objectives. We address two broad categories of management strategies. *Species approaches* result in conservation actions focused on providing for individual species or groups of species with common needs or common ecological characteristics. Species approaches we consider here include those focused on managing for the viability of individual species, those that use surrogate species or groups of species, plus geographic approaches focused on species occurrences or ranges. *System approaches* focus conservation actions on providing for ecosystem composition, structure, or function. System approaches fall into two general categories: those that focus on managing to restore or maintain ecosystem composition and structure and those that focus on restoring or maintaining ecosystem function. These approaches are described in more detail in chapters 6 and 7. For this chapter, we analyzed approaches for which we could find studies or discussions illustrating how the approach might be used to meet either specific needs of a species or a group of species, or to meet more general biodiversity goals.

We evaluated each approach for relative effectiveness in meeting three general categories of conservation objectives: species diversity, genetic diversity, and ecosystem diversity (see chap. 2). *Species diversity* objectives included maintaining or restoring viable populations of (1) RLK species, (2) other species, and (3) other species in natural patterns of abundance and distribution. The objective of *genetic diversity* is to maintain or restore the natural genetic variation of species. *Ecosystem diversity* objectives included (1) providing for native ecosystem types and seral stages within their natural range of variation; (2) maintaining or restoring ecological processes such as disturbance regimes, trophic structures, nutrient cycling, and key functional roles of organisms; and (3) maintaining the capacity of landscapes to provide for resiliency of species in the face of long-term environmental change.

Our evaluation of the various approaches was complicated by a number of factors. Many of these approaches share some degree of overlap, making it difficult to place them in distinct categories. In some cases authors described and evaluated very similar approaches but gave them different names. In these situations, we discussed each author's work under the category where it seemed to fit best but also mentioned it under the category assigned to it by the author. In other papers, the management goals and evaluation criteria were unclear. For example, some papers described the use of a surrogate approach for developing a comprehensive ecosystem management strategy but did not clearly define the species goals for that strategy. Some focused on proximate rather than ultimate goals. For example, the likelihood that surrogate species measures would conserve other species was frequently shorthanded as "protection" of those other species and measured simply as the overlap of the surrogate's locations with locations of other species. This can be misleading because simple overlap with one or more locations of another species does not necessarily ensure adequate conservation measures for that species. We also found that different authors used different evaluation criteria for measuring the effectiveness of the same approaches. These complications were largely because the published research was intended to answer different questions than the ones that we were asking. As a consequence, we were often forced to make assumptions and interpretations beyond those given by the original authors.

We found that variations in the application of an approach could lead to different assessments of its utility. For example, a guild approach based on general species information from the literature might perform differently from one based on site-specific data. For some approaches, we were uncertain if the approach would be applied directly to RLK species or applied to other species with secondary inference to RLK species. For example, the focal species concept might be applied based on the needs of the rare species themselves. Alternatively, it might be applied to a broader array of species, in which case the evaluation becomes a determination of how well the conservation of that broader array of species provides for the rare species. In these cases, we described the assumptions that went into our evaluations.

We used four ranks to rate the effectiveness of each approach in meeting each of the conservation objectives (see table 8.1):

- A—likely to be effective. We assigned this rank for cases where a well-designed application of an approach would explicitly address the conservation objective in a way that would likely be effective in meeting it. For example, strategies that address the viability of individual species would likely be effective in providing for viability of those species.
- B—could be designed to be effective in some circumstances. We assigned this rating to approaches that would be effective in meeting a conservation objective when designed in a particular way but which might not be effective in other applications. For example, the biodiversity indicator approach would be effective in the conservation of rare species when applied directly to rare species and based on site-specific data. Other more generalized applications of this approach might not be effective in meeting this objective.
- C—effectiveness nearly completely dependent on other factors. We assigned this rating when the effectiveness of an approach would be more directly tied to circumstances rather than to the design of the approach itself. For example, the umbrella species approach, which focuses on species with large area requirements, could be effective in conserving rare species if there was significant co-occurrence between the selected umbrella and the rare species. In other circumstances, effectiveness would be low.
- D—unlikely to be effective. This rating was assigned where an

Table 8.1. *Relative effectiveness of various approaches in meeting species, genetic, and ecosystem diversity conservation objectives (see chap. 3)*

SPECIES OR SYSTEM OF APPROACH	CONSERVATION OBJECTIVE									
	SPECIES DIVERSITY — *Maintain viable populations of*			GENETIC DIVERSITY	ECOSYSTEM DIVERSITY					
					STRUCTURE	PROCESSES				RESILIENCY
	Rare and little-known species[1,2]	Other species	Natural patterns-abundance/distribution	Maintain natural genetic variation[3]	Maintain ecosystem types and seral stages within their RNV	Disturbance regimes	Trophic structures	Nutrient cycling	Other processes	Maintain species in face of environmental change
Individual species population viability	A/D	C	C	B/C	C	C	C	C	C	C
Surrogate species										
Umbrella species	C	C	C	B/C	C	C	C	C	C	C
Focal species	B/C	B	B	B/C	B	B	B	B	B	B
Biodiversity indicator species	B/D	B	D	B/D	D	D	D	D	D	D
Flagship species	C	C	D	B/D	C	C	C	C	C	C
Guilds	D	C	C	C/C	D	D	D	D	D	D
Habitat assemblages	C	C	C	D/D	D	D	D	D	D	D
Geographically based approaches										
Locations of target species	B/C	C	C	B/C	D	D	D	D	D	D
Species hot spots	C/C	C	C	D	D	D	D	D	D	D
Reserves or protected areas	B/C	B	B	B/C	B	B	B	B	B	C
System structure/composition										
Natural range of variability	C	B	B	B	A	B	B	C	C	B
Key habitat conditions	C	C	C	C	C	C	C	C	C	C
Species with critical roles	C	C	C	C	C	C	C	C	C	C
System function										
Maintaining disturbance regimes	C	B	B	C	B	A	B	B	B	B
Maintaining other ecosystem functions	C	C	C	C	C	C	C	C	C	C

These rankings are based on using a well-designed example of the approach, and assume that the approach is used alone. Brief rank definitions are as follows: A—likely to be effective; B—could be designed to be effective in some circumstances; C—effectiveness will be nearly completely dependent on other factors; D—unlikely to be effective. See the text for further information on rankings.

[1]In some cases, the approach was assumed to apply directly to rare and little-known species. In other cases a more general application was assumed. See text for additional clarifications.

[2]Where two ranks are shown, the first applies to rare species and the second to little-known species.

[3]For species approaches two ranks are shown. The first applies to the target species of the approach and the second to the other species. In the case of surrogate approaches, the first rank is for the surrogate, and the scond is for the spcies it is intended to represent.

approach, by its nature, would not be effective in meeting a conservation objective. For example, development of individual species strategies based on the concepts of population viability analysis would not be effective if applied to little-known species as there would be inadequate information on which to base the strategies.

Where we felt a particular approach ranked differently for rare versus little-known species, we assigned a rank for each (e.g., A/B). In ranking species approaches for meeting genetic diversity objectives, we also used separate rankings when we felt the approach ranked differently for the species targeted by the approach versus other species. In each section, we briefly reiterate the major tenets of each approach and provide specific details of the examples of the application that we evaluated for the approach. Our intent was to use published, peer-reviewed papers wherever possible, but we also used reports by various agencies and nongovernmental organizations. In some cases we found no examples of application of the approach. In a few of these situations, we attempted to logically evaluate the likely effectiveness of the approach.

Conservation of Individual Species at Risk Based on Concepts of Population Viability

This approach entails developing a conservation strategy for a target species designed to sustain the species in a given geographic area. The most rigorous application of this approach involves incorporating population viability analyses (PVA) that test the likelihood of a species persisting under alternative strategies. Conservation strategies typically provide for conservation of the variety of habitats required for the species and its obligate symbionts (such as pollinators), including mutualist species. They may also provide for management of conditions and threats or stressors that might place a species at risk, such as management of human disturbance factors. The most rigorous approaches have an objective of maintaining the long-term evolutionary potential of the species. This is usually accomplished by maintaining the maximum range of conditions under which the species, including peripheral populations, is known to occur to maintain the species' ability to adapt to changing conditions.

Development of strategies for population viability may be based on detailed single-species models that incorporate quantitative data on species' ecology, demography, biotic interactions, and dispersal and colonization dynamics (Murphy and Noon 1992; USDA Forest Service 2000). These models can be used to explore relative population growth rates of the target species among various populations, in response to alternative management strategies, and in varying climatic conditions. Where the information for such models is lacking, strategies may be based on more general conservation principles (Ralls et al. 2002; Samson 2002), including the general life history characteristics of the species (Wu and Botkin 1980). As noted in chapter 6, application of PVA concepts assumes availability of substantial information about species status, distribution, and autecology, so by definition this approach would generally not be applicable to poorly known species.

Species Diversity Objectives

Some validation testing of PVAs has been completed (see chap. 6). However, the effectiveness of management strategies based on the concepts of population viability in meeting species diversity objectives has not been extensively tested. Such strategies have been implemented for a number of species, including grizzly bears (*Ursus arctos horribilis*) (U.S. Fish and Wildlife Service 1993), northern spotted owls (*Strix occidentalis caurina*) and other old-growth-associated species in the Pacific Northwest (USDA and USDI 1994), northern goshawks (*Accipiter gentilis*) (Reynolds et al. 1992), old-growth-associated species on the Tongass National Forest (USDA Forest Service 1997), red-cockaded woodpeckers (*Picoides borealis*) (Conner et al. 2001), and western prairie fringed orchids (*Platanthera praeclara*) (USDA Forest Service 2000). Because these strategies have all been implemented relatively recently, monitoring data are not yet adequate to provide definitive answers about long-term consequences. However, based on information collected to date, it seems reasonable that these strategies, as modified through time based on new information and conditions, will be successful in conserving their target species. In some cases, however, conditions that cannot be controlled at the scale at which the strategies are implemented (e.g., global warming, major shifts in distribution of competing species) may result in species extirpation/extinction despite the measures adopted through such strategies. For example, the success of the strategy for northern spotted owls is currently in question because of increasing competition from barred owls (*Strix varia*), which have recently moved into the range of the spotted owls (Kelly et al. 2003; Peterson and Robins 2003). Other species, such as the cactus ferruginous pygmy-owl (*Glaucidium brasilianum cactorum*) in Arizona, are already suffering from such severe habitat fragmentation that their long-term persistence is in doubt (Cartron and Finch 2000). Thus these strategies do not ensure species resiliency with all potential futures. They are expected to be successful dealing with the dominant stressors that are known at the time of their development, but not in cases where species and the ecosystems on which they depend are too severely degraded for recovery or where threats to the species cannot be controlled at the scale of the strategy.

We conclude that this approach would be effective (rank A) in maintaining or restoring viable populations of rare species where such species are actually the target of a strategy and information is available to develop the strategy. The information needed to conduct PVAs is often difficult to obtain, which can limit the effectiveness of PVA-based strategies. Sensitivity analysis may be used to bracket potential outcomes where information is highly uncertain. However, the approach is unlikely to be effective (rank D) for little-known species as the information necessary to implement the approach would, by definition, be lacking. (See table 8.1 for all rankings discussed in this and subsequent sections).

Effectiveness of this approach in providing for viability of nontarget species would be highly variable. There are situations where management for a single species could be detrimental to conditions needed to support other species. For example, use of a mechanical device to maintain cavities for red-cockaded woodpeckers may restrict the use of those cavities by Sherman's fox squirrel (*Sciurus niger shermani*), which has also declined significantly (Simberloff 1998). In other situations, target species of this approach may make effective umbrellas for other species and so provide conditions necessary to support viable populations of other species. In one of the best examples of this phenomenon, Thomas et al. (1993) assessed a conservation plan developed specifically for northern spotted owls. Spotted owls were prioritized for attention because they were listed under the Endangered Species Act and not because they were thought to be effective umbrellas. Thomas et al. concluded that conditions conducive to viable populations would be provided for more than half of all other vertebrate species associated with old-growth forests. We conclude that the use of this approach for target species at risk might be effective in providing for conservation of other species, but that this benefit will be incidental rather than the result of characteristics of the approach itself (rank C).

Genetic Diversity Objectives

Strategies aimed at maintaining population viability of individual species can be designed to maintain genetic diversity of the target species as well. Although data quantifying the genetic diversity of individuals in various populations rarely exist, by selecting sites for protection that encompass the diversity of circumstances where the species occurs, it is assumed that genetic diversity encompassed in populations within the planning area will also be retained. These principles were applied in developing the management guidelines for the western prairie fringed orchid on the Sheyenne National Grassland in southeastern North Dakota (USDA Forest Service 2000). In selecting populations deemed critical for the persistence of the species, consideration was given to connectivity of habitat and representation of populations on the periphery of the orchid's range on the planning area. Quantitative population viability analyses were conducted to assess the effectiveness of this level of protection on projected population growth rates of the orchid (Sieg et al. 2003). The success of these guidelines in maintaining viable populations (and hopefully genetic diversity) of the western prairie fringed orchid has yet to be verified by subsequent monitoring data. We conclude that guidelines that attempt to provide for viability of individual species have the potential of maintaining genetic variation of the target species if they are appropriately designed and implemented (rank B). Effectiveness of this strategy in providing for genetic diversity of nontarget species would be entirely dependent on other factors (rank C).

Ecosystem Diversity Objectives

In some situations, strategies developed to meet the needs of individual species may be based on management of overall ecosystem characteristics and thus be compatible with ecosystem management objectives. Where such species are widespread, their management may make a significant contribution to overall ecosystem management. For example, the management strategy for northern goshawks in forests of the southwestern United States (Reynolds et al. 1992) is based largely on the diverse habitat needs of species that serve as prey for goshawks. In this case, management for this wide array of habitats could result in landscapes that would fall both structurally and functionally within their natural range of variation, thus accomplishing ecosystem diversity objectives. However, there are other cases where management for an individual species could be antithetical to the accomplishment of ecosystem management objectives. For example, management guidelines for old-growth-associated vertebrate species in the Sierra Nevada Mountains of California called for higher levels of canopy closure than had been recommended for other ecosystem objectives (USDA Forest Service 2003). Also, in many cases the area affected by management for an individual species will simply be too small to have a significant impact on ecosystem diversity objectives. So we conclude that the use of this approach for target species at risk might be effective in providing for ecosystem diversity objectives, but that this benefit will be the result of fortunate circumstances rather than the characteristics of the approach itself (rank C).

Conservation of Surrogate Species

The following approaches, with the exception of habitat assemblages, use a single species to reflect the needs of a larger group of species. The habitat assemblages approach groups species together to represent their common habitat needs.

Umbrella Species

Umbrella species have generally been defined as species that use similar habitats as other sympatric species, and whose area of occupancy on the landscape physically encompasses the occurrence of other species (Wilcox 1984; Caro and O'Doherty 1999; Suter et al. 2002). Because of the large area requirement, the habitat needed for a viable population of an umbrella species may also provide for the habitat needs of sympatric species.

Umbrella species are used to specify the type and amount of habitat to be protected, but not necessarily specific locations (Berger 1997).

Species Diversity Objectives

A recent evaluation of applications of the umbrella species concept suggests that single-species umbrellas cannot ensure the conservation of all co-occurring species (Roberge and Angelstam 2004). Thomas et al. (1993) found that northern spotted owls served as an effective umbrella for about one-third of other vertebrate, invertebrate, and plant species associated with old-growth forests in the Pacific Northwest; a combination of spotted owls and marbled murrelets (*Brachyramphus marmoratus*) did better, lending protection to additional species. Suter et al. (2002) found that the grouse species Capercaillie (*Tetrao urogallus*) was an effective umbrella for other mountain bird species in the Swiss Prealps, but not for overall avian diversity. Rubinoff (2001) found that a bird associated with coastal sage scrub, the California gnatcatcher (*Polioptila californica*), served as a poor umbrella for three species of butterflies also found in this habitat, but it could be argued that the gnatcatcher was an inappropriate choice for an umbrella species because the butterflies apparently had larger area requirements. Martikainen et al. (1998) concluded that white-backed woodpeckers (*Dendrocopos leucotos*) might serve as an umbrella for the decayed wood habitat required by threatened, saproxylic beetles in Finland. Poiani et al. (2001) found that the use of greater prairie chickens (*Tympanuchus cupido pinnatus*) as an umbrella species provided identification of more "biologically important land" than two other conservation approaches. Rowland et al. (2005) reported that greater sage-grouse (*Centrocercus urophasianus*) would serve as a relatively effective umbrella for sagebrush obligates but not for other species that are associated with sagebrush but also use other habitat types. Variability in success of these tests of umbrella species can be attributed to the care with which the umbrellas were chosen, the characteristics of the landscapes in which the tests were conducted, and the response parameters chosen for the tests.

Effectiveness of the umbrella species approach in meeting objectives for RLK species received limited testing by Thomas et al. (1993). In old-growth forests of the Pacific Northwest, the use of northern spotted owls as umbrellas for rare species was judged to be inadequate. This was due in part to the fine-scale spatial pattern of rare species occurrences, which frequently occupied smaller and more isolated patches of old forest than those that were protected for northern spotted owls. This pattern might hold true in other areas, and the umbrella concept would provide for rare species only when the occurrences of the rare species happened to coincide with those of the umbrella species (rank C). The same condition is likely to hold true for little-known species. Also, because the umbrella species concept relies on species with large area requirements, it is unlikely that it could be tailored specifically to address RLK species.

Because umbrella species can be used to develop comprehensive conservation guidelines, we suspect that the concept might be moderately effective in providing for viability of some other species. The foregoing tests tend to demonstrate that effectiveness. However, this success would be limited to some subset of other species because umbrella species address only one dimension of limitation (i.e., largest area requirement). Consequently we assigned it rank C. In most cases, we expect that the more comprehensive focal species approach would be more effective.

Genetic Diversity Objectives

Under the umbrella species approach, management is directed at providing for viability of the identified umbrella species. Such management would deal with both the full range of the species in the planning area and appropriate connectivity of habitat across the area. Such management would be beneficial in providing for genetic variation of the umbrella species. As with the approach based on concepts of population viability of individual species, we assign the umbrella approach rank B for effectiveness in meeting genetic diversity objectives of the umbrella species itself. The umbrella approach would only be partially effective in providing for genetic diversity of other species, and such effectiveness would be coincidental rather than being based on the design of the approach (rank C).

Ecosystem Diversity Objectives

We found no published tests of the utility of the umbrella species concept in meeting ecosystem diversity objectives. However, because the umbrella species concept focuses on species with large area requirements, it could be effective in identifying large habitat areas for protection, and such areas might have the appropriate attributes to accomplish some level of ecosystem diversity objectives. Analysis done by FEMAT (1993) provides one working example. A conservation network designed around the needs of northern spotted owls was projected to be moderately successful in meeting the objective of maintaining an intact, interconnected, functioning old-growth ecosystem. Adding other umbrella species (e.g., marbled murrelets, salmon [*Oncorhynchus* spp.]) to the conservation design increased the likelihood of meeting ecosystem objectives. We expect that conservation designs based on multiple umbrella species would be more effective than single umbrella species in meeting ecosystem objectives. However, the likelihood that conservation efforts based on umbrella species would meet ecosystem diversity objectives is largely dependent on the characteristics of the species chosen as umbrellas (rank C). For example, species that still occur broadly across the landscape, and that are themselves dependent on a broad set of ecosystem characteristics, will be most likely to be effective in meeting ecosystem diversity objectives. Conservation designs based on the focal species concept, which identifies umbrellas for several different dimensions of species limitations, would likely be more effective in meeting ecosystem objectives than designs based on a single umbrella species.

Focal Species

The concept of focal species has been variously defined (chap. 6). For our evaluation, we accept the concept as defined by Lambeck (1997), who suggested that needs of all species could be represented by a set of species whose characteristics make them most sensitive to resource, area, or dispersal limitations.

Species Diversity Objectives

Several authors have used Lambeck's (1997) concepts to develop landscape designs for species conservation (Bani et al. 2002; Snaith and Beazley 2002). Watson et al. (2001) used the concept to develop spatially explicit conservation guidelines for woodland birds in southeastern Australia. They estimated that all woodland bird species would find suitable habitat in a landscape designed according to this set of guidelines. They cautioned that the guidelines might be inadequate since they were based on occupancy data for populations that might not have reached equilibrium in the test landscape, and recommended that their hypothesis be tested through adaptive management. They concluded that the focal species concept was useful for developing spatially explicit guidelines for bird species, but cautioned that they had not tested the concept for other taxa.

Effectiveness of the focal species approach in meeting objectives for RLK species has not been tested. (Kintsch and Urban's [2002] work, focused on rare plant species, is reviewed under the indicator species section despite their use of the term "focal species.") However, the species identified as focal using Lambeck's concept are likely to be those that occur less frequently on the landscape, simply because their occurrence is most limited by area, resources, or dispersal. Therefore, we conclude that application of the focal species concept to rare species could be used to design a landscape that would provide for viability of those species (rank B). This conclusion is based on the assumptions that (1) there is knowledge of the type, size, and distribution of habitats within which rare species occur; and (2) rare species occurrence is predictable based on these factors. Because this information would not be available for little-known species, the success of the focal species approach in meeting objectives for little-known species would be highly variable and difficult to assess or monitor (rank C).

Similarly, because the concept can be used to develop comprehensive and spatially explicit conservation guidelines, we suspect that it can be effectively applied to provide for viability of other species (rank B). We expect that cross-taxonomic sets of focal species would be more likely to accomplish this objective because it has been frequently demonstrated that surrogates from one taxon are not efficient surrogates for other taxa (Lawler et al. 2003). Also, if the selection and use of focal species were based on knowledge of natural patterns of abundance and distribution of species, the concept could be used to meet the objective of managing for species in such patterns (rank B).

Genetic Diversity Objectives

The focal species approach is intended to identify a subset of species that represent all factors that might significantly limit populations of all species and to establish a strategy that provides for viability of those species. A well-developed application of this approach would consider both the overall distribution of the focal species (at least in the area under consideration) and the connectivity of the landscape for them. As with strategies for viability of individual species, the strategies for viability of focal species would include provisions for the diversity of conditions under which the focal species occur. So we conclude that strategies that provide for viability of focal species could also provide for maintaining the genetic variation of those species (rank B). Further, such strategies would provide for genetic variation in some additional species for which the focal species act as surrogates. This is not likely to be the case for all other species, so we conclude that the focal species strategy would provide incidental benefits for genetic variation of other species (rank C).

Ecosystem Diversity Objectives

The focal species concept is designed to look broadly across species on a landscape and to identify factors that could be critical in limiting the distribution of those species. Because it is inherently based on concepts of species interactions throughout a landscape, it should be compatible with meeting both species and ecosystem needs within that landscape. For example, Raphael et al. (2001) proposed that 31 focal species could be used to design ecosystem management guidelines to meet the needs of a broader array of species of concern in the Interior Columbia River basin in the Pacific Northwest. However, several different forms of management could be used to implement guidelines developed under the focal species concept. For example, the size and distribution of habitat patches might be managed through a system of small reserves, or alternately through consideration of their occurrence over time in a dynamic landscape managed to mimic natural disturbance regimes. Since these two strategies could have very different effects on overall ecosystem diversity standards, we conclude that the focal species concept does not inherently meet ecosystem diversity objectives, even though it could be designed to meet such objectives under some circumstances (rank B).

Guild Surrogates

We discuss the use of a single guild member as a surrogate for other members of a guild in a later section (see "Conservation of Multiple-Species Groups—Guilds").

Management Indicator Species

"Management indicator species" (MIS) is a term used by the U.S. Department of Agriculture (USDA) Forest Service to describe a set of species used to judge the effects of management activities (Landres et al. 1988). Under 1982 regulations (*Federal Register*, Vol. 47, No. 190, September 30, 1982, 36 CFR Part 219) that implemented the National Forest Management Act, national forests were required to select a set of MIS, estimate the effects of proposed management alternatives on them, monitor changes in their populations, and relate those changes to changes in habitat. New regulations issued in 2005 (*Federal Register*, Vol. 70, No. 3, January 5, 2005, 36 CFR Part 219) no longer rely on the MIS concept, but national forests still operating under the 1982 regulations continue to use MIS. All MIS must be species that respond to management activities, and may include species of conservation concern; species with special habitat needs; species that are hunted, fished, or trapped; and species whose population changes may indicate the status of other species. Because there is broad latitude in the designation of MIS, any of the other surrogate concepts could be used to aid in their identification. Since the MIS requirement does not represent a unique surrogate approach, it is not possible to evaluate its effectiveness in meeting objectives.

Biodiversity Indicator Species

As described in chapter 6, biodiversity indicator species are used to identify locations of other species or of communities (Caro and O'Doherty 1999; Chase et al. 2000). Conservation of the biodiversity indicator may provide benefits for co-located species.

Species Diversity Objectives

Kintsch and Urban (2002) found common indicator plant species (that they termed focal) successful in predicting occurrences of rare species, with 62 to 100% of rare species locations predicted by the occurrence of the indicator species. Fleishman et al. (2000) identified two butterfly species whose protection would also protect sites of 97% of all other butterfly species found in the same area. Fleishman et al. (2001) identified sets of plant and butterfly species in California and Ohio whose protection would also protect a large proportion of sites for most other species within the same taxonomic groups. However, these species were not effective as cross-taxonomic indicators. (In both the 2000 and 2001 studies, Fleishman et al. used the term "umbrellas." However, since the species were used to indicate other species locations and not types and amounts of habitat, it seems that they better fit the definition of "biodiversity indicators.") In a study of 35 forest sites in Australia, Pharo et al. (1999) found positive correlations between fern species richness and bryophyte species richness and between overstory species richness and bryophyte species richness. In a study of reserves surrounding the city of Chicago, Panzer and Schwartz (1998) found that plant community richness, plant species richness, and plant genus richness explained significant levels of variation in the richness of insect species that were considered to require conservation attention. Chiarucci et al. (2005) investigated the use of vascular plant richness as surrogates for fungal species richness. Their analysis of data from 25 forest plots in Tuscany, Italy, suggested that vascular plants could not be used as surrogates to maximize fungal species diversity. Chase et al. (2000) tested the effectiveness of potential indicators for identifying areas of high species richness, or particular species composition within coastal sage scrub habitats. They found close associations between the presence of particular species and the composition of bird and small mammal communities but cautioned that additional data were needed to determine whether management for protection of the indicators would also result in protection of the associated species.

The foregoing results suggest that, in some situations, biodiversity indicators could be effectively used to identify locations of both rare species and sets of intra- and intertaxonomic species. However, the mix of positive and negative results reaffirms the need for local data in determining whether surrogacy is warranted. Whether managing for viability of the biodiversity indicators would also provide for viability of the other species would depend on the type of management used, the similarity of habitat needs between the indicator species and the species being indicated, and whether the extant populations of those species represented a viable population. However, it appears that the biodiversity indicator approach could be used to develop effective management direction for both rare species and other species (rank B). Since initial identification of biodiversity indicators requires substantial information on distribution of both target taxa and potential indicators, the approach would not be effective when applied to little-known species (rank D). Further, it is unlikely that this approach would provide for maintenance of other species in natural patterns of abundance and distribution (rank D). Since the approach focuses on existing locations of species, it would not provide information on the natural patterns of abundance and distribution.

Genetic Diversity Objectives

Both Kintsch and Urban (2002) and Fleishman et al. (2000, 2001) showed that biodiversity indicators could be used to locate a substantial portion of the occurrences of target taxa. Management of these sites for the target taxa should provide for significant retention of their genetic variability (rank B). There is no indication that this approach would be effective in maintaining the genetic variability of other, nontarget species (rank D).

Ecosystem Diversity Objectives

Biodiversity indicators are used to indicate the presence of individual rare species or of biodiversity hotspots. Management that would be applied to the locations is not specified by the approach but would presumably include some mixture of active management and reserving some sites from management. However, since this approach is targeted at the locations of individual species, or specified groups of species, it is unlikely to result in the accomplishment of overall ecosystem objectives (rank D).

Flagship Species

Flagship species are species of high public interest that are selected to rally conservation support. Caro and O'Doherty (1999) note that they are frequently large, and may also serve as umbrellas, but that there are no strict criteria for their selection. They are not based on an ecological concept, which distinguishes them from other surrogate approaches.

Species Diversity Objectives

To function as flagships, species must be reasonably well known to the public. Consequently, neither very rare species nor little-known species are likely to be selected as flagships, so the flagship concept is not intended to provide for viability of these species. Some coincidental benefits may accrue to the viability of RLK species (rank C). Although flagships may be large-bodied and have the characteristics of umbrella species, they are not intentionally selected to serve as surrogates for viability of other species, so benefits to those species are coincidental (rank C).

Genetic Diversity Objectives

Under the flagship species approach, management is directed at providing for viability of the identified flagship species. Such management would deal with both the full range of the species in the planning area and appropriate connectivity of habitat across the area. Such management could be beneficial in providing for genetic variation of the flagship species (rank B). However, because flagship species are selected with no strong ecological criteria, management for the flagships is unlikely to provide representation of the full range of ecological conditions in which other species occur. Thus this approach is likely to be ineffective in providing for genetic variability of species other than the flagship (rank D).

Ecosystem Diversity Objectives

Some flagships are large-bodied (Caro and O'Doherty 1999) and would likely have large area requirements. Management for such flagships could have a significant influence on overall ecosystem condition. However, many types of species could be designated as flagships and their management could take many forms, so effects on ecosystem objectives will be variable and coincidental rather than the result of intentional design (rank C).

Conservation of Multiple-Species Groups

Multiple-species approaches include management based on the requirements of species that are identified to be part of a guild or a habitat assemblage.

Guilds

As described in chapter 6, a guild is a group of species with common traits that may include behavior, diet, resource selection, and so forth. Most commonly, guilds are sets of species that use the same or similar resources for the same life function. Guilds have almost always been defined within a single taxonomic group (but see Roberts 1987 for an example of guilds containing birds and mammals).

We discuss two possible approaches. The guild surrogate approach is one in which one member of a guild is chosen to act as a surrogate for all other species in the guild. In the entire guild approach, the focus (e.g., monitoring) is on all members of a guild. Both approaches share common assumptions that the population trends of all members of the guild are closely correlated, and that changes in environmental conditions affect all members of the guild in a similar way. Most research on guilds has focused on tests of these assumptions, with the general finding that species within a guild demonstrate different responses to environmental conditions and changes in those conditions (see chap. 6) and thus violate guild assumptions. Block et al. (1986) concluded that management for an indicator species within a guild might not meet the needs of other guild members and that monitoring of either a guild indicator or of an entire guild might not reflect population fluctuations of individual species within a guild. As an alternative, Szaro (1986) proposed that "response" guilds be based on multiyear data of species' responses to environmental change by simply grouping together species that showed similar responses.

There has only been limited use of guilds to develop management direction. Federal managers have used guilds of cavity excavators to develop guidance for snags and down dead wood, but there has been little testing of the effectiveness of those guidelines (Bull et al. 1997).

Species Diversity Objectives

Use of the guild approach to provide for viable populations of species has not been widely explored. Tests of the guild approach have investigated the underlying assumptions or explored the utility of guilds as monitoring tools. Guilds have rarely been used in developing conservation strategies. Properly defined guilds could potentially be effective in developing conservation strategies, even if population trends of guild members are not closely correlated. In particular, the response guilds proposed by Szaro could be useful in the development of conservation strategies. However, this application of guilds requires significant data on population trends, and multiple guilds would probably be needed to provide for conservation of a broad array of species. Roberts (1987) provided a conceptual example of multiple guilds that could be used to represent key structural features of deciduous forest habitats. However, one shortcoming in the guild concept has been the failure to consider spatial needs of species. Guilds have focused primarily on needs of species within particular habitats, with little attention given to amounts or spatial arrangements of habitats. Thus the guild concept, in applications to date, has been weaker than either the umbrella or the focal species concepts in consideration of spatial needs of species. We conclude that the guild concept would be effective in providing for species viability only if species were not spatially limited, and we assign it rank C.

We did not find any reported attempts to develop guilds of RLK species, or to test the effectiveness of guilds developed for other species in providing for RLK species. Adapting the guild approach to RLK species would be difficult because (1) development of guilds requires significant knowledge of species' life histories (Block et al. 1986), and (2) guild indicators are most effective if they have large populations (Caro and O'Doherty 1999). We conclude that it is unlikely that the guild approach would be effective in the conservation of RLK species (rank D).

Genetic Diversity Objectives

The guild approach has been focused on requirements for specific habitats and not on the spatial distribution of populations. Thus, in application of the guild concept to date, it does not inherently maintain the variety of ecological conditions in which a species occurs. Consequently, benefits to genetic variability under the guild approach would be incidental and not a direct result of the guild approach itself (rank of C for both the guild indicator and the species that it represents within the guild).

Ecosystem Diversity Objectives

The utility of the guild approach for accomplishing ecosystem diversity objectives has not been demonstrated. Guilds constructed to represent key structural features of habitats, as proposed by Roberts (1987), might be used to develop conservation strategies to accomplish ecosystem objectives. However, as noted earlier, there has been little consideration of landscape issues in the identification of guilds. Thus guilds as defined to date would likely perform poorly in accomplishing ecosystem objectives (rank D).

Habitat Assemblages

This approach involves identifying sets of species that use common macrohabitats. The scale and detail of macrohabitat definition may vary considerably in different applications. This approach is widely used by resource managers to develop a basic classification of fauna within a management unit but is not well represented in the scientific literature. Two variants are possible. First, one or more indicators represent the larger group of species in the habitat group and second, the entire group of species (see "Entire Habitat Assemblages" in chap. 6) is the focus.

Species Diversity Objectives

Canterbury et al. (2000) tested correlations of the occurrences of bird species within four habitat groups: mature forest, shrubland, forest edge, and generalists, and by diet, foraging, and nesting guilds. Correlations of occurrences between individual species and the groups in which they were included were highest for habitat assemblages. Wisdom et al. (2000) used habitat groups to assess changes in habitat availability for species of concern in the Columbia River basin. They subsequently (Raphael et al. 2001) used indicators from each group to focus the assessment of potential management alternatives for federal lands in the basin. However, since these indicators were selected as focal species within each of the habitat groups, this was not an example of using habitat groups as a stand-alone approach. Wisdom et al. (2005) placed 40 species of conservation concern into five habitat groups in an assessment of Great Basin habitats and then used those groups to characterize habitat conditions within watersheds across the Great Basin.

The use of habitat assemblages as a stand-alone approach for conservation of species is likely to provide only coincidental benefit (rank C).

Use of macrohabitat groups does not take into consideration either the details of habitat use considered in the guild approach or the various types of spatial and other limitations considered in the umbrella and focal species approaches. Therefore, in developing management strategies, more detailed needs of individual species within each habitat group need to be considered. The groups themselves would only be useful for very coarse direction on maintaining macrohabitats within a landscape.

Genetic Diversity Objectives

Because the use of habitat assemblages as a stand-alone approach would not be effective in providing for viability of species, we conclude that it would also be unlikely to provide for genetic variability of species (rank D).

Ecosystem Diversity Objectives

We found no published reports of the utility of the habitat assemblages approach in meeting ecosystem diversity objectives. It is unlikely that the development of habitat groups as a stand-alone approach would achieve ecosystem diversity objectives. Identifying the habitat groups would only be a first step in the development of conservation strategies. Unless combined with another approach, it probably would be ineffective in achieving ecosystem diversity objectives (rank D).

Geographically Based Approaches

Geographically based approaches include (1) managing for locations of individual target species at risk, (2) delineating hot spots or concentration areas of rare species, and (3) establishing reserves or protected areas designed to conserve biodiversity. These geographic approaches are based on the assumption that appropriate management of the delineated areas will provide for conservation of the species, their biophysical environments, and/or the ecological system's characteristic processes (Flather et al. 1997; Lubchenco et al. 2003).

Locations of Target Species at Risk

Under this approach, documented locations of target species are protected and managed based on specific guidelines designed to sustain the species (see chap. 6). The difference between this approach and those described under "Conservation of Individual Species Based on Concepts of Population Viability" (chap. 6) is that it is most often based on presence of the species of interest, does not incorporate population estimates for known locations, does not attempt to quantify population viability of the target species, and does not generally provide for conservation of suitable but unoccupied areas or of habitat utilized for dispersal between the known locations.

Species Diversity Objectives

The effectiveness of this approach in sustaining viable populations of rare species has not been rigorously tested. This approach is the basis for the Survey and Manage program of the Northwest Forest Plan (Molina et al. 2006, also see chap. 4), as well as a number of Bureau of Land Management and Forest Service management programs intended to sustain sensitive species. Short-term monitoring data have verified continued occupation of known locations by target species in some regions but in most cases data span only a few years. We conclude that in cases where known locations represent quality habitat, and management guidelines are adequate for sustaining this habitat as well as ameliorating threats, this approach has the potential for sustaining the presence, but not necessarily the viability, of populations of the target species (see table 8.1, rank of B for rare species). This approach is less likely to provide for the persistence of viable populations of little-known species, as location data are usually incomplete and information required for quantifying habitat needs and developing management plans is not available (rank of C).

This strategy, when applied to RLK species, would not likely be effective in providing for population viability of other species. In one study, sites selected based on occurrence of rare species in five eastern U.S. states afforded some level of protection for 84% of all other species (Lawler et al. 2003). Based on these findings, species that share particular habitat requirements with the target species (springs, seeps, etc.) might be benefited. However, despite the co-occurrence of rare species with other species, sites protected for rare species would frequently be too small to provide significant benefit to many other species. For example, carnivores with large home ranges may not be protected well under this approach because known locations protected for rare species would be too small to encompass home ranges of the carnivores. We conclude that some benefits for other species might be realized, especially for those that co-occur with the target species and have similar habitat needs, but these benefits would not be the result of the inherent characteristics of the approach (rank of C).

Genetic Diversity Objectives

This approach can be tailored to some degree to retain genetic diversity of some target species (rank B), but contributions to genetic diversity of nontarget species would be dependent on other factors (rank C). The target species most likely to benefit are those for which quality location data are available and either all known locations are protected or, if only a subset will be protected, the goal of maximizing genetic diversity is used in identifying those locations to protect. An example of this approach was used on the Black Hills National Forest in selecting a subset of known locations of bloodroot (*Sanguinaria canadensis*) for monitoring and management (Hornbeck et al. 2003). The selected subset included sites that represented the variety of plant communities where the species occurred, and also considered geographic distribution as well as different types of threats. Such approaches that attempt to conserve genetic diversity represented in the area of interest may assist in meeting genetic diversity objectives of the target species, but whether benefits would also be realized for nontarget species is dependent on other factors.

Ecosystem Diversity Objectives

This approach is not designed to meet general ecosystem diversity objectives but might produce coincidental benefits. For example, managing for target locations of the Kirtland's warbler (*Dendroica kirtlandii*) would entail providing early successional jack pine (*Pinus banksiana*) forest (Probst and Weinrich 1993). If this seral stage were underrepresented relative to long-term structural diversity goals, increasing this community type might also help meet more general ecosystem diversity objectives. Further, using prescribed burning instead of cutting to provide early seral stage jack pine would provide some benefits along the lines of restoring disturbance regimes. However, in many situations, management for locations of target species would be ineffective in meeting ecosystem diversity objectives (rank D).

Species Hot Spots or Concentrations of Biodiversity

The concept of identifying hot spots (Myers 1989) has been used to delineate concentrations of total species richness or various subsets of species of conservation concern, such as endemics and various rare species (chap. 6).

Species Diversity Objectives

Approaches that identify and manage hot spots of species richness may be effective in maintaining the occurrence of rare species for which high-quality distribution data are available but may not assure the viability of these species. This approach is more likely to maintain this subset of species if co-occurrence among target species is high and location data of just target species (as opposed to the entire species pool) are used in identifying hot spots. For species that rarely co-occur with other species, protection of hot spots will not be effective in their conservation. Further, by designating only species-rich sites for protection, sites with fewer species would not be identified for protection. Failing to protect sites with fewer numbers of species may inadvertently result in an inadequate number and quality of sites to sustain some species (e.g., see Kareiva and Marvier 2003). Thus the effectiveness of this approach in maintaining viability of RLK species is strongly dependent on factors that are not accounted for in the approach (rank C).

Preserving species-rich hotspots is unlikely to preserve the overall species pool. Wide-ranging species such as carnivores or species associated with ephemeral habitats that shift on the landscape may not be inherently protected through delineation of species-rich areas. Benefits of this approach to the entire species pool will be dependent on a variety of factors not considered in the approach itself (rank C).

Genetic Diversity Objectives

Hot spot approaches are unlikely to provide for genetic diversity of component species (rank D) because focusing solely on species-rich sites may eliminate less rich sites that are important to protect to sustain genetic variability of the component species. The only exception would be for species whose locations coincide with the boundaries of the delineated hot spots, for which the hot spots represent the maximum geographic distribution of the species in the management area.

Ecosystem Diversity Objectives

Hot spot approaches, whether based solely on locations of RLK species or on the entire species pool, are not, by design, effective in meeting ecosystem diversity goals (rank D). Species-poor ecosystem and community types may be neglected, and species distributions and richness are poor surrogates for ecosystem processes (Conroy and Noon 1996). Sustaining both species-rich sites and those with only a few species may be important components of a landscape that has a capacity to sustain species in light of environmental changes.

Reserves or Other Protected Areas Designed to Conserve Biodiversity

The most sophisticated reserve designs seek to maximize efficiency; that is, contain the greatest number of biodiversity elements at the least cost; and consider complementarity, or the relative gain in conserving biodiversity elements as additional reserve units are added (chap. 6). Other important considerations in reserve designs include "representativeness," where a desired set of biodiversity elements are at least represented, as well as considerations of vulnerability, persistence, and replaceability of candidate sites.

Species Diversity Objectives

Reserve designs can be adapted to increase their effectiveness in meeting species diversity goals. For example, reserves can be designed to maximize the number of known locations of rare species and provide additional suitable habitat. Noss et al. (2002) described an algorithm for prioritizing sites to protect in the Greater Yellowstone Ecosystem. Their analysis showed that it was possible to protect all known locations of highly imperiled species by adding 43 "megasites" to the reserve. Such an approach would improve the likelihood that reserves would provide for viability of rare species (rank B).

The degree to which little known and other species would be protected in reserves would depend on a number of factors. Reserve designs that attempt to incorporate representation of all natural communities in an appropriate landscape configuration while also protecting locations of rare species could be effective in providing for other, nontarget species (rank B). However, even such designs could only provide coincidental benefits to little-known species (rank C).

Genetic Diversity Objectives

Reserve designs can be adapted to sustain genetic diversity of some species. Reserve designs that protect all known locations of imperiled species (e.g., Noss et al. 2002) or focus on protecting endemic species with restricted distributions (e.g., Ceballos et al. 1998) are more likely to sustain genetic diversity of the targeted species (rank B). The degree to which such approaches help meet genetic diversity goals of other species depends on how well the reserve design captures the extremes in the distribution of other species (rank C).

Ecosystem Diversity Objectives

Many existing reserves were not designed to be effective in meeting ecosystem diversity objectives. As pointed out in chapter 6, a number of existing reserves were designed around a single species or around scenic attractions or roadless areas. In recent years a broader set of biodiversity objectives have been proposed in developing reserve designs. Approaches that emphasize representativeness tend to be more effective in maintaining ecosystem types and seral stages and offer a better opportunity for maintaining more resilient landscapes. Noss et al. (2002) estimated that the addition of 15 "megasites" to the Greater Yellowstone Ecosystem would increase the representation of geoclimatic classes, which could potentially result in greater representation of ecosystem types and seral stages. Whether ecosystem types and processes would be sustained would depend on the degree to which disturbance regimes could be restored and maintained. Ideally, reserves could be designed to be large enough to maintain both the taxa and the ecological processes native to a bioregion over ecological time. This critical size was identified as the minimum dynamic area by Pickett and Thompson (1978). Reserves designed using these concepts could sustain ecological processes (rank B).

Maintaining System Structure and Composition

We classified approaches that focus on maintaining system structure and composition into two general categories: those based on an understanding of the range of natural variability (RNV), and those based on concepts other than RNV, including the use of key habitat conditions and managing for species that play critical ecological roles.

Range of Natural Variability

These approaches focus on maintaining the mix of ecosystems and seral stages across the landscape within the best approximation of a historic range of natural variability (RNV) (see chap. 7). It is frequently asserted that maintaining the mix of communities within RNV will provide for the current needs of associated species (Morgan et al. 1994). The approach assumes that adequate amounts of the ecosystem remain or can be restored and that threats such as introduction of exotic invasive species can be ameliorated.

Species Diversity Objectives

A variety of approaches that incorporate an understanding of RNV in restoring the mix of ecosystems and successional stages across the landscape are described in the literature. For the most part, these approaches have only recently been attempted, and therefore long-term data on their effectiveness in meeting species diversity objectives are limited.

Three management alternatives proposed for the Columbia River basin (USDA and USDI 2000) incorporated the concepts of RNV. Raphael et al. (2001) estimated the effects of these alternatives on 31 vertebrate species considered to be of conservation concern using a Bayesian belief network model. The model incorporated attributes of the quantity and quality of habitat for each species, including a composite measure of their departure from the historical range of variability. Their results suggested that all alternatives would improve the abundance and distribution of most forest-associated species but would provide few improvements for species associated with rangelands. Problems for rangeland species stemmed from a lack of available techniques to restore communities seriously affected by exotic plant invasion; such communities may be in an "altered state" whereby restoration of native plant communities may not be possible (Hemstrom et al. 2001).

Two proposed alternatives in the Sierra Nevada Forest Plan amendment were designed to promote ecosystem conditions expected under RNV. One alternative emphasized restoration of ecosystem conditions and processes within RNV under prevailing climate; another was designed to actively manage entire landscapes to establish and maintain a mosaic of forest conditions approximating patterns expected under RNV (USDA Forest Service 2003). Analyses indicated that neither of these alternatives ranked higher than other proposed alternatives for meeting plant and animal diversity objectives. Instead, the alternative that incorporated species-specific guidelines for conserving rare species was judged to rank higher in meeting species diversity objectives.

These examples suggest that, in some circumstances, approaches designed to maintain ecosystems and successional stages based on an understanding of RNV could be designed to provide for viability of many species in the species pool (rank B). However, to be successful in providing for viability, such approaches would have to consider RNV of macrohabitats (e.g., major vegetation types and successional stages and their arrangement on the landscape), microhabitats (e.g., snags and logs), special habitat features (e.g., seeps and springs), and human disturbances. For example, Litvaitis (2003) proposed that maintenance of "thicket-dependent" species in the northeastern United States will require more than simply maintaining pre-settlement levels of early successional habitat. Instead, for conservation efforts to be successful in maintaining this group of species, creative solutions that address other aspects such as high road densities, loss of special habitat features such as beaver dams, and habitat parcelization are needed. Such approaches may also provide habitat for some RLK species, but the degree to which viability for individual species is supported will be dependent on other factors such as the treatment of individual sites occupied by the species (rank C). Since the concept of RNV does not consider individual species locations, many RLK species sites might not be protected under such approaches.

Genetic Diversity Objectives

Approaches designed to retain ecosystem types and successional stages based on their RNV would likely provide for genetic diversity of many common species, especially in ecosystems that have not been greatly altered since settlement (rank B). Genetic diversity of some rare species may be retained, as well. But, for many RLK species, the success of such approaches in retaining genetic diversity would be dependent on other factors (rank C).

Ecosystem Diversity Objectives

Assuming that it is feasible to restore native ecosystem types and seral stages within RNV, such approaches have a high probability of meeting a number of ecosystem diversity objectives. One alternative of the Sierra Nevada Forest Plan amendment aims to establish and maintain a diversity of forest ages and structures over the landscape in a mosaic approximating patterns that would be expected under natural conditions, that is, conditions characterized by current and expected future climates, biota and natural processes (USDA Forest Service 2003). Ecosystems and ecological processes would be actively managed to maintain and restore them to desired conditions. Analyses indicated that some improvements in ecosystem diversity (i.e., providing for ecosystem types and seral stages within RNV) were achieved by this alternative. Verification of these projected results requires long-term monitoring, but the theoretical basis for accomplishing the objective for ecosystem types and seral stages within RNV is sound, assuming that it is possible to mimic natural community mosaics using available tools (rank A).

In regard to other ecosystem diversity objectives, if management were designed so that vegetation types and successional stages were created and maintained through the introduction of natural disturbance regimes, these approaches would be likely to achieve the objective of restoring disturbance regimes (rank B). It could also be assumed that by providing habitat for many species in the species pool, the restoration of trophic structures would be accomplished to some degree as well (rank B). The objectives of restoring nutrient cycles and key functional roles of organisms would be promoted by approaches that maintain ecosystem types and seral stages within RNV using natural disturbance regimes. However, since these functions may be partly dependent on individual species whose fate is uncertain under such approaches, we assigned rank C to the objectives of nutrient cycling and other processes. Finally, by providing for RNV of native ecosystems and successional stages, landscapes would likely retain a high level of resiliency in the face of long-term environmental changes (rank B). RNV-based approaches would be most successful in accomplishing these objectives if locations of various successional stages represented the entire geographic range available. We conclude that approaches that attempt to restore component ecosystem types and successional stages may have merit in meeting ecosystem diversity objectives, especially if resource managers can overcome political and legal constraints required to restore historic disturbances and consideration is given to maximizing the geographic range over which a given type occurs.

Diversity of Habitat Conditions

We evaluated two types of approaches that emphasize diversity of habitat conditions using concepts other than RNV: providing key habitat conditions, and managing for species that play critical ecological roles.

Key Habitat Conditions

This approach focuses on maintaining a mix of habitat conditions. The goal is not necessarily to provide these conditions based on RNV. Instead, the intent is to include the mix of habitat conditions, including their extremes, in an effort to provide habitat for a variety of species (Haufler et al. 2002; see chap. 7).

Species Diversity Objectives

A number of conservation plans incorporate the provision of key habitat conditions with the goal of sustaining species associated with this range in conditions. Examples include the residual cover guidelines proposed in the U.S. Forest Service's Northern Great Plains plan that are designed to provide a range of grazing intensities (USDA Forest Service 2001) and those proposed by Taft et al. (2002) for managing wetlands in California by providing the extreme ends of the water depth spectrum to provide habitat for both diving birds and shorebirds. We are not aware of any empirical tests of the effectiveness of these approaches in meeting species diversity objectives, but such approaches seem likely to provide for the persistence of species associated with the range of conditions provided. For rare species associated with the habitat conditions provided, their occurrence would be likely, but whether populations were viable would be dependent on other factors (rank C). Rare species dependent on entirely different key conditions would not be benefited. Benefits for little-known species would depend on the degree to which they were dependent on the conditions provided (rank C). Whether such approaches provided for the viability of other species or natural patterns of abundances of all native species would be dependent on other factors (rank C).

Genetic Diversity Objectives

Approaches designed to provide key habitat conditions might, in some applications, contribute to meeting genetic diversity objectives of some species (rank C). However, genetic benefits would usually be the result of factors that are not an inherent part of the approach such as whether these key conditions are geographically well distributed and consider the distribution of rare species. Benefits to little-known species would be dependent on their occurrence in, and dependence on, the habitat conditions provided in the approach.

Ecosystem Diversity Objectives

Approaches designed to provide key habitat conditions may contribute to meeting ecosystem diversity objectives, but such benefits are usually restricted to the ecosystems affected by the approach. For example, management guidelines proposed by Taft et al. (2002) that provide a range in water depth of wetlands would not, by design, provide for other ecosystem types and their successional stages. Similarly, such approaches may restore ecosystem processes to some degree in the affected ecosystem, but any benefits to other ecosystem types would not be inherent in the design. In summary, the effectiveness of approaches designed to provide key habitat conditions in meeting ecosystem diversity objectives depends on the approach being used, the diversity of resulting key conditions and affected ecosystems, and how widely the management guidelines are applied (rank C).

Species That Play Critical Ecological Roles

These approaches focus on restoring species that have large effects on community structure or ecosystem function that are disproportionate to their abundance, and that perform roles not performed by other species or processes (Power et al. 1996). Included in this group of species are *keystone species,* which have a large effect on species diversity and competition, and *ecosystem engineers* that modulate physical habitat.

Species Diversity Objectives

There is evidence that restoration of species with critical ecological roles helps meet overall species diversity objectives as well. For example, for the nine vertebrate species that depend on prairie dog (*Cynomys* spp.) colonies and 20 species that opportunistically use these habitats (Kotliar et al. 1999), maintenance of prairie dog colonies would be beneficial (Sharps and Uresk 1990; Mulhern and Knowles 1997). Other taxa, including nematodes (Ingham and Detling 1984) and birds (Agnew et al. 1986), may also benefit from maintenance of prairie dog colonies. However, the degree to which prairie dog colonies would provide for viable populations of both rare and common species would depend greatly on factors such as the number, size, and juxtaposition of the prairie dog colonies (Hof et al. 2002), as well as the effect of threats from other factors such as diseases and predation. Therefore, the benefit provided to other species by such approaches would depend on other factors (rank C).

Restoration of species with critical ecological roles may result in increases of overall species diversity. American bison (*Bison bison*) were reintroduced to a tallgrass prairie preserve in Kansas in 1987. Within a few years, increases in the abundances of forbs and an overall increase in plant species richness and diversity due to bison grazing and nongrazing activities were observed (Hartnett et al. 1996; Knapp et al. 1999). However, the degree to which the natural abundances and distributions of species could be restored through such an approach would be dependent on other factors (rank C).

Genetic Diversity Objectives

Reintroduction and maintenance of species with critical ecological roles may assist in retaining the natural genetic variation of both the introduced species and the associated species, but benefits would depend greatly on how well the species is established across the range of environmental conditions where the species naturally occurs (rank C).

Ecosystem Diversity Objectives

Restoration of species with critical ecological functions has the potential to meet some ecosystem diversity objectives, although effects along these lines are unlikely to be generalizable. Restoration and maintenance of prairie dog colonies can promote lower seral stage prairie habitat and assist in maintaining some ecosystem processes (Hansen and Gold 1977; Uresk and Bjugstad 1983; Knowles 1986). The degree to which the capacity of landscapes supporting prairie dog colonies would provide for diversity of species in the face of long-term environmental change would depend on a number of factors, including how well resulting colonies represent the maximum geographic range of the prairie dog.

We conclude that the degree to which ecosystem diversity objectives are met through the restoration and maintenance of a species with critical ecological function such as the prairie dog would depend largely on the characteristics of the species, the manner in which it affects ecosystem diversity, and how extensively it is reintroduced and maintained (rank C).

Maintaining System Processes and Functions

We evaluated two approaches that focus on maintaining system processes and functions. The first focuses on maintaining disturbance regimes, and the second focuses on maintaining other ecosystem functions.

Maintaining Disturbance Regimes

Maintaining disturbance regimes is considered an important aspect of maintaining the structure and function of many, if not most, ecosystems (see chap. 7). Many approaches that attempt to restore disturbance processes base management recommendations on an estimation of the range of natural variability of the disturbances under which the system evolved. This approach assumes that by restoring historic disturbance regimes (including frequency, intensity, timing, and spatial attributes), the functional and structural attributes will be restored as well. In reality, we found few examples where success in restoring all aspects of historic disturbances could be proclaimed.

Species Diversity Objectives

Conceptually, providing for disturbance regimes under which a given system evolved should provide a sound basis for sustaining associated species. In actuality, long-term studies that evaluate the effectiveness of disturbance-based approaches in maintaining species diversity are rare. Even for the well-studied Konza Prairie Research Natural Area in Kansas, where efforts to restore historic disturbances began in the 1970s (Hulbert 1973), there is uncertainty about the long-term effects on species diversity. Initial research was limited by burning treatments that encompassed only a few hundred hectares in only one season (Hulbert 1985). Despite recent increases in the scale of treatments and studies (Knapp et al. 1999), information is still limited by the small scale of treatments relative to historical scales and by the near absence of data on the response of some taxa.

The effectiveness of such strategies for meeting overall species diversity goals may be tied to how well historic disturbance regimes can be emulated. A high level of what Jordan et al. (2003) term "pyrodiversity," including varied fire regimes across landscapes and over many decades that include a range in fire intensities and severities and variable seasonality in accordance with natural rhythms, is expected to best enhance overall species diversity. Morrison et al. (1995) found that increasing the variability of the length of time between fires in sandstone communities in Australia was associated with an increase in species richness of both fire-tolerant and fire-sensitive species. In some situations it is possible that viable populations of some other species could be maintained by restoring and maintaining disturbance regimes, but this would not be guaranteed. Andersen et al. (2005) found that most taxa associated with tropical savannas in Australia were very resilient to fires, regardless of seasonality; yet riparian vegetation and associated stream biota, as well as small mammals, were mostly associated with unburned areas. Therefore, management designed to provide some less frequently burned areas might be required to maintain viable populations of fire-sensitive species. Parr and Andersen (2006) concluded that there is a need for more critical consideration of the levels of pyrodiversity needed to maintain diversity and greater attention to developing management guidelines for such approaches.

We conclude, therefore, that strategies that incorporate historic disturbance regimes have the potential to be designed to provide for viability of some species (rank B), and may lead to natural patterns of abundances (rank B). However, the degree to which such approaches result in viable populations of RLK species would depend on other factors (rank C), and viability of populations would not be guaranteed without viability assessments.

Genetic Diversity Objectives

Approaches designed to restore disturbance regimes based on their natural range of variation would potentially provide for genetic diversity of species that benefit from such disturbances. For rare species, some tailoring might be required to provide habitat conditions the species require across as broad a geographic area as possible. For little-known species, success in retaining genetic diversity would be dependent on other factors. Overall, we assign approaches designed to restore disturbance regimes a rank of C.

Ecosystem Diversity Objectives

As discussed in chapter 7, strategies aimed at restoring disturbance processes are fraught with a number of practical and logistical issues that would need to be overcome to meet other ecosystem diversity objectives. Accounting for the rarity of some ecosystems and the small size of remaining patches of many vegetation types are not trivial problems to overcome. For example, remaining patches need to be large enough to allow for disturbances to occur at an appropriate scale and intensity (Pickett and Thompson 1978). Restoring disturbance regimes would have a good chance of maintaining nutrient cycles, but the degree to which trophic structures and key functional roles of organisms are maintained would depend on the degree to which component species are restored. To the degree that restoring historic disturbances results in a heterogeneous landscape that spans the maximum geographic range available, such approaches have the potential to provide for resiliency in the face of long-term environmental changes.

We conclude that approaches that focus on restoring disturbance regimes could be designed to be effective in meeting some ecosystem diversity objectives in some circumstances (rank B). Especially in relatively intact ecosystems, restoration of disturbances has the potential to lead to restorations of species patterns and system processes.

Maintaining Other Ecosystem Processes

This section reviews other management strategies that seek to restore and maintain other ecosystem processes. We found only two examples where management strategies designed to restore key ecological functions were attempted. The first example was a functional groups approach proposed by the Forest Ecosystem Management Assessment Team (1993), whereby arthropods associated with late-successional forests in the Pacific Northwest were aggregated into 11 functional groups based on their ecological roles. The team then rated seven land management options as to their sufficiency in providing adequate habitat on federal lands to provide for well-distributed populations of the various functional groups. This approach focused on the functional aspects of the species and was necessitated by the fact that a great number of arthropods have not been identified, and distributional information was not adequate to conduct individual species viability assessments. The ultimate utility of this approach has not been tested.

The second example is the management strategy for northern goshawks in southwestern U.S. forests, which was based on a food web strategy, whereby the habitat needs of the primary prey species of the goshawk were used in developing management guidelines (Reynolds et al. 1992).

Species Diversity Objectives

Either of the approaches already described might provide for the occurrence of some RLK species but would not necessarily provide for the viability of those populations. Given that the functional groups proposed for late-successional forests in the Pacific Northwest encompass a variety of trophic levels (predators, herbivores, pollinators, and decomposers) as well as community types (aquatic, riparian, coarse wood, litter, understory, forest gap, canopy, and epizootic forest species), there is a potential that some subset of species in other taxa might be captured as well.

Because the goshawk management guidelines result in a high diversity of habitat conditions of southwestern forests, their implementation would be expected to provide for the occurrence of species associated with the resulting range of conditions. Whether the patterns of species occurrences that result from implementing the goshawk guidelines emulate their natural patterns depends on how well the habitat needs of these specific prey species represent natural habitat patterns in southwestern forests. The degree to which little-known species are accommodated by implementing these guidelines would be unknown. We conclude that approaches that attempt to provide for ecosystem functional attributes such as food webs or functional groups might provide habitat for rare, little-known, and common species, but benefits are not predictable (rank C).

Genetic Diversity Objectives

The degree to which approaches that focus on functional aspects of ecosystems such as functional groups of arthropods or food webs meet the objective of restoring or maintaining the natural genetic variation of species would depend on attributes other than the approach itself (rank C).

Ecosystem Diversity Objectives

The functional groups approach proposed for arthropods associated with late-successional forests in the Pacific Northwest would rank highly in regard to restoring key functional roles of arthropods, but whether this benefit extended to other ecosystem diversity objectives would depend on other factors. As discussed earlier, since the management guidelines for goshawks in southwestern U.S. forests result in a great variety of habitat conditions and are based on the subset of goshawk prey species, implementation of these guidelines has the potential of restoring some portion of the natural variation of native ecosystem types and seral stages.

We conclude that approaches that focus on restoring some functional aspect of ecosystems, such as functional roles of arthropods or food webs, have the potential to meet some ecosystem diversity objectives, especially related to those ecosystem structures and functions associated with the species or species groups that are featured in the approach. Ecosystem diversity attributes not addressed in the approach would not likely be provided, and therefore we assigned a rank of C for ecosystem processes. Whether or not landscapes with the capacity for resiliency of species in the face of long-term environmental change would result would depend on other factors (rank C).

Summary of Effectiveness in Conserving RLK

The primary focus of this book is conservation of RLK species. Successful conservation efforts must deal with both short-term and long-term sources of risk. In an attempt to consider all potential sources of risk, this chapter has taken a comprehensive view of conservation approaches and objectives. Some of those approaches tend to focus on short-term risks, such as protection of known species sites, while others incorporate longer-term risk and system resiliency. In this section we summarize success of the approaches in dealing with both short- and long-term risks associated with the conservation of RLK species.

Our evaluation of the various species-based and system-based approaches (see table 8.1) indicates that none is adequate to reliably meet all conservation objectives and thus meet both short- and long-term needs for conservation of RLK species. For most, effectiveness in meeting specific conservation goals depends on how they are implemented (ranks B and C). Some approaches are better designed for short-term conservation of RLK species. Others may perform well in the conservation of a broad array of more common species and habitats. Still others have their primary strength in addressing ecosystem-level objectives, including the long-term resiliency of those ecosystems.

To better understand underlying causes for the performance of the approaches, we examined how well each approach addresses seven of the elements that are key to species conservation (table 8.2). These are use of macrohabitats, use of fine-scale habitat features, spatial arrangement of habitat, habitat dynamics and resiliency, current species locations, biotic interactions of the species, species demographics, and species genetics. We rated the degree to which each approach addresses each of these elements (see table 8.2). An approach was rated 1 for an element if that element is a primary consideration of the approach and 2 if the element is a secondary consideration of the approach or addressed in at least some applications of the approach. A dash in the rating table indicates that the element is not addressed through the approach.

Ratings for table 8.2 were based on the descriptions of the conservation approaches presented in this chapter and information on those approaches from both this chapter and chapters 6 and 7. The ratings reflect an informed judgment about the elements of conservation that are addressed by each approach, but they are subjective and cannot represent nuances of every possible application of an approach.

As with the ratings for effectiveness of the various approaches (see table 8.1), the evaluation of species conservation elements (see table 8.2) was complicated by the fact that different approaches are targeted to different sets of species, or to overall systems rather than species. Consequently, the question of how the approaches address species-related elements takes on different meanings for different sets of approaches. For the individual species approaches, we evaluated how well the approach addresses the individual species under consideration. For the surrogate approaches, we evaluated how well the approach addresses the set of species being represented by the surrogate. For the geographic approaches, we evaluated how well the approach addresses the species or set of species targeted by the approach. For the system approaches, the term "habitat" was treated as a major vegetation type rather than habitat specific to a species.

Based on our evaluation of species conservation elements (table 8.2), the approach that comes closest to addressing all seven elements associated with species conservation is the development of conservation strategies for individual species based on concepts of species viability. Its long-term effectiveness may be limited, however, by lack of consideration of habitat dynamics. This shortcoming could be managed through an adaptive management approach, but other shortcomings such as the high cost, particularly when applied to many species, and lack of the necessary data, are more difficult to overcome.

Table 8.2. *The degree to which each approach addresses seven elements important to species conservation*

Approach	Macrohabitat	Fine-scale habitat features	Spatial habitat arrangement	Habitat dynamics and resiliency	Current species locations	Biotic interactions	Demographics	Genetics
Viability of individual species	Ratings reflect the individual species under consideration							
Strategy based on concepts of population viability	1	1	1	-	1	1	1	2
Surrogate species	Ratings reflect the set of species being represented by the surrogate							
Focal species	1	1	1	-	-	-	-	-
Umbrella species	1	-	2	-	-	-	-	-
Guilds	2	1	-	-	-	-	-	-
Habitat assemblages	1	-	-	-	-	-	-	-
Biodiversity indicator species	2	-	-	-	1	-	-	-
Flagship species	-	-	-	-	-	-	-	-
Geographic approaches	Ratings reflect the species that are the target of the technique							
Management for locations of target species	2	-	-	-	1	-	-	2
Hot spots	2	-	-	-	1	-	-	-
Reserves or protected areas	2	-	-	-	1	-	-	-

(continues)

Table 8.2. Continued

Approach	Macrohabitat	Fine-scale habitat features	Spatial habitat arrangement	Habitat dynamics and resiliency	Current species locations	Biotic interactions	Demographics	Genetics
Maintaining system structure and composition	Term "habitat" is interpreted as major vegetation type rather than habitat specific to a species							
Managing for RNV	1	2	2	1	-	-	-	-
Managing for diversity of habitats	1	2	2	2	-	-	-	-
Strongly interacting species	2	-	-	1	-	-	-	-
Maintaining system function	Term "habitat" is interpreted as major vegetation type rather than habitat specific to a species							
Maintaining disturbance regimes	2	2	2	1	-	-	-	-
Maintaining other ecosystem functions	-	-	-	2	-	-	-	-

Ratings are defined as follows: 1—the element is a primary consideration of the approach; 2—the element may be considered in some applications of the approach; dash—the element is not considered unless this approach is combined with another.

Elements of conservation are defined as follows:

Macrohabitats—species use of habitats mappable at fairly large-scale (e.g., 10 acre) resolution. Includes major vegetation types and structural stages. May also include specific topographic features and positions (e.g., cliffs or streamside zones).

Fine-scale habitat features—species use of habitat features that are not detectable in large-scale (e.g., 10 acre resolution) mapping. May include structural features such as logs or snags, or the presence of particular tree species that serve as habitat for some RLK species and are not predictable through knowledge of the major vegetation type.

Spatial habitat arrangement—reflects the species' spatial use of habitat, including habitat patch size, interpatch distances, juxtaposition of various habitats used by the species (e.g., breeding habitat, dispersal habitat, winter habitat), and so forth.

Habitat dynamics—deals with changes in habitat over time and ability of habitat to persist during disturbance and/or restore itself after disturbance.

Current species locations—sites that are currently occupied by populations of the species.

Demographics—vital population rates of the species and factors that influence those rates.

Genetics—genetic composition of the species' populations, and factors that influence that genetic composition.

Four of the surrogate approaches (focal species, umbrella species, guilds, and habitat assemblages) tend to focus on habitat use by the surrogate and the set of species represented by the surrogate. There is nothing inherent to these approaches that leads to consideration of the current locations or demographics of species or of habitat dynamics and system resiliency. The biodiversity indicator approach has its strength in denoting the locations of one or more species but does not address the other factors. The flagship approach does poorly in addressing any of the seven elements.

The system approaches tend to focus on habitats and dynamics of those habitats. Note however that under these approaches there is no knowledge or intentional manipulation of habitats associated with any particular species. Rather, they provide for the major compositional and structural components of vegetation that serve as habitat, and for the underlying processes that lend resiliency to systems. These approaches fail to explicitly consider species locations or demographics.

Conclusion

It is clear that no single approach acts as a comprehensive conservation strategy. This mirrors the findings of authors (Caro and O'Doherty 1999; Carignan and Villard 2002; Kintsch and Urban 2002; Kareiva and Marvier 2003; Hess et al. 2006) who have cited the need to combine approaches to provide for effective conservation. For example, RLK species that are small bodied, and therefore are expected to have small home ranges, might be most efficiently managed through approaches that focus on current species locations. Rare species for which habitat requirements are well understood and landscape configuration of habitat is important could be managed through the focal species approach. Species for which requirements and risks are complex and involve significant biotic interactions may need individual consideration that involves the concepts of population viability. Systems approaches may be needed in addition to species approaches to provide for overall function and resiliency of the ecosystem. Finally, it is important to remember that management strategies developed through any of these approaches are hypotheses that remain to be tested, and that validation of those hypotheses requires collection of local data.

References

Agnew, W., D. W. Uresk, and R. M. Hansen. 1986. Flora and fauna associated with prairie dog colonies and adjacent mixed-grass prairie in western South Dakota. *Journal of Range Management* 39:135–39.

Andersen, A. N., G. D. Cook, L. K. Corbett, M. M. Douglas, R. W. Enger, J. Russell-Smith, S. A. Setterfield, R. J. Williams, and J. C. Z. Weinarski. 2005. Fire frequency and biodiversity conservation in Australian tropical savannas: Implications from the Kapalga fire experiment. *Austral Ecology* 30:155–67.

Bani, L., M. Baietto, L. Bottoni, and R. Massa. 2002. The use of focal species in designing a habitat network for a lowland area of Lombardy, Italy. *Conservation Biology* 16:826–31.

Berger, J. 1997. Population constraints associated with the use of black rhinos as an umbrella species for desert herbivores. *Conservation Biology* 11:69–78.

Block, W. M., L. A. Brennan, and R. J. Gutierrez. 1986. The use of guilds and guild-indicator species for assessing habitat suitability. Pp. 109–13 in *Wildlife 2000: Modeling habitat relationships of terrestrial vertebrates*, ed. J. Verner, M. L. Morrison, and C. J. Ralph. Madison: University of Wisconsin Press.

Bull, E. L., C. G. Parks, and T. R. Torgersen. 1997. Trees and logs important to wildlife in the interior Columbia River Basin. General Technical Report PNW-GTR-391. Portland, OR: U.S. Department of Agriculture, Forest Service, Pacific Northwest Research Station.

Canterbury, G. E., T. E. Martin, D. R. Petit, L. J. Petit, and D. F. Bradford. 2000. Bird communities and habitat as ecological indicators of forest condition in regional monitoring. *Conservation Biology* 14:544–58.

Carignan, V., and M. Villard. 2002. Selecting indicator species to monitor ecological integrity: A review. *Environmental Monitoring and Assessment* 78:45–61.

Caro, T. M., and G. O'Doherty. 1999. On the use of surrogate species in conservation biology. *Conservation Biology* 13:805–14.

Cartron, J. E., and D. M. Finch. 2000. Ecology and conservation of the cactus ferruginous pygmy-owl in Arizona. General Technical Report RM-GTR-43. Fort Collins, CO: U.S. Department of Agriculture, Forest Service, Rocky Mountain Research Station.

Ceballos, G., P. Rodríguez, and R. A. Medellín. 1998. Assessing conservation priorities in megadiverse Mexico: Mammalian diversity, endemicity, and endangerment. *Ecological Applications* 8:8–17.

Chase, M. K., W. B. Kristan III, A. J. Lynam, M. V. Price, and J. T. Rotenberry. 2000. Single species as indicators of species richness and composition in California coastal sage scrub birds and small mammals. *Conservation Biology* 14:474–87.

Chiarucci, A., F. D'Auria, V. De Dominicis, A. Langana, C. Perini, and E. Salerni. 2005. Using vascular plants as a surrogate taxon to maximize fungal species richness in reserve design. *Conservation Biology* 19:1644–52.

Conner, R. N., D. G. Randolph, and J. R. Walters. 2001. The red-cockaded woodpecker: Surviving in a fire-maintained ecosystem. Austin: University of Texas Press.

Conroy, M. J., and B. R. Noon. 1996. Mapping of species richness for conservation of biological diversity: Conceptual and methodological issues. *Ecological Applications* 6:763–73.

FEMAT (Forest Ecosystem Management Assessment Team). 1993. Forest ecosystem management: An ecological, economic, and social assessment. Washington, DC:

U.S. Government Printing Office 1993-793-071. Available at: Regional Ecosystem Office, P.O. Box 3623, Portland, OR 97208.

Flather, C. H., K. R. Wilson, D. J. Dean, and W. C. McComb. 1997. Identifying gaps in conservation networks: Of indicators and uncertainty in geographic-based analyses. *Ecological Applications* 7:531–42.

Fleishman, E., D. D. Murphy, and P. F. Brussard. 2000. A new method for selection of umbrella species for conservation planning. *Ecological Applications* 10:569–79.

Fleishman, E., R. B. Blair, and D. D. Murphy. 2001. Empirical validation of a method for umbrella species selection. *Ecological Applications* 11:1489–1501.

Hansen, R. M., and I. K. Gold. 1977. Black-tailed prairie dogs, desert cottontails and cattle trophic relations on shortgrass range. *Journal of Range Management* 30:210–13.

Hartnett, D. C., K. R. Hickman, and L. E. Fischer-Walter. 1996. Effects of bison grazing, fire, and topography on floristic diversity in tallgrass prairie. *Journal of Range Management* 49:413–20.

Haufler, J. B., R. K. Baydack, H. Campa III, B. J. Kernohan, C. Miller, L. J. O'Neil, and L. Waits. 2002. Performance measures for ecosystem management and ecological sustainability. Technical Review 02-1. Bethesda, MD: Wildlife Society.

Hemstrom, M. A., J. J. Korol, W. J. Hann. 2001. Trends in terrestrial plant communities and landscape health indicate the effects of alternative management strategies in the interior Columbia River basin. *Forest Ecology and Management* 153:1–3.

Hess, G. R., F. H. Koch, M. J. Rubino, K. A. Eschelbach, C. A. Drew, J. M. Favreau. 2006. Comparing the potential effectiveness of conservation planning approaches in central North Carolina, USA. *Biological Conservation* 128:358–68.

Hof, J., M. Bevers, D. W. Uresk, G. L. Schenbeck. 2002. Optimizing habitat location for black-tailed prairie dogs in southwestern South Dakota. *Ecological Modelling* 147:11–21.

Hornbeck, J. H., C. H. Sieg, and D. J. Reyher. 2003. Conservation assessment for bloodroot in the Black Hills National Forest, South Dakota and Wyoming. Custer, SD: U.S. Department of Agriculture, Forest Service. http://www.fs.fed.us/r2/blackhills/projects/planning/assessments/bloodroot.pdf.

Hulbert, L. C. 1973. Management of Konza Prairie to approximate prewhiteman fire influences. Pp. 14–16 in *Proceedings of the Third Midwest Prairie Conference*, ed. L. C. Hulbert. Manhattan: Kansas State University.

———. 1985. History and use of Konza Prairie Research Natural Area. *Prairie Scout* 5:63–93.

Ingham, R. E., and J. K. Detling. 1984. Plant–herbivore interactions in a North American mixed-grass prairie, III: Soil nematode populations and root biomass on *Cynomys ludovicianus* colonies and adjacent colonies and adjacent uncolonized areas. *Oecologia* 63:307–13.

Jordan, M. J., W. A. Patterson III, and A. G. Windisch. 2003. Conceptual ecological models for the Long Island pitch pine barrens. *Forest Ecology and Management* 185:151–68.

Kareiva, P., and M. Marvier. 2003. Conserving biodiversity coldspots. *American Scientist* 91:344–51.

Kelly, E. B, E. D. Forsman, and R. G. Anthony. 2003. Are barred owls displacing spotted owls? *Condor* 105:45–53.

Kintsch, J. A., and D. L. Urban. 2002. Focal species, community representation, and physical proxies as conservation strategies: A case study in the Amphibolite Mountains, North Carolina, U.S.A. *Conservation Biology* 16:936–47.

Knapp, A. K., J. M. Blair, J. M. Briggs, S. L. Collins, D. C. Hartnett, L. C. Johnson, and E. G. Towne. 1999. The keystone role of bison in North American tallgrass prairie. *BioScience* 49:39–50.

Knowles, C. J. 1986. Some relationships of black-tailed prairie dogs to livestock grazing. *Great Basin Naturalist* 46:198–203.

Kotliar, N. B., B. W. Baker, and A. D. Whicker. 1999. A critical review of assumptions about the prairie dog as a keystone species. *Environmental Management* 24:177–92.

Lambeck, R. J. 1997. Focal species: A multispecies umbrella for nature conservation. *Conservation Biology* 11:849–56.

Landres, P. B., J. Verner, and J. W. Thomas. 1988. Ecological uses of vertebrate indicator species: A critique. *Conservation Biology* 2:316–28.

Lawler, J. J., D. White, J. C. Sifneos, and L. L. Master. 2003. Rare species and the use of indicator groups for conservation planning. *Conservation Biology* 17:875–82.

Litvaitis, J. A. 2003. Are pre-Columbian conditions relevant baselines for managed forests in the northeastern United States? *Forest Ecology and Management* 185:113–26.

Lubchenco, J., S. R. Palumbi, S. D. Gaines, and S. Andelman. 2003. Plugging a hole in the ocean: The emerging science of marine reserves. *Ecological Applications* 13:S3–S7.

Martikainen, P., L. Kaila, and Y. Haila. 1998. Threatened beetles in white-backed woodpecker habitats. *Conservation Biology* 12:293–301.

Molina, R., B. G. Marcot, and R. Lesher. 2006. Protecting rare, old-growth forest-associated species under the survey and manage program guidelines of the Northwest Forest Plan. *Conservation Biology* 20:306–18.

Morgan, P. G., H. Aplet, J. B. Haufler, H. C. Humphries, M. M. Moore, and W. D. Wilson. 1994. Historical range of variability: A useful tool for evaluating ecosystem change. *Journal of Sustainable Forestry* 2:87–111.

Morrison, D. A., G. J. Cary, S. M. Pengelly, D. G. Ross, B. J. Mullins, C. R. Thomas, and T. S. Anderson. 1995. Effects of fire frequency on plant species composition of sandstone communities in the Sydney region: Interfire interval and time since fire. *Australian Journal of Ecology* 20:239–47.

Mulhern, D. W., and C. J. Knowles. 1997. Black-tailed prairie dog status and future conservation planning. Pp. 19–30 in *Conserving biodiversity on native rangelands: Symposium Proceedings*, D. W. Uresk, G. L. Schenbeck, and J. T. O'Rourke, tech. coords. General Technical Report RM-GTR-298. Fort Collins, CO: U.S. Department of Agriculture, Forest Service, Rocky Mountain Research Station.

Murphy, D. D., and B. R. Noon. 1992. Integrating scientific methods with habitat conservation planning: Reserve design for northern spotted owls. *Ecological Applications* 2:3–17.

Myers, N. 1989. Threatened biotas: "Hotspots" in tropical forests. *Environmentalist* 8:1–20.

Noss, R. F., C. Carroll, K. Vance-Borland, and G. Wuerthner. 2002. A multicriteria assessment of the irreplaceability and vulnerability of sites in the Greater Yellowstone Ecosystem. *Conservation Biology* 16:895–908.

Panzer, R., and M. W. Schwartz. 1998. Effectiveness of a vegetation-based approach to insect conservation. *Conservation Biology* 12:693–702.

Parr, C. L., and A. N. Andersen. 2006. Patch mosaic burning for biodiversity conservation: A critique of the pyrodiversity paradigm. *Conservation Biology* 20:1610–19.

Peterson, A. T., and C. R. Robins. 2003. Using ecological-niche modeling to predict barred owl invasions with implications for spotted owl conservation. *Conservation Biology* 17:1161–65.

Pharo, E. J., A. J Beattie, and D. Binns. 1999. Vascular plant diversity as a surrogate for bryophyte and lichen diversity. *Conservation Biology* 13:282–92.

Pickett, S. T. A., and J. N. Thompson. 1978. Patch dynamics and the design of reserves. *Biological Conservation* 13:27–37.

Poiani, K. A., M. D. Merrill, and K. A. Chapman. 2001. Identifying conservation-priority areas in a fragmented Minnesota landscape based on the umbrella species concept and selection of large patches of natural vegetation. *Conservation Biology* 15:513–22.

Power, M. E., D. Tilman, J. A. Estes, B. A. Menge, W. J. Bond, L. S. Mills, G. Daily, J. C. Castilla, J. Lubchenco, and R. T. Paine. 1996. Challenges in the quest for keystones. *BioScience* 609–20.

Probst, J. R., and J. Weinrich. 1993. Relating Kirtland's warbler population to changing landscape composition and structure. *Landscape Ecology* 8:257–71.

Ralls, K., S. R. Beissinger, and J. F. Cochrane. 2002. Guidelines for using population viability analysis in endangered-species management. Pp. 521–50 in *Population Viability Analysis*, ed. S. R. Beissinger and D. R. McCullough. Chicago: University of Chicago Press.

Raphael, M. G., M. J. Wisdom, M. M. Rowland, R. S. Holthausen, B. C. Wales, B. C. Marcot, and T. D. Rich. 2001. Status and trends of habitats of terrestrial vertebrates in relation to land management in the interior Columbia River basin. *Forest Ecology and Management* 153:63–88.

Reynolds, R. T., R. T. Graham, M. H. Reiser, R. L. Basset, P. L. Kennedy, D. A. Boyce Jr., G. Goodwin, R. Smith, and E. L. Fisher. 1992. Management recommendations for the northern goshawk in the southwestern United States. General Technical Report GTR-RM-217. Fort Collins, CO: U.S. Department of Agriculture, Forest Service, Rocky Mountain Forest and Range Experiment Station.

Roberge, J.-M., and P. Angelstam. 2004. Usefulness of the umbrella species concept as a conservation tool. *Conservation Biology* 18:76–85.

Roberts, T. H. 1987. Construction of guilds for habitat assessment. *Environmental Management* 11:473–77.

Rowland, M. M., M. J. Wisdom, L. H. Suring, and C. W. Meinke. 2005. Greater sage-grouse as an umbrella species for sagebrush-associated vertebrates. Pp. 232–49 in *Habitat threats in the sagebrush ecosystem: Methods of regional assessment and applications in the Great Basin*, ed. M. J. Wisdom, M. M. Rowland, and L. H. Suring. Lawrence, KS: Alliance Communications Group.

Rubinoff, D. 2001. Evaluating the California gnatcatcher as an umbrella species for conservation of southern California coastal sage scrub. *Conservation Biology* 15:1374–83.

Samson, F. B. 2002. Population viability analysis, management, and conservation planning at large scales. Pp. 425–41 in *Population Viability Analysis*, ed. S. R. Beissinger and D. R. McCullough. Chicago: University of Chicago Press.

Sharps, J. C., and D. W. Uresk. 1990. Ecological review of black-tailed prairie dogs and associated species in western South Dakota. *Great Basin Naturalist* 50:339–45.

Sieg, C. H., R. M. King, and F. Van Dyke. 2003. Conservation and management issues and applications in population viability analysis. Pp. 115–22, Exercise 15, in *A workbook of practical exercises in conservation biology*, ed. F. Van Dyke and contributors. New York: McGraw-Hill.

Simberloff, D. 1998. Flagships, umbrellas, and keystones: Is single-species management passe in the landscape era? *Biological Conservation* 83:247–57.

Snaith, T. V., and K. F. Beazley. 2002. Moose (*Alces alces americana* [Gray Linnaeus Clinton] Peterson) as a focal species for reserve design in Nova Scotia, Canada. *Natural Areas Journal* 22:235–40.

Suter, W., R. F. Graf, and R. Hess. 2002. Capercaillie (*Tetrao urogallus*) and avian biodiversity: Testing the umbrella-species concept. *Conservation Biology* 16:778–88.

Szaro, R. C. 1986. Guild management: An evaluation of avian guilds as a predictive tool. *Environmental Management* 10:681–88.

Taft, O. W., M. A. Colwell, C. R. Isola, and R. J. Safran. 2002. Waterbird responses to experimental drawdown: Implications for the multispecies management of wetland mosaics. *Journal of Applied Ecology* 39:987–1001.

Thomas, J. W., M. G. Raphael, R. G. Anthony, E. D. Forsman, A. G. Gunderson, R. S. Holthausen, B. G. Marcot, G. H. Reeves, J. R. Sedell, and D. M. Solis. 1993. Viability assessments and management considerations for species associated with late-successional and old-growth forests of the Pacific Northwest. Washington, DC: U.S. Government Printing Office.

Uresk, D. W., and A. J. Bjugstad. 1983. Prairie dogs as ecosystem regulators on the northern High Plains. Pp. 91–94 in *Seventh North American prairie conference proceedings*. Springfield: Southwest Missouri State University.

USDA (U.S. Department of Agriculture), Forest Service. 1997. Tongass land management plan revision. R10-MB-338e. Juneau: U.S. Department of Agriculture, Forest Service, Alaska Region.

———. 2000. Management guidelines for the western prairie fringed orchid on the Sheyenne National Grassland. Bismarck, ND: U.S. Department of Agriculture, Forest Service, Dakota Prairie Grasslands.

———. 2001. Land and resource management plan for the Nebraska National Forest and Associated Units. Denver: U.S. Department of Agriculture, Forest Service, Rocky Mountain Region.

———. 2003. Sierra Nevada forest plan amendment draft supplemental environmental impact statement. R5-MB-019. Vallejo, CA: U.S. Department of Agriculture, Forest Service, Pacific Southwest Region.

USDA (U.S. Department of Agriculture) and USDI (U.S. Department of the Interior). 1994. Final supplemental environmental impact statement on management of habitat for late successional species and old-growth-forest-related species within the range of the northern spotted owl, vols. 1 and 2. Portland, OR: U.S. Department of Agriculture, Forest Service, Regional Office.

——-. 2000. Interior Columbia Basin supplemental draft environmental impact statement. Report No. BLM/OR/WA/Pt-00/019 +1792. Portland, OR: U.S. Department of the Interior, Bureau of Land Management.

U.S. Fish and Wildlife Service. 1993. Grizzly bear recovery plan (revised). Denver, CO: U.S. Department of Agriculture, Fish and Wildlife Service.

Watson, J., D. Freudenberger, and D. Paull. 2001. An assessment of the focal-species approach for conserving birds in variegated landscapes in southeastern Australia. *Conservation Biology* 15:1364–73.

Wilcox, B. A. 1984. In situ conservation of genetic resources: Determinants of minimum area requirements. Pp. 639–47 in *National Parks: Conservation and Development*, ed. J. A. McNeely and K. R. Miller. Washington, DC: Smithsonian Institution Press.

Wisdom, M. J., R. S. Holthausen, B. C. Wales, C. D. Hargis, V. A. Saab, D. C. Lee, W. J. Hann, et al. 2000. Source habitats for terrestrial vertebrates of focus in the Interior Columbia Basin: Broad-scale trends and management implications. 3 vols. General Technical Report PNW-GTR-485. Portland, OR: U.S. Department of Agriculture, Forest Service, Pacific Northwest Research Station.

Wisdom, M. J., M. M. Rowland, L. H. Suring, L. Schueck, C. W. Meinke, S. T. Knick, and B. C. Wales. 2005. Habitat for groups of species. Pp. 205–31 in *Habitat threats in the sagebrush ecosystem: Methods of regional assessment and applications in the Great Basin*, ed. M. J. Wisdon, M. M. Rowland, and L. H. Suring. Lawrence, KS: Alliance Communications Group.

Wu, L. S., and D. B. Botkin. 1980. Of elephants and men: A discrete, stochastic model for long-lived species with complex life histories. *American Naturalist* 116:831–49.

9

Social Considerations

John D. Peine

This chapter describes social considerations and consequences to include in the analysis of alternative approaches to conserve rare or little-known (RLK) species. A comparison of social and ecological systems is helpful. Ecological ecosystems can be referred to in the context of structure, process, and functions. "Structure" refers to species composition within communities and their distribution across attributes such as biomass, age class, reproduction, and mortality. "Process" describes what drives the system, such as climatology, nutrient cycling, and species evolution and succession. "Functional" refers to elements that include habitat characterization and the role of the species and community in the food chain. At various stages nutrient cycling processes such as uptake, decomposition, and mineralization are considered functional elements of ecosystems (Odum 1956).

As a corollary to biological systems, consider navigating social systems in the context of structure, process, and functions. *Social structure* may be expressed in the context of units of social interaction and association, such as family and friends, commercial enterprise, voluntary association, public agencies, communities and neighborhoods, and governmental units. Typical parameters used to describe structure include age, gender, ethnicity, occupation, and income. *Social processes* may be defined in terms of institutional systems directing social knowledge and values; order in terms of policy and politics; and the production of goods and services, ergo capital and wealth. Typical parameters to describe process include the interaction of social, cultural, and economic positions related to any given issue of concern. The degree of social orientation and economic dependency on natural resources is a direct link between human and natural resource dynamics. Elements of *social function* include raising and providing for a family, entertainment and leisure, and intellectual and spiritual enlightenment. Typical parameters reflecting social function include income and educational level, orientation of values, and lifestyles and leisure activities. These social dynamics may be greatly influenced by and reflect the context of the natural environment within which society functions (Peine et al. 1999). The challenge is to identify those social elements relevant to the biological analysis of alternative conservation strategies for RLK species. It is instructive to identify social considerations for any given case in the context of social structure, process, and function.

This chapter describes relevant social considerations and how to incorporate them into the analysis of alternative conservation strategies being considered for RLK species. The array of social dimensions included is admittedly broadly cast so as to provide a holistic perspective on the social challenges of environmental conservation. Human values are at the center of social considerations in evaluating alternative strategies for conservation of RLK species. Understanding the values at play in any given conservation alternative is the social considerations equivalent to knowing the biological functions necessary to support suitable habitat for an RLK species of concern. The difference is that these social values are not absolutes but rather starting points that allow us to define where common and conflicting interests exist and the potential for collaboration.

Other social considerations included in this chapter focus on factors influencing the decision-making process, such as alternative land protection strategies, institutional dynamics and risk taking, and the role of laws and statutes, politics, and religion. Several case examples are included to illustrate how decision-making processes have been applied in various circumstances, what values were in play, what were the influencing social factors, what social indicators were used and measured, and whether the conservation goal for the RLK species of concern was achieved. The fundamental social building blocks to achieve that ultimate goal via the communications, analysis, and decision-making process include the following objectives: (1) cultivating inclusiveness, (2) maintaining a holistic social perspective, (3) maximizing stakeholder benefits, and (4) institutionalizing the implementation of the preferred conservation strategy. This social considerations foundation will enhance the potential to establish long-term conservation buy-in by the parties involved—the ultimate goal. The challenges are formidable as the drumbeat of land use conversion from open space to development accelerates at a current rate of 2 million acres per year in the United States (Dowling 2000; Land Trust Alliance 2005). But there is also a growing social consensus that environmental stewardship values are pervasive. Environmental conservation has become a bipartisan issue of concern by a majority of Americans (Lubchenco 1998; DiPeso 2002).

A litany of metrics is routinely measured to gauge the social dimensions of species conservation (Glesne and Peskin 1992; Taylor 1994). Citizen surveys are commonly used to measure social dimensions but are limited in their application to understanding the complexity of social dimensions surrounding species conservation (Fiorino 1990). Public forums may also be problematic if they become dominated by a few vocal individuals with strong opinions (Lukensmeyer and Brigham 2003). The tenor of dialogue can easily become adversarial and confrontational. Content analysis is a key social science tool that can be applied to a variety of social dimensions: literature review, public policy documents, news articles, organizational literature, Web sites, video and audio tapes of meetings, and meeting documentation (Krippendorff 1980). To achieve a deeper perspective on the social dimensions of RLK species conservation strategies, focus groups and personal interviews are effective tools. Interactive Web sites are a relatively new strategy being applied to solicit personalized feedback as well (Higdon 2003).

And on a final introductory note, just how common are little-known species? The preponderance of species has not been identified (Wilson 1999). For instance, the first of its kind All-Taxa Biodiversity Inventory being conducted since 1999 in the approximately 500,000 acre Great Smoky Mountains National Park has to date identified about 12,000 species, 4666 of which are newly documented in the park, and of those, 651 are new to science. There are an estimated 50,000 to 100,000 nonmicrobial species in the park (ATBI 2006). One of the greatest social considerations challenges is how to convey to the global human population the value of these little-known and unknown species and their habitat as indicators of environmental health and species richness.

Social Goals and Objectives

As referenced in chapter 2, the challenging social goal is to establish a long-term commitment by all sectors of society to maintain sustainable populations of RLK species. It is one thing to support the principal of environmental stewardship in general but quite another to support a specific conservation strategy for an RLK species that might directly impact individuals, their families, and their jobs. The following objectives provide the building blocks to achieve this ambitious goal:

- Insure equitable inclusiveness among stakeholder participation in the planning, analysis, and decision-making process that evaluates alternative conservation strategies to sustain RLK species. The importance of achieving this objective is documented in several case examples of RLK species conservation.
- Maintain a holistic and nonpartisan social perspective of the social considerations and consequences of alternative RLK species conservation strategies under consideration. These include the range of social values, social structure and welfare, cultural heritage and values, natural resource utilization, institutional dynamics and risk taking, political influence, and the social dimensions of decision making.
- Maximize an array of social and economic benefits to the degree practical without compromising the RLK species conservation goal. Conservation strategies that focus on this objective, particularly on privately owned land, are more likely to succeed in the long run.
- Establish and maintain an institutional framework for sustaining the chosen conservation strategy. This critical social consideration is illustrated in several case examples of successful RLK species conservation practices.

Values

The sections on social, structural and welfare, cultural, and natural resource values following here include discussion of the components of those values, factors that contribute to the formulation of those values, and suggested indicators to measure them. A series of case examples discuss conservation strategies to sustaining RLK species and illustrate how indicators of various values were measured, analyzed, and applied to decision making.

Social Values

According to Merriam-Webster (www.m-w.com), the term "social values" as used in this section refers to individuals, families, and like-minded cooperatives and communities. Identifying relevant social values, how deeply they are held, and in what context is a social process. The human values listed below summarize a broad spectrum. The relative importance of these values will likely vary as to the context of the social process and relevancy to any given RLK species conservation strategy under consideration. Personal values can be quite divergent from collective values represented by families, communities, businesses, and political interests. The components of social values briefly described below offer some context on this topic, but they are invariably interrelated to some degree for any given circumstance associated with alternative conservation strategies under consideration. Stewardship and aesthetics tend to be at the top of the list of social values espoused by advocates of species conservation. The following social values are listed in order of probable social consideration relevancy to assessing alternative conservation strategies for RLK species.

Stewardship

Since the publishing in 1962 of *Silent Spring* by Rachel Carson, followed by the first Earth Day on April 22, 1970, there has been an emerging social awareness and endorsement of a stewardship ethic for the natural environment (Hand and Van Liere 1984; Kempton et al. 1995; DePeso 2002). Various indigenous cultures have held such values for generations, but society drifted away from such a close relationship to the earth with the industrial revolution and the urbanization of society (Jung 1964; Nolt and Peine 1999).

A basic understanding of how the public perceives environmental issues has become increasingly important in the development of policy and planning related to the conservation of RLK species. Over the years, sociologists and natural resource managers alike have attempted to discover a scale by which to measure environmental values and attitudes. Several paradigms with accompanying measurement scales have been developed to distinguish a spectrum of environmental beliefs. As environmental issues came to the forefront in public policy during the late 1970s, the new environmental paradigm (NEP) scale became an important tool in the measurement of environmental attitudes. Developed by Dunlap and Van Liere (1984), the NEP scale was intended to measure the proenvironmental stewardship shift that was becoming more evident in the study of environmental beliefs. A positive reaction to the NEP scale emphasized a shift from more traditional understanding of the environment as a tool for human use to cognition of the environment as something that must be protected and respected for its own sake. The NEP has important implications for the natural resource manager. The use of the NEP to isolate demographic variables that correlate with a positive environmental worldview allows planners and land managers to better understand the human dimensions of the areas in which they work. Research findings indicate that environmental stewardship is more likely to be embraced by young people, the more highly educated, those with higher incomes, those with a liberal political philosophy, those living in an urban area, and those employed outside of primary industries (Scott and Willits 1994). See the introductory section on social structure.

Privately owned lands adjacent to protected areas are focus areas for new development (Freyfogle 1998). In-migrants into rural areas are more knowledgeable on environmental issues, more concerned about the environment, place higher priority on environmental protection, and are more engaged in activities that promote environmental values than nonmigrants (Jones et al. 2003). The propensity for environmental stewardship can be assessed from public surveys, focus groups and personal interviews, assessment of environmental policy and governance practices, and/or content analysis of literature generated by the media or special interest groups. Stewardship values are likely to be more prevalent in the following circumstances.

The distinction of public land ownership is suggested as a key perspective on stewardship, aesthetics, and sense of place (Kessler et al. 1992). People tend to focus on "their parks" when expressing environmental concerns. Public surveys are standard tools to measure stewardship values. For instance, a survey of visitors to Great Smoky Mountains National Park found that the park attribute contributing most to the quality of the park experience was visitor knowledge that the natural environment was healthy and well cared for. Park facilities and services consistently ranked relatively lower (Morse 1988). This is a classic example of how important environmental stewardship can be to users of public lands.

The Siouxon Valley case example illustrates how stewardship values prevailed when the public had a voice in decision making. The Siouxon is one of the few remaining pristine lowland valleys located in the Gifford Pinchot National Forest near Portland, Oregon. In 1990, the U.S. Department of Agriculture (USDA) Forest Service initiated an integrated resource analysis of the area. Public input was initially solicited during five scoping meetings. A social values assessment was initiated by first conducting a limited number of personal interviews using a draft questionnaire to gain a more in-depth reflection of opinions on the area and to pretest a visitor and mail-back survey instrument (Hansis 1997). Visitors to the areas were interviewed and mail-back surveys sent to a sample of households in the Oregon–Washington region. Results of the analysis indicated that most of the public preferred that the late-successional forest area remain roadless and be managed for primitive forms of recreation (e.g., allowing footpaths and primitive campsites but no vehicular access). The majority also supported wild river designation for Siouxon Creek.

However, national versus local perspectives on stewardship of public lands can be contentious. Local people who rely directly on federal lands for their livelihood and/or recreation are very likely to have "selective" perspectives on stewardship that are protective of their special interests (Karp 1996). An example illustrating this dichotomy is the controversy related to the northern spotted owl habitat conservation plan in the Pacific Northwest, which is discussed later in this chapter (Noon and Blakesley 2006). Examples of public and commercial interests that can jeopardize the habitat of RLK species include overgrazing, poaching of wildlife, off-road vehicle use, forest clear-cutting, oil and gas drilling, or open pit mining.

Aesthetics

Hargrove (1989) defines aesthetics as traditionally beautiful, sublime, picturesque and the scientifically interesting. The old cliché "beauty is in the eyes of the beholder" suggests that this social value is subjective and difficult to quantify. The focus in the context of RLK species conservation is to assess whether or not the setting of critical habitat necessary to sustain the RLK species of concern includes the social value associated with a sense of beauty inherent in nature and picturesque landscapes. Individuals react differently in defining their sense of aesthetics so it is difficult to identify indicators of a common denominator. Indicators reflecting a degree of naturalness summarized below are suggested to imply a sense of aesthetic values.

- *Natural landscapes.* The degree of naturalness of the landscape can be a useful indicator for quantifying aesthetic values. Description of land cover types provides useful indicators such as native vegetation; topographic relief; and the presence of water, forest or grasslands, open space, and/or agricultural fields. These types of landscapes hold aesthetic values for many people (Bell 1999). More than one-half of the population living near wilderness areas do so because of the presence of these lands (Rudzitis and Johansen 1991). Analysis of satellite imagery to categorize landscapes can provide a means to quantify this value (Bell 1999). The accelerating movement of the U.S. population toward coastal areas is in part a reflection of the aesthetic value of being near the coastline (Frey and Speare 1992).
- *Solitude.* The opportunity to become one with nature with minimal distraction is an important human dimension of the aesthetics of the environment. One of the principal values derived from natural/wilderness areas is escape—an opportunity for temporary release from the routine and pressures of everyday life (Hendee et al. 1968; Stankey 1973). The degree of solitude may be an indicator of the likelihood that a sense of beauty and/or spirituality associated with nature will be strongly felt. Delineation of roadless areas of a given size, such as 2000 hectares or more, is a useful parameter to measure this attribute. These areas also serve as sanctuaries for RLK species, an attribute rarely discussed in the ongoing debate about reversing the federal policy of setting aside roadless areas (Falk and Holsinger 1991).
- *Spirituality.* Possibly the least-recognized social value associated with the environment and species conservation is that stemming from the spirituality that some segments of society associate with the natural world. Such perspectives are not necessarily associated with any formal religion (Kellert 2002). The organization Creation Care is an evangelical environmental network that raises awareness of climate change and other environmental

issues (www.creationcare.org). There is a kind of primal heritage association that some people make when experiencing the natural environment. Native American peoples invariably have a rich heritage of environmental attitudes and values about place, which originate from a much different cultural conditioning and mindset (Swan 1989). The built environment of modern society tends to be very place-generic, reflecting a growing sameness in social, cultural, and environmental context (Calthorpe 2000).

Structure and Welfare

This section focuses on values related to personal and family structure and welfare in the context of analysis of alternative strategies to conserve RLK species. The structure and welfare values are listed following here in the order of their probable relevancy to the conservation of RLK species. Economic considerations associated with social structure and welfare are addressed in chapter 10. Natural resource managers invariably support the need to understand the condition of the natural resources and habitat required to sustain populations of RLK species. There is just as great a need to understand the social structure and welfare conditions of people, particularly those with vested interests in the alternative conservation strategies under consideration (Wondolleck and Yaffee 2000). They are most likely to be directly affected by natural resource management decisions and therefore are usually the ones whose support is most important to obtain. This information is central to achieving the third social considerations objective for the conservation of RLK species mentioned earlier, to maximize social and cultural benefits. Understanding the social structure and welfare context helps to navigate the building of collaboration and consensus. It is insightful for leaders in natural resource management to follow the old adage, "walk a mile in my moccasins."

Implications of the social demographics related to the NEP theory on environmental stewardship discussed earlier apply to the parameters expressed here. As described by psychologist Abraham Maslow, human needs can be portrayed in a hierarchical range from basic survival and security needs at the bottom of a triangle to the need for self-realization at the apex. Higher-order needs emerge as lower-order needs are satisfied

(Mehta and Ouellet 1995). People tend to be most fundamentally concerned with livelihood, health care, safety, housing, education, and financial security. They are likely to be interested in government programs that support those basic needs, particularly within the family context. Environmental concerns are not likely to be ranked above those societal needs previously mentioned unless environmental conditions directly threaten those fundamental individual needs.

On the other hand, the characteristics and condition of the natural environment have significant influence on the perceived sense of place and quality of life (Rutzitis and Johansen 1991). Environmental stewardship values are likely to be related to social structure and welfare values. Certainly the assertion that human welfare is a key social consideration in RLK species conservation was vigorously supported by a panel convened at the Innovations in Species Conservation Symposium (ISCS) held in 2003 (ISCS 2003). Their charge was to address the question: How does society perceive, support, challenge, or interact with various conservation approaches? For instance, panelist Tom Story from the San Diego mayor's office referred to Maslow's scale of needs indicating that family well-being comes first before environmental conservation. To paraphrase Tom Story, when approached concerning RLK species habitat conservation, the first question on a landowner's mind is: How will this impact my family welfare? Suggested indicators of social structure and welfare values follow in order of probable relevancy to RLK species conservation.

Population Dynamics

Key human population indicators include community size, population density, population change, and the degree of rurality. Trends in community size and population density are sometimes related to expressed satisfaction with social conditions. Stinner and Van Loon (1992) evaluated the relationship between community size and satisfaction using data from a statewide telephone survey in Utah. They found that the natural environment was one factor determining community satisfaction. Population density and migration patterns are indicators of social conditions. High density of people is correlated with greater likelihood of increased crime and degraded environmental conditions. Rapidly expanding populations strain community infrastructure (Howe 1987). Beginning in the 1960s, increasing numbers of people in the United States have escaped the urban setting to the suburbs. Such out-migration is frequently associated with deteriorating social conditions such as a loss of available jobs, increased crime, and a decline in educational services (Wilson 1987). Urban and exurban sprawl continues today as some people are now escaping the sameness of the built suburban landscape to the more distinctive character of small satellite towns (Davis et al. 1994). As far back as 1989, Raines (1989) found that nonmetropolitan counties in the United States were experiencing greater average rates of population growth than metropolitan areas. These dynamics are of concern when evaluating alternative conservation strategies for sustaining populations of RLK species located on privately owned lands in a projected urban–wildlands interface. In those cases, long-term projections of development growth patterns and their impacts on the environment should be a first order of business. The Environmental Protection Agency Region 4 has developed the Southeastern Ecological Framework for just that purpose (Durbrow et al. 2002). Other EPA regional offices are now conducting similar analyses (Rick Durbrow, pers. comm., 2006).

Commute time to and from work, a parameter included in the U.S. census, can be an indicator of a lack of available preferred jobs close to home (Davis et al. 1994). Greater commute time suggests an inadequate job market but in many cases may also be in response to a preferred quality of home life that people are willing to make compromises to sustain. More and more, regions with scenic landscapes in sparsely populated areas are becoming fast growth regions as more people desire to live there (Jones et al. 2003). Natural resource managers are becoming more cognizant of current and projected future land use conversion patterns triggered by sprawl, particularly in the context of habitat fragmentation for RLK species (Durbrow et al. 2002). Urban planners continually struggle to stem the tide of sprawl and entice people to rediscover the benefits associated with an urban lifestyle. Exurban transportation planners are now beginning to evaluate the implications of highway development on the environment (Howe 1987; Davis et al. 1994; Durbrow et al. 2002). Land use conversion driven by sprawl is the greatest threat to forests in the southeastern United States with implications for RLK species (Wear and Bolstad 2004).

Family Structure

Family structure relates to social structure, welfare, and cultural values. With the dramatic social shift from the traditional nuclear family to a rapidly growing variety of personal relationships living under one roof, the contemporary family structure is an important indicator of social condition (Vosler 1996; Botts 2003). Single-parent households, female heads of households, and family size are likely to be related to levels of income, education, and ethnicity. In the United States, almost half of today's children are not living on a regular basis with both birth parents (Dawson 1991; Botts 2003). As a result, one can conjecture that orientation to the natural environment for children in their formative years, which has traditionally been passed on from one generation to the next within families, may be significantly altered for those living in nonnuclear families. Reduction in awareness and utilization of natural resources within family groups may have implications for the conservation of RLK species of concern. There are new campaigns designed to get children outdoors. The National Park Service's version is "Leave no child indoors" (Soukup 2006).

Education

Formal education provides empowerment to escape an unsatisfactory social condition. Parameters selected to profile education should reflect the quality of the education services available and the level of achievement (Downey 1994). Providing for a quality education should be a priority government service at all levels. Too often, there is significant disparity between urban versus rural and rich versus poor communities because public education has traditionally been funded with local property taxes. The level of education achieved tends to be positively correlated with the degree of endorsement of environmental stewardship values, and these become relevant to the conservation of RLK species (Kanagy et al. 1994: Tindall 1994).

Safety

There are many dimensions of safety and security ranging from risks of personal assault and property damage in the home and neighborhood to risks while driving to school and the workplace. The potential disparity of security conditions within a geographic area is of particular concern. Natural resource managers need to be particularly cognizant of this fundamental dimension of the quality of life. In addition, public land managing agencies spend significant resources dealing with crime in national parks, forests, wildlife reserves, and natural areas so they are familiar with security issues (Pendleton 1996).

Safety has also become a major issue at the urban–wildlands interface where humans and carnivores coexist. In the rapidly growing region of south Florida, for example, the remnant Florida panther population (*Puma concolor coryi*) has increased from 30 to approximately 100 with the release of eight female cougars (*P. c. stanleyana*) in 1995 from southern Texas (Maehr et al. 2002). The 2.5 million acre Florida Panther National Wildlife Refuge where panthers are primarily located is adjacent to the Big Cypress National Reserve and Florida Everglades National Park. A recent panther encounter by a local resident has raised concerns about human safety (Scoloff 2006). The U.S. Fish and Wildlife Service has devised a landscape model of panther movement patterns and predictions of where new development might be located to minimize the potential for wildlife–human encounters (Maehr et al. 2004). Developers are being counseled on how to minimize the potential for panther encounters with humans (Larry Richardson, USFWS, pers. comm., 2006). This is a case where environmental stewardship and human safety values conflict (Defenders of Wildlife in Scoloff 2006). There is no resolution at this point to insure survival of this high-profile endangered species.

Poverty

The concern for poverty is rightfully a major component of most public social agendas. Natural resource managers should be aware of the potential for inequity of environmental health among socioeconomic classes (Szasz 1997). This concern is an important consideration for the third social considerations objective mentioned earlier: maximize benefits associated with RLK species conservation. Personal observation suggests that in rural areas in close proximity to federal lands in the Appalachian Mountains, low-income families are being displaced by wealthy in-migrants, thus changing the social, cultural, and economic dynamics of the region. These low-income families have occupied the region for generations and to some degree have lived off the land via firewood, hunting, fishing, and small-scale agriculture. Land purchases by new in-migrants have greatly increased land values and significantly increased the size of the environmental footprint per household.

Cultural Values

Cultural dimensions expressed here are those most likely to have relevancy to the conservation of RLK species. According to Merriam-Webster (www.m-w.com) "cultural values" reflect the pattern of social behavior and the state of being cultured reflecting intellectual and moral faculties. Cultural behavior reflects the anthropological context in which people live. The past is seen to symbolize the familiar and known, and hence is reassuring and conducive to a feeling of security and continuity. The past is often of particular interest because it relates to previous events and places relevant to our own lives and those of our families (Taylor and Konrad 1980). History, ancestry, and a sense of cultural cohesion have the ability to cause emotional and symbolic links with specific places in the environment. The influence of the environment on culture has been shown in stories passed down through Native American tribal custom and culture. Specific areas are considered to be sacred because of their linkages with the tribe (Swan 1989). Howell (2002) proposes that there should be an expanded role for the cultural sciences in environmental planning and stewardship. The intent of this section is to provide a brief characterization of key cultural indicators that can be incorporated in the process of assessing alternative conservation strategies for RLK species. Suggested indicators of cultural values follow in order of probable relevancy to species conservation.

Cultural Traditions Related to the Environment

The U.S. census includes information on ancestry, though cultural orientation has become diluted over time (Farley 1991). Family traditions concerning resource utilization and stewardship values can be reflective of the heritage of cultural ancestry from native lands dating back several generations (Taylor and Konrad 1980). It is insightful to be aware of these dynamics when determining vested interests of various constituencies in the conservation of RLK species. Local and regional cultural traditions sometimes reflect the natural landscape setting and utilization of natural resources (Howell 2002; Noon and Blakesley 2006). Relationships with nature are manifest in innumerable ways, some of which may be of particular importance for any given management issue. Relevant cultural traditions can be cataloged via oral history (Thompson 2000), a procedure commonly practiced by the U.S. National Park Service. State Historic Preservation Officers have access to county cultural resource surveys, which can be very helpful (www.cr.nps.gov/nr/shpolist.htm). Frequently, these traditions center on utilization of native species in family traditions. For instance, a literature search in ethnobiology conducted for Great Smoky Mountains National Park found that over two-thirds of the species present were known to be used by people (Campell et al. 1985).

Time in residence, an indicator recorded by the U.S. Census Bureau, builds emotional bonds to place, particularly during the adolescent stage of life (Williams et al. 1992). Length of residency can be a good indicator of public attitude toward stewardship and utilization of natural resources (Peine et al. 1999). For an area growing rapidly in a setting rich in scenic beauty and other natural resource amenities, growth is likely to be in part a response to the landscape setting (Jones et al. 2003). Jones et al. indicate that newcomers are more likely to be proponents for managing growth so as to protect the resource conditions that attracted them to the area in the first place. This is particularly true of new retirees who harbor specific expectations in their lifestyle and landscape setting. Long-time residents revere the resources as well but are more likely to be reluctant to accept zoning or other means to control development (Fortmann and Kusel 1990; Peine and Stephens 1997).

Natural Resources Utilization

People consume goods and services both directly and indirectly from the natural environment. Examples of direct consumption include extraction of minerals and petroleum, timber, wildlife, and plant materials. Some of these practices are renewable and others not. These natural resources are used either as end-use goods or as inputs into the production of other goods and services. Indirect consumption occurs with nonconsumptive activities that use the natural environment to produce services (e.g., recreational activities and natural insurance). Natural insurance occurs when ecosystems act to buffer the market economy from external shocks to production and consumption. For example, forest systems provide protection, thus insurance from flooding due to weather events (Crocker et al. 1998). In addition, ecosystem attributes or services such as scenic beauty, air quality, water quality, and climate indirectly influence a community's economic health through their influence on business location, job creation, and income levels in a region (Johnson and Rasker 1995). This exemplifies the interrelationships of social and economic values. The economic values associated with renewable and nonrenewable natural resource utilization is discussed in detail in chapter 10 but is briefly mentioned here to emphasize the close relationship between social and economic values within the context of assessing alternative conservation strategies for RLK species. Several of the case examples of conservation strategies applied to sustain RLK species in this chapter are focused on competing social and economic interests on federally owned lands.

Conservation Strategies

An often overlooked social consideration is the evolution of creative strategies to protect open space and critical habitat for RLK species. Land use change reflects generational evolution. For instance, the baby boomer generation is selling off family farms to liquidate equity, which results in more land to meet the growing demand for exurban sprawl development (Briassoulis 2000). As land use conversion accelerates in the United States at a current rate of over 2 million acres per year (Land Trust Alliance 2005), ecosystems become more fragmented and open space parcel sizes become smaller. This trend will accelerate into the future.

To address this extraordinary challenge, the Region 4 office of the U.S. Environmental Protection Agency has created the Southeastern Ecological Framework, a tool to ascertain critical regional ecosystems supporting species richness and diversity as well as a system of environmental corridors to sustain a cohesive network (Durbrow et al. 2002). The assemblage of georeferenced databases will soon be web-enabled to increase access for planning and conservation of RLK species. Other EPA regional offices are initiating similar projects. A related analysis, initiated at the state level in Oregon, conducted a statewide analysis of biological diversity and sustainability risk assessment. A state model was then created to define and implement stewardship incentives to conserve biodiversity and RLK species (Vickerman 1999). Their incentives program is a robust example of the third social considerations objective discussed earlier in this chapter—maximize benefits to stakeholders. Highlights include offering something for everyone, flexibility to meet diverse requirements of landowners, focusing incentives in a regional context (including incentives to promote cluster development), and conducting environmental policy analysis to minimize inconsistency among all levels of government.

Because the federal estate is finite, in the future other stakeholders and landowners will become increasingly important in protecting habitat for RLK species. For example, regional and local governments working in partnership with local land trust nongovernment organizations have been remarkably successful in protecting habit of RLK species. The total acres conserved by 1667 private land trusts in the United States increased by 54% in the last 5 years to 37 million acres (Land Trust Alliance 2005). Conservation tax incentives for making a charitable donation of a conservation easement are expanding.

Press et al. (2006) make the case that local and regional agencies in concert with land trusts are well suited to conserve habitat of RLK species. They advocate a convergence of scale among land trusts, community and regional open space districts, greenways, and parks. They focused on a case study encompassing four counties in the central coastal region of California. The co-occurrence of endemic species in relatively small areas made the ambitious plans compelling. The specialized habitats are serpentine outcropping and coastal dune–coastal scrub areas. Alternative conservation strategies evaluated included special-area zoning, transferable development rights, conservation easements, and purchase of development rights (Wright 1994, cited in Press et al. 2006). The authors concluded that outright purchase was the conservation strategy of choice due to some uncertainty with the longevity of other options and the high value of development rights. A combination of land trusts and regional and local governments are collaborating to complete the purchase of these critical habits with a variety of fund-raising strategies.

Institutional Dynamics and Risk Taking

The complexity associated with the conservation of RLK species can be exacerbated by the context of institutional groupthink and tradition. Effective leadership can entail risk in thinking outside the box.

Institutional Influence on Decision Making

According to Ostermeier (1999), there are several important considerations regarding decision-making processes and the evolution of institutional positions associated with use of natural resources and species conservation. First, decision-making processes have become increasingly adversarial, where participants have learned to stand firm on positions they believe to be their "rights" concerning the pursuit of economic opportunity or protecting environmental integrity. Shannon (1991) states, "The discourse of divisiveness created by our current social, economic and political institutions threatens our capacity to address sustainability."

Second, natural resource decision making has become exceedingly procedural. Both managerial and judicial checks on administrative and legislative processes are heavily process oriented. Adversaries will seek to uncover even the smallest procedural deficiencies. "The adversarial, win-lose nature of judicial and administrative appeals promotes strong, one-sided argumentation from each of the affected parties, with little incentive to think of creative solutions that bridge diverse interests" (Yaffee 1994).

Third, natural resource and environmental decision making has become increasingly complex and technical. A key driver of this dynamic is the compunction to avoid legal challenge to policy and resources management practice. The court system has become the "elephant in the room" for environmental conservation. Primary participants have become representatives of government and special interest groups. The role of the citizens has been marginalized. By empowering our institutions, such as governments, interest groups, and courts, conflict resolution is less likely to involve communities and their members. In so doing, alternative community values compete with each other without engagement by the most affected parties.

Fourth, this adversarial and interest-driven decision-making system has become entrenched in the political system. Although legislative decision making often results in "splitting the difference" among interest groups, rarely does the process result in creativity or new approaches. As a result, local institutions have effectively been preempted.

And fifth, these institutional arrangements have become ends in themselves. Institutions are systematic, and self-preservation is an ultimate force. Exerting their power is likely to marginalize the significance of the role of individual citizens in the decision-making process. Citizens lose, or never develop both the ability and desire to represent their interests relative to others.

Federal Land Management Agencies

Federal land management agencies have diverse congressional mandates and policies that have developed over decades. Policies may conflict between different agencies concerning a particular resource issue. Each agency has a distinct culture that to some degree reflects social dynamics. Sometimes different factions within institutions have different perspectives on natural resource utilization and/or environmental conservation practices. In addition, there has consistently been inadequate funding for environmental stewardship activities (Alkire 2003). Striking a balance between natural resource utilization versus species conservation is an ongoing challenge. Vining (1992) found that the general public and environmental groups are more inclined to change forest management plans to protect the environment than natural resource managers.

Conservation of RLK species is challenging and often requires taking risk due to uncertainty of the science (see chaps. 4 and 5), the openness of stakeholders to work collaboratively, and who holds political sway in the decision-making process (Peterman and Anderson 1999). Courage is an attribute that allows leaders to take necessary risks in order to work toward a solution on a critical problem. Because of the difficulty in calculating when the risk is warranted, experience and knowledge become of paramount importance. A resource manager will need to define the implications of inaction in the context of stakeholder values and sustainability of the RLK species of concern.

Administering the Endangered Species Act has become an institutional risk-laden challenge (Lieben 1997). For a variety of reasons, the U.S. Fish and Wildlife Service has been reluctant to designate required Conservation Habitat Plans for listed threatened and endangered species (Noss et al. 1997; Wilcox and Elderd 2003). These studies suggest that, as a result, court-mediated implementation of critical habitat designation has greatly impeded full implementation of the law. Part of the problem is a lack of adequate funds to devise the recovery plans, but it goes beyond that. Political controversy is at the heart of the issue. One study found Democrats are generally more supportive of listed species conservation and Republicans are more likely to focus on private property rights (Czech and Borkhataria 2001). Hagen and Hodges (2006) call for change of critical-habitat guidelines to a decision-analysis framework that will make critical habitat strategies scientifically and legally workable as a conservation tool.

Finally, arguably one of the most contentious examples of designating critical habitat centered on the northern spotted owl (*Strix occidentalis caurina*). In 1993, President Bill Clinton placed approximately 9.7 million ha of federal lands in Washington, Oregon, and Northern California under ecosystem management, seeking to protect old-growth forests and over 1000 RLK species associated with them (Noon and Blakesley 2006). This policy shift also reduced federal timber sales by more than 75%, adversely affecting about 300 rural communities (Swedlow 2002). The ensuing policy decision was in response to a federal mandate by Congress for several federal agencies to collaborate in conducting a scientifically sound analysis centered on conservation strategies for the northern spotted owl, a charismatic megafauna (Noon and Blakesley 2006). Studies showed that the number of territorial nesting females had declined significantly and that old-growth forests are preferred for nest sites (Thomas et al. 1990). The conservation plan to conserve old-growth forests became controversial due to the plan's economic consequences. Litigation focused on scientific uncertainty. Timber industry lawyers labeled scientists as conservation advocates (Noon and Blakesley 2006). As a result, decision makers were placed in a highly politically charged position.

Yaffee (1995) maintains that the northern spotted owl controversy in the Pacific Northwest old-growth forests is an example where federal agency leaders lacked courage and "effectively turned over their role as agents of managed organizational and policy change to other voices, both inside and outside the agency." He further states "their response to the changes underlying the issue was inadequate in either resolving the controversy or building an agency less likely to find itself in similar situations in the future." He asserts that this lack of courage and abdication of responsibility further exacerbated and extended the conflicts and set poor precedent for future controversies and the future in general.

Political Influence

The political stage is where the collective manifestations of conflicting social values and institutional dynamics can influence policy decisions concerning conservation of RLK species. Political changes in perspective on environmental policy are inevitable and have varied among federal administrations. Robbins (2004) argues that political agendas, the structure of the decision-making process, and financial resources used to aggrandize a position have great influence on whether the dynamics of a process for devising a habitat conservation plan for a listed species is based on conflict or compromise. The number of new listings of endangered species varies among presidential administrations and reflects political positions due in part to the resources applied to the process (Stinchcombe 2000). All of these dimensions are interrelated but not necessarily of comparative influence.

Voting records by the U.S. Congress provide a good mechanism to quantify political perspective. Project Vote Smart (2003) has thoroughly documented U.S. congressional voting records on environmental issues. Such an analysis is conducted periodically by groups representing different philosophical perspectives, such as the League of Conservation Voters and the Competitive Enterprise Institute, which represent advocates and adversaries of environmental conservation, respectively. On environmental legislative issues, members of Congress are frequently polarized in their positions and provide graphic portrayal of diversity among leadership values. Dunlap (1992) found that members of Congress representing constituencies that were young, white, urban, and above average in socioeconomic status had higher than average proenvironment voting scores.

Social Dimensions of Decision Making

Collaboration is essential to sustain environmental health and populations of RLK species (Wondolleck and Yaffee 2000). The three social consideration objectives expressed earlier of inclusiveness, a holistic nonpartisan perspective, and maximizing benefits to stakeholders, are the building blocks for a successful decision-making process. It is important to recognize, however, that decision making evolves as dynamics change. Flexibility is necessary but so are structure and neutrality of perspective by those leading the process. The inevitable complexity of assessing alternative conservation strategies for RLK species conservation requires diligence in seeking out all aspects of concern and making appropriate adjustments over space and time as necessary (Peine et al. 1999).

George Stankey provided an informative presentation entitled "People, politics, and the protection of rare and at-risk species" at the Innovations in Species Conservation Symposium held April 28–30, 2003, that addressed many important factors influencing the decision-making process (http://outreach.forestry.oregonstate.edu/isc/agenda.htm). He emphasized there and in a later article (Stankey and Shindler 2006) the increasing diversity of public values and uses, the growing mistrust of institutions of governance, the growing recognition of complexity and uncertainty, and the dynamic changes of human populations on the landscape. These topics collectively represent the enormity of contemporary challenges to create an effective and inclusive decision-making process that leads to effectively defining strategies to sustain populations of RLK species. Peterson et al. (2004) documented the complexity of this challenge. They used an ethnographic approach to critically review the habitat conservation planning processes employed to develop regional Conservation Habitat Plans for the Houston toad (*Bufo houstonensis*) and the Florida Key deer (*Odocoileus virginianus clavium*). In both cases, the process focused on a search for the optimum solution through collaboration and consensus building, and in neither case was a solution achieved. The authors concluded that, "The paradoxical nature of liberal democracy precluded the possibility of a single, ideal solution." Other case examples included below illustrate that success can be achieved by incorporating the first social consideration objective mentioned earlier, inclusiveness of the decision-making process. Providing equitable inclusiveness of stakeholders in the decision-making process is arguably the greatest social challenge to species conservation.

Robbins (2004) found that limited access to decision making leads to increased use of confrontational activities, whereas increased information, particularly specialized knowledge, decreases the probability of confrontation. Yaffee and Wondolleck (2003) studied trends in natural resource planning in the United States and identified a marked shift toward collaboration in the last 30 years. They assert that this shift has been prompted by changes in the perceived legitimacy of agencies as expert decision makers, an increase in the availability of information, the perceived nature of the problems facing land managers, and a significant broadening of political power in the United States, combined with legal tools that gave outside groups access to decision making.

Gregory (2000) states that "extensive involvement is not synonymous with meaningful public input." He argues that this is a result of a number of factors, including a lack of specific criteria for accessing stakeholder interests and incorporating their input into the process. He proposes that structured public involvement should engage stakeholders to be part of the process of framing the decision, defining objectives, establishing alternatives, and identifying consequences and tradeoffs. He concludes, "In order to increase the probability of successful outcomes . . . communication gaps among stakeholders need to be narrowed and overall communication levels need to be improved."

Perspectives of Stakeholders

Divergent perspectives are difficult to reconcile. For example, Ruud and Sprague (2000) examined the discourse between environmentalists and loggers in the context of old-growth redwood forests in California. They believed that, "through language humans produce meanings and values that are available for others to hear and interpret." Thus they felt it important to understand both what is spoken and what is heard on opposing sides to establish a true understanding of the issues. They found that the divergent perspectives on what was heard on the topic resulted in a lack of acceptance and meaningful dialogue. The environmentalists focused on a personhood perspective that emphasized stewardship and aesthetic environmental values, whereas the loggers focused on a corporate perspective of commercial value and economic benefits. The temporal perspective diverged dramatically, with the environmentalists focused on long-term ecosystem sustainability and the loggers focused on short-term profits. The environmentalists also tried to bring attention to global issues of timber harvest, whereas loggers steadily focused on jobs in their Northern California communities. Though the study does not provide a conclusion of how to rectify the gap between modes of discourse, it does indicate that, in order for the subject to be broached, the language must speak for and to both sides.

Consensus Building

Engaging decision makers in a process of consensus building is another key aspect of the social dimension concerning species conservation. Isenhart and Spangle (2000) challenge traditional assumptions on conflict resolution. Moving from "rights" to "interests" and from "forcing" to "negotiation" is particularly important, given the litany of conflicting interests invariably associated with species conservation. "Joint Fact Finding" provides a model to build consensus among stakeholders (Susskind and McCreary 1985; Ehrmann and Stinson 1999). This technique is centered on the theory of joint fact finding whereby stakeholders gain perspectives from others through shared participation in the decision-making process (Susskind et al. 2000). The discourse starts by asking stakeholders around the table to describe their interests in the issue at hand. That is followed by a discussion of shared interests and then divergent interests. Participants next discuss strategies to accommodate those divergent issues. This is likely to be a lengthy process.

The multidimensional conservation strategy devised for the lesser prairie-chicken (LPC) (*Tympanuchus pallidicinctus*) and the sand dune lizard (SDL) (*Sceloporus arenicolus*), two candidate species for federal listing, provides an example of successfully engaging divergent stakeholder interests in join fact finding (BLM 2005). The New Mexico LPC-SDL Working Group met for two years to devise a complex conservation strategy. Members included representatives from seven federal and state agencies; conservation groups; three representatives of the oil and gas industry that supports 28,000 jobs in the region; and one representative of the cattle industry. The first task was to become familiar with habit requirements in the "sand shinnery" community to sustain the two RLK species. The habitat has become greatly reduced and fragmented due to agriculture and mining activity. The group devised nine pathways for sustaining LPC populations related to the variety of threats: maintain quality range for breeding and nesting, address fragmentation of habitat, develop strategies to maintain and improve LPC habitat, focus on long-range planning to extend habitat, institute additional measures to address causes of mortality and low nesting success, conduct research and monitoring, provide education and outreach for stakeholders, and fund a permanent position. A similar set of strategies were devised for the SDL. These strategies include numerous details about land stewardship policy and practice. Federal and state agencies sat down and worked with industry to protect the RLK species while allowing continuation of the natural resource utilization. The congenial nature of the group allowed them to work together to fix a problem (http://bpappalachia.nbii.gov/portal/server.pt) (BLM 2005).

Another successful example of collaboration in decision making to protect an RLK species concerns a watershed in northern Georgia, as documented on the Best Sustainability Practices Web site (http://bpappalchia.nbii.gov/portal/server.pt) (2007). Although the environmental setting was different, the challenge was equally as difficult as that described in the California example. The Etowah River is a major headwater tributary of the Coosa River system of the Mobile River drainage. Lying entirely within Georgia at the foothills of the Southern Appalachians, it originates in the Blue Ridge physiographic province but also drains Piedmont and Valley and Ridge provinces. There are 10 imperiled aquatic species known to inhabit this watershed, and five others are believed to have been extirpated. The Etowah basin also lies on the north edge of the Atlanta metropolitan area. The suburban counties that make up the lower portion of the system have been among the fastest-growing counties in the nation over the last decade. Based on these factors, local governments; state, regional, and federal agencies; universities; nongovernmental organizations; and other stakeholders began a planning process to develop a comprehensive habitat conservation plan (HCP) for the Etowah watershed.

This initiative was established to protect the 10 RLK species known to inhabit the area while at the same time allowing growth and development in the watershed. The planning process is overseen by a steering committee composed of representatives from each of the counties and municipalities within the watershed. The steering committee, funded by a grant from the U.S. Fish and Wildlife Service, is assisted by a team of scientists, policy analysts, and educators from the University of Georgia, Kennesaw State University, and the Georgia Conservancy. The HCP advisory committee members at the University of Georgia and Kennesaw State University have conducted extensive research to support the development of the conservation strategy. The HCP consists of a set of development policies adopted by the local governments of the Etowah basin. These policies have been developed by technical committees made up of experts from within the watershed and staffed by researchers from the University of Georgia. The steering committee decided what policies were included in the plan, based on a variety of factors, including recommendations from a number of stakeholder and technical committees with expertise in each policy area. In 2003, the steering committee developed an extensive outreach and public participation program to ensure that the HCP fits into local community contexts and addresses local variations in policy needs across the basin. Ordinances and policies covering issues such as storm water, best site design, runoff limits, erosion, sedimentation, mass grading, stream buffers, road and utility crossings, conservation subdivisions, and water supply management continue to be investigated. Each local government in the watershed will adopt a set of policies to limit the impacts of development on the imperiled fish. This will allow them to receive an incidental take permit that covers development activities within their jurisdiction.

The conservation strategies in both of these cases were complex, and their implementation required extraordinary cooperation, but the goals were the same—sustaining viable populations of RLK species while at the same time sustaining traditional land-use activity. In both cases, success was achieved by adhering to the basic building blocks emphasized in this chapter: (1) participation in the process was inclusive, and all stakeholders as well as scientists were involved from the beginning of the process; (2) the analysis of alternative conservation strategies under consideration was holistic as to implications on demands for use of the natural resources and habitat requirements to sustain the RLK species in question; (3) desired stakeholder benefits were maximized, the RLK species populations were sustained while the development permit process was efficiently expedited; and (4) a framework was established to systematize compliance to agreed-to conservation strategy via a partnership among federal agencies, university faculty, community planners/regulators, and developers.

Decision Support System Capabilities and Limitations

Technology applied to decision making should not be overlooked as a social consideration. New technology is becoming more accepted as a means to improve communication and illustrate alternatives in the decision-making process. A powerful application of the technology is to portray alternative futures based on divergent management strategies in a geospatial context. Social inequity among stakeholders may occur due to a lack of universal access to or skills in applying the technology. The more technically skilled stakeholders can hold a distinct advantage in the decision-making process. It is important to measure the influence that decision support technology has on the decision-making process (Nedovic-Budic 1999). Understanding these impacts is a critical area for future research as the impact from information technology becomes more important and pervasive to decision making (Nedovic-Budic 1998). Delone and McLean (1990) identify six categories to evaluate information systems: system quality, information quality, information use, user satisfaction, individual impact, and organizational impact. Nedovic-Budic (1999) includes societal impact as well. This emerging technology will ultimately revolutionize the decision-making process.

Role of Science

A less than obvious social dimension of species conservation is the role of science in the decision-making process (Adler et al. 2000). Decision makers may not be all that persuaded by the biological science presented, possibly due to the scientific uncertainty associated with the species of concern. Climate change is a good case in point: there is growing scientific evidence that human-caused global warming is taking place, but much less is known as to specific ecological, social, and economic ramifications. That is in part due to divergent predictions from different models (Rastetter 1996). That situation is changing rapidly via the release of a series of reports in 2007 from the Intergovernmental Panel on Climate Change. There is a high degree of consensus and more specificity of reported adverse effects predicted (IPCC 2007). The prescribed actions required to reverse this human influence on global climatic processes require, for the short term, significant changes away from "business as usual." Dueling scientists is a familiar scene in courtroom litigation (Edwards 1999). Scientific information is not always taken at face value. Nedovic-Budic (1999) uses data accuracy, availability, collection time, accessibility, currency, and format to determine information quality. For scientific information to be useful to decision making, accompanying metadata must thoroughly describe methodology, limitations of the data, and its relevance to the issue of concern.

Applying All Dimensions of Decision Making

In the 1990s, an attempt to combine the principles discussed above in a holistic fashion was made by Utah's governor Michael Leavitt and Oregon governor John Kitzhaber. They coauthored a new philosophy for confronting contentious environmental issues and called it Enlibra, which is intended to serve as "an evolving set of new principles for environmental management . . . based upon principles that have proven effective in resolving environmental and natural resource debates in a more inclusive, faster and less expensive fashion" (Western Governors' Association 2002 in Higdon 2003). There are eight guiding principles of Enlibra: (1) national standards, neighborhood solutions—assign responsibilities at the right level; (2) collaboration, not polarization—use collaborative processes to break down barriers and find solutions; (3) reward results, not programs—move to a performance-based system; (4) science for facts, process for priorities—separate subjective choices from objective data gathering; (5) markets before mandates—pursue economic incentives whenever appropriate; (6) change a heart, change a nation—environmental understanding is crucial; (7) recognition of costs and benefits—make sure all decisions affecting infrastructure, development, and environment are fully informed; and (8) solutions transcend political boundaries—use appropriate geographic boundaries to address environmental problems.

Governor Leavitt has repeatedly stated that Enlibra is a philosophy and not a policy, and has a primary tenet: bringing all stakeholders involved in environmental disputes together to work out differences. The philosophy was applied in the study of the San Rafael Swell area in Emery County, Utah, with the goal of accommodating appropriate resource utilization while insuring cultural and environmental resource conservation. Several RLK species are located in the area. The U.S. Bureau of Land Management created a progressive Web site to detail natural resource features and utilization activities in the area and alternative management objectives and strategies. Things were going well until Governor Leavitt inadvertently referred in a speech to the potential goal of local interests designing a resource management strategy that could potentially lead to designating the area as a national monument. That generated a very negative response from a well-organized off-road-vehicle interest group that lobbied successfully for ending the planning effort (Higdon 2003). The point here is a cautionary one: it is important for all parties to maintain an inclusive and transparent decision-making process.

Conclusion

Social sciences provide a mechanism to measure social considerations associated with the process of assessing alternative strategies to sustain populations of RLK species. As described in this chapter, there are numerous dimensions of social science vital to crafting these strategies. There is a growing awareness, as acknowledged by many participants of the Innovations in Species Conservation Symposium (ISCS 2003), that viable conservation strategies for RLK species should be based on the application of interdisciplinary science. Mainstreaming this theory into practice remains a challenge for the future. A centerpiece of this challenge is developing effective communication and dialogue among all stakeholders based on a transparent decision-making process (Wondolleck and Yaffee 2000). The case example discussed earlier of conserving the listed threatened and endangered aquatic species in the Etowah River watershed is a robust example of the role of interdisciplinary science (http://bpappalachia.nbii.gov/portal/server.pt). University faculty working in conjunction with local community officials devised a complex yet viable conservation strategy. The National Environmental Policy Act permitting process was streamlined, and the comprehensive conservation strategy was consistently applied.

The emerging science of information technology is a new innovation that has the potential to revolutionize the decision-making process and make it more transparent, inclusive, and collaborative (Adler et al. 2000). To integrate social and environmental science, there is a need to focus on how best to capture the capabilities of this technology and better communicate and embed science in the decision-making process. Effectiveness is dependent on designing inclusiveness into the process, having a holistic nonpartisan perspective of the social considerations, and working to maximize, to the extent practical, benefits to the stakeholders.

The case examples discussed earlier provide encouragement that innovative thinking and collaboration among stakeholders can successfully meet the formidable challenge of sustaining populations of RLK species.

The strategy to sustain a viable population of the lesser prairie-chicken in New Mexico included devising complex management and research strategies to maintain and restore viable habitat and rebuild populations (BLM 2005). At the same time, strategies were devised to maintain ongoing cattle ranching and the economically vital gas and minerals industries. Another remarkable example is the conservation of RLK species along the central California coastal region. In this case, a robust collaboration of public agencies, private industries, and land trusts are purchasing critical habitat where the rare species assemblages are concentrated (Press et al. 2006).

Conserving RLK species is dependent on long-term commitments to the devised conservation strategy. The mission of the Land Trust Alliance (2005) is to grow and refine land conservation practices and continue the rate of protection. The regional and state-level ecological sustainability analyses case examples conducted by EPA Region 4 (Durbrow et al. 2002) and the State of Oregon (Vickerman 1999), respectively, provide robust scientific rationales for designing a conservation strategy, but no guarantee that their implementation will be sustained over the long term. The case studies for the lesser prairie chicken in New Mexico (BLM 2005), the aquatic RLK species of the Etowah River watershed (BSP 2006), and the coastal RLK species in central California (Press et al. 2006) exemplify the importance of institutional commitment to complex conservation strategies. In each case, financial and institutional incentives to stay the course were included as integral parts of the chosen strategies.

There are commonalities in social considerations associated with successful and failed applications of conservation practices as illustrated in the case examples included in this chapter and elsewhere. There is a need to apply insight gained from this book to plan for the future conservation of RLK species. The recent dramatic increase of protected lands via conservation easements and regional and local planning provides an extraordinary opportunity to build awareness of RLK species and encourage biological inventories and habitat assessments and the adoption of stewardship strategies when warranted. The use of emerging information technology as an agent for information dispersal is promising as a tool to facilitate inclusive, holistic, and transparent assessment of alternative conservation strategies for RLK species. There is a need to routinely apply comprehensive and quantitative strategies to conduct integrated science to improve the conservation of RLK species. And the last word is: Focus on institutionalizing long-term stewardship commitment to sustain viable populations of RLK species.

Acknowledgments

I thank Joe Clark of the USGS Southern Appalachian Field Lab for his insightful review comments and facilitating administrative support, and I thank Darryl Johnson of the University of Washington for his suggestions to organize elements of the topic and his insightful review comments.

References

Adler, P. S., C. Barrett, M. C. Bean, J. E. Birkhoff, C. P. Ozawa, and E. B. Rudin. 2000. Managing scientific and technical information in environmental cases: Principles and practices for mediators and facilitators. Pp. 1–80 in *Managing scientific-intensive public policy disputes*. Tucson, AZ: Resolve Inc., U.S. Institute for Environmental Conflict Resolution, Water Justice Center.

Alkire, C. 2003. Supplemental funding for environmental stewardship. PhD diss., George Washington University, Washington, DC.

ATBI (All taxa biological inventory). 2006. http://www.dlia.org/.

Bell, S. 1999. *Landscape pattern, perception and process*. London: Spon Press.

Blahna, D. and S. Yonts-Shepard. 1989. Public involvment in resource planning: Toward bridging the gap between policy and implementation. *Society and Natural Resources* 2:222–24.

BLM (Bureau of Land Management). 2005. Collaborative conservation strategies for the lesser prairie chicken and sand dune lizard in New Mexico: Findings and recommendations of the New Mexico LPC/SDL Working Group. www.nm.blm.gov/misc/conservation_strategies/docs/LPC_SDL_conservation_strategies/CD.PDF.

Botts, E. 2003. *Family and social network: Roles, norms and social networks in ordinary urban families*. London: Routledge.

Briassoulis, H. 2000. *Analysis of land use change: Theoretical and modeling approaches*. www.rri.wvu.edu/WebBook/Briassoulis/contents.htm.

BSP (Best Sustainability Practices). 2007. http://bpappalachia.nbii.gov/portal/server.pt (accessed July 20, 2007).

Calthorpe, P. 2000. New urbanism and the apologists for sprawl. *Places* 13:68.

Campbell, J. A., A. Bailey, G. Darugh, W. Gregg, and P. S. White. 1985. *Ethnobiological study of Great Smoky Mountains*. Atlanta: National Park Service. 300 pp.

Carson, R. 1962. *Silent spring*. Boston: Houghton Mifflin.

Crocker, T. D., S. B. Kask, and J. F. Shogren. 1998. Valuing ecosystems as natural insurance. http://uwacadweb.edu/Shogren/jaysho/ecosystem.pdf.

Czech, B., and R. Borkhataria. 2001. The relationship of political party affiliation to wildlife conservation attitudes. *Politics and the Life Sciences* 20:3–12.

Davis, J. S., Nelson, A. C., and K. J. Duecker. 1994. The new "burbs": The exurbs and their implications for planning policy. *Economics* 67:15–29.

Dawson, D. 1991. Family structure and children's health and well-being: Data from the 1988 national health interview survey on child health. *Journal of Marriage and the Family* 53:573–84.

DeLone, H., and E. R. McLean. 1990. Information system success: The quest for the dependent variable. *Information Systems Research* 3:60–95.

DiPeso, J. 2002. The environment is bipartisan. *Environmental Quality Management* 11:43–51.

Dowling, T. J. 2000. Reflections on urban sprawl, smart growth, and the fifth amendment. *University of Pennsylvania Law Review* 148:873–87.

Downey, C. J. 1994. *The quality education challenge*. Thousand Oaks, CA: Corwin Press.

Dunlap, E. 1992. *American environmentalism: The US environmental movement 1970–1990*. New York: Taylor & Francis.

Dunlap, R. E., and K. D. Van Liere. 1984. Commitment to the dominant social paradigm and the concern for environmental quality. *Social Science Quarterly* 65:1013–28.

Durbrow, B. R., C. W. Berish, N. B. Burns, J. R. Richardson, N. B. Burns, and T. S. Hoctor. 2002. Protection of ecological function using the southeast ecological framework. Pecora 15/land satellite information IV/ISPRS Commission I/FIEOS Conference Proceedings. http://www.isprs.org/commission1/proceedings02/paper/00073.pdf.

Edwards, P. N. 1999. Global climate science, uncertainty and politics: Data-laden models, model-filtered data. *Science as Culture* 8:437–72.

Ehrmann, J. R., and B. L. Stinson. 1999. Joint fact-finding and the use of technical experts. Pp. 271–94 in *The consensus building handbook: A comprehensive guide to reaching agreement*, ed. L. Susskind, S. McKearn, and J. Thomas-Larmer. London: Sage.

Falk, D. A., and K. E. Holsinger. 1991. *Genetics and conservation of rare plants*. New York: Oxford University Press.

Farley, R. 1991. The new census question about ancestry: What did it tell us? *Demography* 28:411–29

Fiorino, D. J. 1990. Citizen participation and environmental risk: A survey of institutional mechanisms. *Science, Technology and Human Values* 15:226–43.

Fortmann, L., and J. Kusel. 1990. New voices, old beliefs: Forest environmentalism among new and long-standing rural residents. *Rural Sociology* 55:214–32.

Frey, W. H., and A. Speare Jr. 1992. The revival of metropolitan population growth in the United States: An assessment of findings from the 1990 Census. *Population and Development Review* 18:129–46.

Freyfogle, E. T. 1998. Bounded people, boundless land. Pp 15–38 in *Stewardship across boundaries*, ed. R. Knight and P. B. Landres. Washington, DC: Island Press.

Glesne, C., and A. Peskin. 1992. *Becoming qualitative researchers: An introduction*. White Plains, NY: Longman.

Gregory, R. 2000. Using stakeholder values to make smarter environmental decisions. *Environment* 42:34–44.

Hagen, A. N., and K. E. Hodges. 2006. Resolving critical habitat designation failures: Reconciling law, policy, and biology. *Conservation Biology* 20:399–407.
Hand, C., and K. Van Liere. 1984. Religion, mastery-over-nature and environmental concern. *Social Forces* 63:555–75.
Hansis, R. 1997. The Siouxon Valley in the Gifford Pinchot National Forest. Pp. 153–64 in *Public lands management in the west: Citizens, interest groups, and values*, ed. B. S. Steel. Westport, CT: Praeger.
Hargrove, E. C. 1989. *Foundations of environmental ethics*. Englewood Cliffs, NJ: Prentice-Hall.
Hendee, J. C., W. R. Catten, L. D. Marlow, and C. F. Brockman. 1968. Wilderness users in the pacific northwest: Their characteristics, values and management preferences. Research Paper PNW-61. Portland, Oregon: U.S. Department of Agriculture, Forest Service, Pacific Northwest Research Station.
Higdon, M. 2003. Consulting local interests first in planning for a national monument. MS thesis, University of Tennessee, Knoxville.
Howe, C. W. 1987. On the theory of optimal regional development based on an exhaustible resource. *Growth and Change: A Journal of Urban and Regional Policy* 18:53.
Howell, B. 2002. Appalachian culture and environmental planning: Expanding the role of the cultural sciences. Pp. 1–16 in *Culture, environment and conservation in the southern Appalachians*. ed. R. Howell. Urbana: University of Illinois Press.
IPCC (Intergovernmental Panel on Climate Change). 2007. Working Group II: Impacts, adaptation and vulnerability. www.ipcc.ch.
ISCS (Innovations in species conservation: Integrative approaches to address rarity and risk. Innovations in Species Conservation Symposium). 2003. http://outreach.cof.orst.edu/isc/index.htm.
Isenhart, M. W., and M. Spangle. 2000. *Collaborative approaches to resolving conflict*. London: Sage.
Johnson, J., and R. Rasker. 1995. The role of economic and quality of life values in rural business location. *Journal of Rural Studies* 11:405–16.
Jones, R. E., J. M. Fly, J. Tally, and H. K. Cordell. 2003. Green migration into rural America: The new frontier of environmentalism? *Society and Natural Resources* 16:221–38.
Jung, C. 1964. *Civilization in transition*. Princeton: Princeton University Press.
Kanagy, C., C. Humphrey, and G. Firebaugh. 1994. Surging environmentalism: Changing public opinion or changing publics? *Social Science Quarterly* 75:804–19.
Karp, D. G. 1996. Values and their effect on proenvironmental behavior. *Environment and Behavior* 28:111–33.
Kellert, S. R. 2002. Values, ethics, and spiritual and scientific relations to nature. Pp. 29–48 in *The good in nature and humanity: Connecting science, religion, and spirituality with the natural world*, ed. S. R. Kellert and T. J. Farnham. Washington, DC: Island Press.
Kempton, W. M., J. S. Boster, and J. A. Hartley. 1995. *Environmental values in American culture*. Boston: MIT Press.
Kessler, W. B., H. Salwasser, C. W. Cartwright Jr., and J. A. Caplan. 1992. New per-

spectives for sustainable natural resources management. *Ecological Applications* 2:221–25.

Krippendorff, K. 1980. *Content analysis: An introduction to its methodology*. 2nd ed. London: Sage.

Land Trust Alliance. 2005. The conservation campaign in national land trust census report. www.lta.org.

Lieben, I. J. 1997. Political influences on USFWS listing decisions under the ESA: Time to rethink priorities. *Environmental Law* 27:1323–71.

Lubchenco, J. 1998. Entering the century of the environment: A new social contract for science. *Science* 279:491–97.

Lukensmeyer, C., and S. Brigham. 2003. Taking democracy to scale: Creating a town hall meeting for the twenty-first century. *National Civic Review* 91:351–66.

Maehr, D. S, E. D. Land, D. B. Shandle, O. L. Bass, and T. S. Hoctor. 2002. Florida panther dispersal and conservation. *Biological Conservation* 106:187–97.

Maehr, D. S., J. L. Larkin, and J. J. Cox. 2004. Shopping centers as panther habitat: Inferring animal locations from models. *Ecology and Society* 9:art9. www.ecologyandsociety.org/iss2/art9.

Mehta, M. D., and E. Ouellet. 1995. *Environmental sociology*. Concord, Ontario, Canada: Captus Press.

Morse, D. 1988. Air quality in national parks. Natural Resources Report 88-1, no. 3. Denver: U.S. Department of the Interior, National Park Service, Air Quality Division, Natural Resources Program.13 pp.

Nedovic-Budic, Z. 1998. The impact of GIS technology. *Environment and Planning B: Planning and Design* 25:681–92.

———. 1999. Evaluating the effects of GIS technology: Review of methods. *Journal of Planning Literature* 13:284–95.

Nolt, J. E., and J. D. Peine. 1999. The evolution of land use ethics and resource management: Coming full circle. Pp. 41–62 in *Ecosystem management for sustainability: principles and practices illustrated by a regional biosphere reserve cooperative*, ed. J. D. Peine. New York: Lewis.

Noon, B. R., and J. A. Blakesley. 2006. Conservation of the northern spotted owl under the northwest forest plan. *Conservation Biology* 20:288–96.

Noss, R. F., D. D. Murphy, and M. A. O'Connell. 1997. *The science of conservation planning: Habitat conservation under the endangered species act*. Washington, DC: Island Press.

Odum, E. P. 1956. *Fundamentals of ecology*. Philadelphia: Saunders.

Ostermeier, D. M. 1999. The role of institutions in ecosystem management. Pp. 457–74 in *Ecosystem management for sustainability: Principles and practices illustrated by a regional biosphere reserve cooperative*, ed. J. D. Peine. New York: Lewis.

Peine, J., and B. Stephens. 1997. Tuckaleechee cove values questionnaire: Technical report to the Tuckaleechee Cove Advisory Board. Knoxville: University of Tennessee, Graduate School of Planning.

Peine, J. D., R. E. Jones, M. English, and S. W. Wallace. 1999. Contributions of sociology to ecosystem management. Pp. 73–99 in *Integrating social sciences with ecosystem management: Human dimensions in assessment, policy and management*, ed. H. K. Cordell and J. C. Bergstrom. Champaign, IL: Sagamore.

Pendleton, M. R. 1996. Crime, criminals and guns in "natural settings": Exploring the basis for disarming federal rangers. *American Journal of Police* 15:3–25.

Peterman, R., and J. Anderson. 1999. Decision analysis: A method for taking uncertainties into account for risk-based decision-making. *Human and Ecological Risk Assessment* 5:231–44.

Peterson, M. N., S. A. Allison, M. J. Peterson, and R. L. Roel. 2004. A tale of two species: Habitat conservation plans as bounded conflict. *Journal of Wildlife Management* 68:743–61.

Press, D., D. F. Doak, and P. Steinberg. 2006. The role of local government in the conservation of rare species. *Conservation Biology* 10:1538–48.

Project Vote Smart. 2003. *Voter's self-defense manual: 2003 election edition.* Corvallis, OR: Center for National Independence in Politics.

Raines, G. A. 1989. Nonmetropolitan county net migration and industrial differentiation, 1960–1985. Ann Arbor: University Microfilms International.

Rastetter, E. B. 1996. Validating models of ecosystem response to global change. *BioScience* 46:190–98.

Robbins, S. M. 2004. Consensus and conflict: Interest group strategy in the policy process. Diss., State University of New York at Stony Brook.

Rudzitis, G., and H. E. Johansen. 1991. How important is wilderness? Results from a United States survey. *Environmental Management* 15:227–33.

Ruud, G., and J. Sprague. 2000. Can't see the [old growth] forest for the logs: Dialectical tensions in the interpretive practices of environmentalists and loggers. *Communication Reports* 13:55–65.

Scoloff, B. 2006. Humans and panthers not mixing well in South Florida. *USA Today*, December 28.

Scott, D., and F. Willits. 1994. Environmental attitudes and behavior: A Pennsylvania survey. *Environment and Behavior* 26:239–60.

Shannon, M. A. 1991. Is American society organized to sustain forest ecosystems? Proceedings of the 1991 National Society of American Foresters Convention. Bethesda, MD.

Soukup, M. 2006. Remarks at the Appalachian Trail MEGA-transect symposium. http://apptrail.nbii.gov/portal/server.pt (requires membership to this Web site).

Stankey, G. H. 1973. Visitor perception of wilderness recreation carrying capacity. Research Paper INT-142. U.S. Department of Agriculture, Forest Service.

Stankey, G. H., and B. Schindler. 2006. Formation of social acceptability judgments and their implications for management of rare and little-known species. *Conservation Biology* 20:28–37.

Stinchcombe, J. 2000. U.S. endangered species management: The influence of politics. *Endangered Species Update* 17:118–21.

Stinner, W. F., and M. Van Loon. 1992. Community size preference status, community satisfaction and migration intentions. *Population and Environment* 14:177–95.

Susskind, L., and S. T. McCreary. 1985. Techniques for resolving coastal resource management disputes through mediation. *Journal of the American Planning Association* 51:365–74.

Susskind, L., S. McKearnan, and J. Thoms-Larmer. 2000. *The consensus building handbook: A comprehensive guide to reaching agreement*. Thousand Oaks, CA: Sage.

Swan, J. 1989. Sacred sites: Cultural values and management issues. *Humanity and Society* 13:442–56.

Swedlow, H. B. 2002. Scientists, judges, and spotted owls: Policymakers in the Pacific Northwest. PhD diss., University of California, Berkeley. *Dissertation Abstracts International*, publ. nr. AAT 3063567, DAI-A, 63/09:3343.

Szasz, A. 1997. Environmental inequalities: Literature review and proposals for new directions in research and theory. *Current Sociology* 45:99–120.

Taylor, R. 1994. Qualitative research. Pp. 265–79 in *Mass Communication Research*, ed. M. Singletary. New York: Longman.

Taylor, S. M., and V. A. Konrad. 1980. Scaling dispositions toward the past. *Environment and Behavior* 12:283–307.

Thomas, T. W. , E. D. Forsman, J. B. Lint, E. C. Meslow, B. R. Noon, and J. Verner. 1990. A conservation strategy for the northern spotted owl. Interagency Scientific Committee to address the conservation of the northern spotted owl. Portland, OR: U.S. Department of Agriculture, Forest Service, Pacific Northwest Research Station.

Thompson, P. 2000. *The voice of the past: Oral history*. New York: Oxford University Press.

Tindall, D. 1994. Collective action in the rain forest: Personal networks, identity and participation in the Vancouver Island wilderness preservation movement. PhD diss., Department of Sociology, University of Toronto.

Vickerman, S. 1999. A state model for implementing stewardship incentives to conserve biodiversity and endangered species. *Science of the Total Environment* 24:41–50.

Vining, J. 1992. Environmental emotions and decisions: A comparison of the responses of forest managers, an environmental group and the public. *Environment and Behavior* 24:3–34.

Vosler, N. R. 1996. *New approaches to family practice*. Thousand Oaks, CA: Sage.

Wear, D. N., and P. Bolstad. 2004. Land-use changes in southern Appalachian landscapes: Spatial analysis and forecast evaluation. *Ecosystems* 1:575–94.

Western Governors' Association. 2002. Enlibra. www.westgov.org/wga/initiatives/enlibra/default.htm.

Wilcox, C. V. and B. D. Elderd. 2003. The endangered species act petitioning process: Successes and failures. *Society and Natural Resources* 16:551–59.

Williams, D. R., M. E. Patterson, J. W. Roggenbuck, and A. E. Watson. 1992. Beyond the commodity metaphor: Examining emotional and symbolic attachment to place. *Leisure Science* 14:29–46.

Wilson, E. O. 1999. *The diversity of life*. New York: W. W. Norton.

Wilson, W. V. 1987. *The truly disadvantaged: The inner city, the underclass and public policy*. Chicago: University of Chicago Press.

Wondolleck, J. M., and S. L. Yaffee. 2000. *Making collaboration work*. Washington DC: Island Press.

Wright, J. H. 1994. Designing and applying conservation easements. *Journal of the American Planning Association* 60:330–38.
Yaffee, S. L. 1994. *The wisdom of the spotted owl: Policy lessons for a new century*. Washington, DC: Island Press.
———. 1995. Lessons about leadership from the spotted owl controversy. *Natural Resources Journal* 35:381–412.
Yaffee, S. L., and J. M. Wondolleck. 2003. Collaborative ecosystem planning processes in the United States: Evolution and challenges. *Environments* 31:59–72.

10

Economic Considerations

Richard L. Johnson, Cindy S. Swanson, and Aaron J. Douglas

This chapter describes methods for and gives examples of estimating the economic consequences and values of conserving rare or little-known (RLK) species. In the first section, we describe the discipline of economics in terms of its goals and self-imposed limits. Economics does not intend to impose moral or ethical judgments. Positive economics describes "what is," whereas normative economics shows how alternate economic policies can affect outcomes. Next, we describe the evolving political economic framework within which decisions are made in the United States. Finally, we discuss economic measurement tools and their application to RLK species.

The Essential Economic Problem

Economic growth, as defined in standard economic textbooks, is an increase in the production of goods and services. Economic growth occurs when there is an increase in the multiplied product of population and per capita consumption. As human populations increase through fertility and immigration, the economy grows; and as natural resource use increases through agriculture, extraction, and the manufacturing and service sectors, the economy grows. At the same time, limited natural resource supplies are diminished and by-products that can degrade the remaining resources are produced (Meadows et al. 1972).

Economic growth has been a primary, perennial goal of American government and society. Some economists believe there is no limit to economic growth, particularly given that technological progress allows use of fewer resources for greater returns. However, based on established principles of ecology and physics, we believe there is a limit to economic growth. Though technological progress can reduce the ecological impacts of economic growth, it can also exacerbate ecological problems and therefore is not sound reasoning to support unlimited economic growth (Krutilla 1967; Costanza et al. 1997).

The mechanisms of economic growth may impact the environment through pollution, overexploitation, habitat loss and fragmentation, and climate change. Excessive consumption results in degradation of resources as evidenced by the decline of some fisheries, the growing number of threatened and endangered species, and the loss of open space. Decline in environmental quality and loss of biological diversity are not easily assessed in economic terms. Current economic policies do not usually take into account the value of natural resources in contributing to biological processes such as flood control by wetlands, climate regulation by forests and open space, and pollination by insects, birds, and mammals (Constanza et al. 1997). Furthermore, economic health is not measured accurately by gross domestic product (GDP) alone since GDP does not incorporate the condition and sustainability of natural resources.

As a result of the protection of personal property rights and allowing a competitive free market to operate unregulated, negative impacts on the environment can result. The concept of a competitive free market operating without interference should result in meeting consumer needs at the lowest price possible, and firms receive reasonable return or profit from their investment. If markets work perfectly, well-being is maximized because no reallocation exists that could yield a net gain to society (consumers or producers). Unfortunately, a private, competitive free market can only result in the efficient allocation of resources if the production of goods and services does not produce unwanted effects on others (economists refer to this as an externality; see Bator 1958; Arrow 1965; Just et al. 1982). Examples include water/air pollution downstream/downwind from a factory, noise pollution resulting from air traffic over a wilderness, or loss of critical fish and wildlife habitat from the construction of dams and roads.

The concept of a private, competitive free market is also based on the belief that the market price is a reflection of the value of a good. It shows how much consumers are willing to pay for another unit and how much sellers are willing to sell it for. Resources that are nonrival (consumption by one individual does not effect the amount available for consumption by others, e.g., scenic view, hiking in an uncongested wilderness/park/forest) and nonexclusive (property rights cannot be enforced, e.g., ocean fisheries, migrating wildlife) are not efficiently allocated in a market, and therefore a market failure results. We are again faced with an allocation of resources that does not result in societal benefit maximization since nonrival/nonexclusive goods do not have a market price (because they cannot be owned or sold), and therefore society has inferred a value of zero (Walsh et al. 1984; Rubin et al. 1991).

These sources of nonmarket failure are not fatal, but the structure of the current concept of private property rights and a free, competitive market will result in continued high levels of pollution and the value of "public goods," including RLK species, not being recognized.

Economic Criteria Are Not the Only Decision Criteria

Economic issues are often a critical element in species conservation because preservation of a species often requires some people, somewhere to constrain their behavior (i.e., clearing forests for farming, building houses, draining wetlands), incur additional costs (clean up discharge water, put in smokestack scrubbers), or forgo activities they view as beneficial (building along a river, building privacy fencing).

Of course, there are many other important criteria that society uses when making important policy decisions concerning species conservation. Ethical considerations about what we as a society feel is right or just correctly play a legitimate role. The U.S. Endangered Species Act requires agencies to ignore costs in making listing decisions up to the limits set by the Endangered Species Committee (Freeman 2003), but costs are considered when critical habitat designation decisions are made (Schamberger et al. 1992).

Althought there are clearly costs associated with species preservation, there are also substantial benefits. Many of these benefits are from the public-good nature of the resources and thus are nonmarket in nature, and these benefits often do not show up in the company's bottom-line profit (Arrow 1965; Just et al. 1982). A thesis of this chapter is that we should explicitly recognize the economic costs and benefits of species preservation and use that as one criterion in our decision making. Economic benefits and costs are certainly not the only thing that matters, but neither can they be ignored.

Developing a Framework for Analysis

A description of the measurement problem for the economic consequences and benefits of conservation of RLK species requires specification of a framework for analysis. At issue is the organization of human activity for the purpose of production, distribution, and consumption of goods and services given unlimited wants and limited means. This essential economic problem exists across all types of cultures, political organizations, or economic systems. The exchange of goods and services among individuals increases people's material welfare. Often, however, virtually all developed nations, including the United States, have chosen some form of constrained market capitalism to address concerns of externalities and nonrival/nonexclusive goods and services. Constrained market capitalism eschews both laissez-faire and centralized planning for a market-based system that is modified to achieve public goals. The framework for analysis is the solution to the essential economic problem provided by market capitalism with government intervention (Schumpeter 1942). Familiar forms of intervention include fiscal and monetary macroeconomic policy, efforts by government to calculate values and beliefs and modify behavior, direct government operation and control of property, the impositions of subsidies and taxes, government regulation, access charges for the use of public property, property zoning, the institution of tradable pollution permits, other redefining of property rights, and rationing.

The Detached and Human-Centered Nature of Positive Economics

The positive (as opposed to normative) economics methodology is intentionally value-free and anthropocentric (Ehrlich and Ehrlich 1996). Value in economic theory, and in market capitalism in particular, derives from behavior reflecting the cumulative tastes and preferences of human beings

(Henderson and Quandt 1980). If individuals believe certain things have intrinsic value, then those values will be counted, but intrinsic values derived from sources other than human wants are excluded. This stance is in contrast to some other disciplines, which may wish to influence human tastes and preferences or beliefs. Outside the economics discipline, biodiversity, ecosystem health, environmental justice, and sustainability are often discussed both in terms of their effect on humans and as purely environmental issues that humans have a duty to resolve. Economics becomes prescriptive rather than descriptive when it is used to make judgments as are found in feasibility studies and benefit–cost analyses. The key factor is that economic prescription is based on measures of the wants, desires, and beliefs of the citizenry and avoids interposition of the beliefs of economists.

Pure Market Capitalism

Prior to discussing the widely perceived need to constrain pure market capitalism we must define it. The ideal market capitalism model assumes that all business firms and consuming units are sufficiently small and independent and that they have no perceptible influence on market prices. Consumers are assumed to have unlimited wants for goods and services, and firms work to satisfy customers. Firms not making economic profits run out of money and are automatically terminated. Survival of the fittest and most progressive firms ensures maximum consumer satisfaction. At its best, pure capitalism embodies the flexibility and incentives to allow constant growth and change. Only entrepreneurs who consistently satisfy consumers' wants and needs accumulate wealth. The ideal model achieves the highest possible level of economic efficiency from any given set of resources. Market prices measure the marginal benefits to individuals and society. The marginal benefits (prices) from the consumption of goods and services are equal to the marginal costs of production so that any reallocation of production and distribution of consumption would make society worse off in terms of the marketable goods (Just et al. 1982). The coordination of production, distribution, and consumption activities among producers and consumers is accomplished by trading in the impersonal, value-free, market place. The unfathomable number of decisions and resulting transactions in a nation's economy occur without direction from any centralized authority to achieve a maximum standard of living (gross domestic product) for society.

Criticisms of Market Capitalism Leading to Constraints (Government Intervention)

Many of the assumptions embedded in the ideal market capitalism model are inconsistent with reality and with the conservation of RLK species. Once one moves beyond the ideal market capitalism model, the analysis leaves the province of pure economic theory and includes political science, law, and ethics.

A key requirement for market capitalism is private ownership of the means of production, including one's own labor. Private ownership ensures, according to the model, that real property will be enhanced and conserved over time rather than exploited and discarded. Unfortunately, there are situations where exploit and discard may be the superior policy for a firm. Thus minimal acceptable rules of conduct are required. More important, the firm is a single-purpose entity focused on the goal of maximization that ignores holistic management of property. In a market economy, property is conserved and enhanced only for purposes of profit maximization, not for the purpose of retaining and enhancing its many desirable attributes. In particular, the market system fails to recognize that the consequences of individual actions by the few are far different from the collective actions of many (Bator 1958; Arrow and Hahn 1986). The enforcement of minimal acceptable standards of conduct can provide some protection for long-lived, extra-market property, but it must take account of the cumulative effects over individuals, space, and time. For example, international sanctions prohibiting sales and transport of endangered species may be unenforceable because the sellers are in underdeveloped nations with limited law enforcement capabilities, and the buyers are in states that defy a wide array of international regulations (Douglas 1990). Thus prohibitions against trade in endangered species may be effectively enforced in the Western industrial democracies but fail to prevent the extirpation of critical habitats and endangered species.

An obvious weakness in the model is the incorrect assumption of small, competitive firms possessing no market power and the incorrect presumption that wealth accumulation is always an indicator of superior performance in want satisfaction (Henderson and Quandt 1980; Arrow and Hahn 1986). Apprehensive managers have strong incentives to monopolize markets and subvert government policies (such as environmental protection) that threaten the continued existence of their firms. Thus the powerful profit incentive can be counterproductive if antitrust enforcement is weak and regulatory agencies and/or the political systems are susceptible to coercion or bribery. The invisible hand of the market only functions to increase well-being of the citizenry if a centralized authority prevents the accumulation of market power by individual firms, trade associations, unions, consumer groups, and the like. The ability of society to constrain the structure, conduct, and performance of these economic organizations depends on the viability of political institutions. For example, the interplay of political power among the executive, legislative, and judicial branches of government may play a role in limiting the accumulation of economic power. Moreover, certain types of production incur continually declining average production costs with increased output (economies of scale) so that they are inherently huge and monopolistic and cannot fit the atomistic mold. Some method of dealing with natural monopolies, such as regulation or government ownership, must be appended to the market capitalism model.

A second weakness in the market capitalism model is the presumption that increased quantities and qualities of marketed goods and services are sufficient to satisfy consumer wants. The model completely ignores common-pool resources. To be transacted in a market, a good or service must have the attribute of exclusivity. If consumers cannot be excluded from consuming a good, a price cannot be charged and therefore no funds will be made available for production in a market system. Firms emerge to manage resources for the production of profits creating market goods. Few firms emerge to manage common-pool resources. This effect has been called the tragedy of the commons (Hardin 1968). An example of a nonmarket good is the protection afforded by national defense. While some public-spirited citizens might feel obligated to pay for the common defense, many free-riders would withhold their funds. Some method of collecting payment from all members of society is required to adequately and fairly fund national defense. Other nonmarket goods include law enforcement, protection from contagious diseases, unemployment insurance, and environmental protection, including air and water quality and RLK species.

A third weakness in the ideal market capitalism model is the assumption of no negative externalities. A negative externality is said to occur if the physical production or consumption of one good creates physical damage to some other production or consumption process. Almost all consumption or production creates some physical damage to other consumption or production. Population pressure can exacerbate and intensify the loss of forest and rangeland habitat (Spurr and Barnes 1980; Caufield 1986). Squatting on undeveloped land is usually highly regulated in contemporary Western societies, but urban population pressures may generate uncontrolled use of undeveloped land tracts in many parts of tropical regions such as Brazil or Malaysia (Caufield 1986). Unregulated tenancy in fragile rain forest ecosystems may cause the loss of plant cover, in turn causing soil erosion, stream contamination, and the loss of critical habitat and productive cropland (Spurr and Barnes 1980; Caufield 1986; Perrings et al. 1995).

One way to reduce these damages is to internalize the cost by creating a tax on the offending consumer or producer group to recoup damages created. The offending group then has an increased incentive to reduce damages until the efficient level of damages is achieved. Both market and nonmarket goods can be protected from negative externalities through enforcement of environmental laws and regulations. An alternative approach to internalizing costs is provided by the Coase theorem (Coase 1960). Coase showed that if property rights are clearly defined in law, then bargaining between the actors could also internalize externalities through voluntary negotiation. Again, property rights must be established for both market and nonmarket goods, and existing firms have a strong incentive to oppose the creation of new property rights for nonmarket goods. Further, for the Coase theorem to yield an optimum internalization of the externalities, the transaction costs must be low. This only occurs when there are a few polluters and, more importantly, only a few recipients. Thus the Coase theorem is of limited applicability for most environmental problems, including the protection of most RLK species.

The presence of negative externalities and public concern for nonmarket goods tend to increase as population density increases. Increased recognition of the limits of market capitalism means the public sector component of the economy has expanded greatly over time. The danger that the public sector will be legislated into areas best served by the private sector is ever present.

Property Rights for Environmental Resources

Protection of environmental common-pool resources via the legal system in the United States has gained favor since the enactment of the Endangered Species Act (16 U.S.C. § 1531). Federal laws evolved over time from the establishment of national forests/grasslands, parks and monuments, wildlife refuges, and wilderness areas to the inclusion of protection for environmental resources within the requirements of the federal land management agencies. However, this early legislation applied only to public lands. Actual legal standing for RLK species was more directly established by the Endangered Species Act, which applies differentially to both public and private lands. The wild and scenic river designation has value in and of itself (Walsh et al. 1984) but can also affect the use of private lands and thus afford protection of RLK species. Thus a form of property rights does now exist in the United States for certain environmental common-pool resources. Many nonprofit organizations are dedicated to the protection or preservation of various aspects of the natural environment. Thus the actions suggested by the Coase theorem have come into fruition for the environmental resources as a result of property rights having been more clearly defined in law (for example, ocean commercial fisheries now have equipment, harvest, and seasonal restrictions to protect a common property resource).

Of the nearly 2.3 billion acres of land in the United States, approximately 650 million acres (28% of the land base) are owned by the federal government. Four agencies administer 96% of this federal land (the U.S. Department of Agriculture Forest Service and the U.S. Department of the Interior Bureau of Land Management, Fish and Wildlife Service, and National Park Service) (U.S. Bureau of Census 1992). Laws have been enacted to address/assign property rights of environmental resources within these agencies. The Forest Service and the Bureau of Land Management are both multiple-use, sustained-yield management agencies following the National Forest Management Act and Federal Land Policy and Management Act, respectively. The U.S. Fish and Wildlife Service, under the Refuge Improvement Act of 1977, emphasizes preservation of fish, wildlife, and their habitats. The National Park Service has the most preservation-oriented legislation in its Organic Act of 1916. Box 10.1 provides a list of the relevant environmental statutes for the management of federal agencies (also see chap. 2).

Box 10.1. Relevant federal statutes.

A. Forest Service Related Statutes:

Multiple-Use Sustained-Yield Act of 1960: Act of June 12, 1960; P.L. No. 86-517, 75 Stat. 215. 16 U.S.C. A7528, et seq.

Rangeland Renewable Resources Planning Act of 1974: Act of August 17, 1974; P.L. No. 93-378, 88 Stat. 476. 16 U.S.C. A71600, et seq.

National Forest Management Act of 1976: Act of October 22, 1976; P.L. No. 94-588, 90 Stat. 2949.

Forest and Rangeland Renewable Resources Research Act of 1978: Act of June 30, 1978; P.L. No. 95-307, 92 Stat. 353. 16 U.S.C. A71641, et seq.

Cooperative Forestry Assistance Act of 1978: Act of July 1, 1978; P.L. No. 95-313, 92 Stat. 365.16 U.S.C. A72101, et seq.

B. Bureau of Land Management Related Statutes:

Mineral Leasing Act of 1920: Act of February 25, 1920; ch. 85, 41 Stat. 437. 30 U.S.C. A7181, et seq.

Taylor Grazing Act of 1934: Act of June 28, 1934; ch. 865, 48 Stat. 1269. 43 U.S.C. A7315, et seq.

Mineral Leasing Act for Acquired Land: Act of August 7, 1947; ch. 513, 61 Stat. 913; 30 U.S.C. A7351-359.

Materials Disposal Act: Act of July 31, 1947; ch. 406, 61 Stat. 681. 30 U.S.C. A7601, et seq.

Wild Horses and Burros Act of 1971: Act of December 15, 1971; P.L. No. 92-195, 85 Stat. 649. 16 U.S.C. A71331, et seq.

Federal Land Policy and Management Act of 1976: Act of October 21, 1976; P.L. No. 94-579, 90 Stat. 2744. 43 U.S.C. A71701, et seq.

Public Rangelands Improvement Act of 1978: Act of October 25, 1978; P.L 95-514, 92 Stat. 1803.

Alaska National Interest Lands Conservation Act of 1980: Act of December 2, 1980; P.L. No. 96-487, 94 Stat. 2371. 16 U.S.C. A73101, et seq.

C. Fish and Wildlife Service Related Statutes:

National Wildlife Refuge System Administration Act of 1966: Act of October 15, 1966; P.L. No. 90-404, 80 Stat. 927. 16 U.S.C. A7668dd-668ee.

Alaska National Interest Lands Conservation Act of 1980: Act of December 2, 1980; P.L. No. 96-487, 94 Stat. 2371. 16 U.S.C. A73101, et seq.

Fish and Wildlife Act of 1956: Act of August 8, 1956; ch. 1036, 70 Stat. 1120. 16 U.S.C. A7742a, et seq.

San Francisco Bay National Wildlife Refuge (1972): Act of June 30, 1972; P.L. No. 92-330, 86 Stat. 399. This Act is a typical statute establishing a refuge by law. 16 U.S.C.

Box 10.1. *Continued*

D. National Park Service Related Statutes:

Preservation of American Antiquities: Act of June 8, 1906; ch. 3060, 34 Stat. 225. 16 U.S.C. A7431-433.

Rights-of-Way Through Parks and Reservations: Act of March 4, 1911; ch. 238, 36 Stat. 1253, 16 U.S.C. A75.

The National Park Service Organic Act of 1916: Act of August 25, 1916; ch. 408, 39 Stat. 535. 16 U.S.C. A71-4.

Land and Water Conservation Fund Act: Act of Sept. 3, 1964; P.L No. 88-578, 78 Stat. 897. 16U.S.C. A71, et seq.

Concession Policy of the National Park Service: Act of October 9, 1965; P.L. No. 89-249, 79 Stat. 969. 16 U.S.C. A720-20g.

Mining in National Parks: Act of September 28, 1976; P.L. No. 94-429, 90 Stat. 1342. 16 U.S.C. A71901-1912.

In addition, there are a number of laws which establish specific units of the National Park System. For example see: Yellowstone National Park Act: R.S. 2474, derived from Act of March 1, 1872; ch.24, 17 Stat. 32. 16 U.S.C. A721, et seq.

E. Special Management Systems on Public Lands:

Outdoor Recreation Act of 1963; P.L. No. 88-29. 16 U.S.C. A74601

Wilderness Act: Act of September 3, 1964; P.L. No. 88-577, 78 Stat. 890. 16 U.S.C. A71131, et seq.

Wild and Scenic Rivers Act: Act of October 2, 1968; P.L. No. 90-542, 82 Stat. 906. 16 U.S.C.A71271, et seq.

National Trails System Act: Act of October 2, 1968; P.L. No 90-543, 82 Stat. 919. 16 U.S.C. A71241, et seq., as amended August 1992.

National Parks and Recreation Act of 1978: Act of November 10, 1978; P.L. No. 95-625, 92 Stat. 3467.

Alaska National Interest Lands Conservation Act of 1980; P.L. No. 96-487, 94 Stat. 2371. U.S.C. A73210.

California Desert Protection Act of 1994; P.L. No. 103-433, 108 Stat. 4471.

The Economics of Rarity

Rarity of species implies to an economist a relative scarcity, and this leads to high values at the margin for preserving another individual in the population. Extreme cases of scarcity, such as 12 California condors or a handful of returning sockeye salmon to Redfish Lake in Idaho, may imply astronomically high values. This is because, as the quantity gets smaller, the value of those remaining individuals gets larger (Randall and Stoll 1983). As the population approaches the last breeding pair the value could approach infinity depending on the shape of the demand curve for the species.

Unfortunately, because the costs of preservation are often the only economic data available, the decision on species protection may be based on costs rather than benefits. This is an inappropriate use of economics and may result in only "low cost" species being preserved (Swanson et al. 1994).

Rarity may increase the likelihood of extinction (chap. 3). Extinction, being irreversible, may generate a supply-side option value for individuals (Bishop 1982). That is, individuals who are certain they wish to view a species in its native habitat in the future may be willing to pay a premium over and above the species' expected benefits to maintain that opportunity in the face of possible uncertainty about whether the species will still exist. A quasi option value to resource managers arises from the value of new knowledge about the benefits of preservation that can be gained by postponing an irreversible action (Arrow and Fisher 1974; Fisher and Hanemann 1986; Freeman 2003). That is, with extinction being irreversible, the benefits of development must be greater than the benefits of preservation by some additional amount, to ensure that development is truly the optimal action. This requirement arises because preservation in this period maintains the future option to preserve or develop in the next period. The decision to develop in the current time period precludes the opportunity to preserve in the future time period. Thus, to be optimal, society must be extra certain that the benefits of development outweigh the benefits of preservation today and all future time periods. This extra certainty can be thought of as the quasi option value premium.

A great deal has been written about how human well-being depends on ecosystem functions (e.g., nutrient cycling) that provide ecosystem services of value to people (Costanza et al. 1997; Daily 1997). Ecosystem services include clean water, wildlife, and recreation resources. Even though a species may be rare or occur in small numbers it can exert an influence or play a critical role in the function of an ecosystem and therefore indirectly be of high value. Maintaining species helps to maintain ecosystem stability, thus reducing the variance in flow of ecosystem services. Maintaining RLK species acts like an insurance policy to reduce the potential for large losses in the future (see also chaps. 3, 6, and 7).

Whereas anthropocentric values are human based, human values can go beyond just the instrumental use values such as on-site viewing or medicinal use (see also chap. 9). Krutilla (1967) first suggested that people might derive satisfaction or utility from just knowing that a species exists in situ without actual use. He further suggested that people today might also pay to protect a species today from irreversible loss so as to bequeath that species to future generations. That is, people receive satisfaction knowing that preservation today provides the species to their children, grandchildren, and future generations. Collectively, these existence and bequest values are now known as passive use values.

Evaluating the Economic Consequences of Conserving RLK Species

Economists measure the economic impacts, benefits, and costs of policy changes by using market prices if recorded prices are reliable guides to the relevant social costs and benefits. For reasons discussed previously, market prices do not reflect policy-relevant social marginal values for a host of goods and services, including preservation of RLK species. This can lead to the inappropriate conclusion that they have little or no value (Just et al. 1982; Rubin et al. 1991; Hagen et al. 1992). Thus alternate methods and techniques have been and are being developed to measure costs and benefits for nonmarket goods such as RLK species. Below is a discussion of some of the question(s) that the procedures address and the types of methods and models that are utilized by various approaches.

Economic Impact Analysis

An economic impact analysis (also known as a regional economic analysis) traces the flows of spending in order to identify and quantify changes in sales, tax revenues, personal income, and jobs caused by changes in final demand activity. Final demand is the last point of purchase and includes exports from a region such as tourism expenditures, federal spending, or regional investment. Wassily Leontief's (1936, 1941, 1966, 1974) famed input–output (I-O) model is the basic analytical tool used to model these impacts. The basic assumptions of the I-O model are (1) there is a fixed group of final demanders; (2) there is a group of industries that use the outputs of other industries to supply a product or set of products to final demanders; (3) all production processes are linear, hence a fixed positive fraction called a "production coefficient" captures the contribution of the input industry i to the output industry j; (4) all flows are measured in dollars; and (5) supply always equals demand. Although inputs and outputs are measured in dollars, the model can be adapted to show the creation of additional by-products such as smoke or congestion as an economy expands. Or dollars can be replaced by water or some other relevant unit of account (Miller and Blair 1985; Lave, et al. 1995; Hendrickson, et al. 1998).

The principal empirical data sources for economic impact measurement are: business or visitor spending surveys and secondary data from government economic statistics, the economic base model, and I-O models and multipliers. The I-O model is a general equilibrium model of an economy, which takes account of the interrelatedness of production across industries and government. From a national accounting stance, a positive economic impact in one region will usually be offset by losses in other regions so that the impacts wash out (Miller and Blair 1985). Because the economic impact estimate only shows changes in transactions, it does not address the issue of enhanced public welfare. However, I-O analysis can be extended through optimization techniques to forecast the environmental effects of economic changes and the market and nonmarket outcomes of alternative policies.

Hunting- and fishing-related tourism is the best example of how an I-O model can be applied to a nonmarket resource. By tracking expenditures in a local economy, the importance of fish and wildlife can be demonstrated. In the case of RLK species, if a species (like the California condor) attracts travel to a location, then an I-O model can be applied.

Benefit–Cost Analysis

If some allocations of inputs generate a higher social value than alternative uses of the same inputs, then the allocation in question is said to be economically efficient (Samuelson 1954). A benefit–cost analysis estimates the relative economic efficiency of alternative policies by comparing marginal benefits and marginal costs (Mishan 1976). The goal of benefit–cost analysis is to identify the policy with the highest benefit to cost ratio (Mishan 1976). Benefit–cost analysis makes use of a wide range of methods for estimating values of nonmarket goods and services, such as the travel cost method (TCM) and contingent valuation method (CVM) (Just et al. 1982). Benefit–cost analysis compares the present value of the benefits of preservation to the present value of the costs. Of course it is important that all benefits and all costs be included and measured in commensurate monetary terms (e.g., in nominal or constant dollars).

Quantifying the benefits of RLK species preservation with market prices is quite difficult. In part, this is due to our limited biological and ecological knowledge about the role the species play in providing ecosystem services and because of the lack of knowledge the public has about these species. Thus, although benefits are often underestimated, market prices can often be used to estimate the costs of preservation and identify the agents who bear the costs (Just et al. 1982; Boyle and Bishop 1987). Although these costs are frequently overstated by including employment issues—which are simply interregional transfers—the costs are often readily available.

Given the likelihood that benefits will be underestimated and costs will be fully or perhaps overestimated, it is often difficult to justify species protection on benefit–cost grounds. Thus there are few examples of benefit–cost analyses that indicate that species preservation is economically efficient (e.g., see the northern spotted owl, Rubin et al. 1991; Hagen et al. 1992; Douglas and Taylor 1999; see Olsen et al. 1991 for the opposite conclusion). Opponents of preservation advocate benefit–cost analysis in listing decisions because the difficulty of capturing the benefits is well known (Just et al. 1982).

Cost-Effectiveness

One commonly used economic tool is cost-effectiveness analysis (CEA). The CEA approach is to evaluate the cost of alternative preservation scenarios to determine the least expensive way to reach the policy goal. Thus, if maintaining a viable population is the goal, the CEA evaluates and ranks the costs for maintaining and/or purchasing land tracts that provide the requisite habitat. Resource managers might prefer to purchase the cheapest land, or to purchase expensive land that yields more habitat value per dollar. There are a variety of optimization routines that can be used to search for the combinations of habitats or policies that reach a specific level of species protection at minimum cost. This technique has been applied to trace out the minimum costs of increasing the certainty of survival of northern spotted owls in the Pacific Northwest (Montgomery and Brown 1992, Montgomery et al. 1994; Hof and Raphael 1997).

Safe Minimum Standard

The safe minimum standard (SMS) takes cost-effectiveness analysis a step further by providing a guide to decision making (Bishop 1978; Ready and Bishop 1991). SMS states that species protection is warranted unless a cost-effectiveness study indicates that the social costs are unacceptably large. Of course, the numerical values that result in an unacceptably large cost depend on one's perspective. Note that it is possible to compensate the agents incurring the costs. For example, under the *Forest Plan for a Sustainable Economy and a Sustainable Environment* (Clinton and Gore 1993), the Northwest Economic Adjustment Initiative (NEAI) made approximately $1.2 billion available in Oregon, Washington, and Northern California to assist with economic transitions caused by reducing timber harvest to preserve northern spotted owl habitat. The NEAI was intended to assist workers and their families, businesses, and communities with adjusting and preparing for a more sustainable future (Tuchmann et al. 1996). In this way, some of the costs of protecting spotted owls and their ecosystem-conserving habitat were spread across many beneficiaries.

Total Economic Value as an Economic Framework

An attempt to tie together the direct and passive use values into an overall economic framework was first proposed by Randall and Stoll (1983) and later refined by Randall (1991). We adopt a broad definition for passive use value for RLK species where individuals who do not intend to make immediate use of such resources would nevertheless feel a loss if they were to disappear. Our definition includes existence value, option value, quasi option value, and bequest value, previously defined (also see Perrings et al. 1995). The total social benefit is the sum of the recreation use value and passive use. However, from a policy perspective, all that matters is the magnitude of "total economic value," not the magnitudes of the constituents. Thus Randall (1991) suggested that total economic value be elicited directly from respondents.

For economists, the maximum amount of other goods and services an individual is willing (and able) to sacrifice to gain another unit of something else is a fundamental measure of economic value. The preservation value of a species to an individual who has no contractual ownership or proprietary rights to the species is the individual's maximum willingness to pay (Hicks 1943; Just et al. 1982; Mitchell and Carson 1989). If, however, the individual has contractual ownership rights with respect to a species, then the appropriate value construct is the minimum amount of additional goods and services they would accept to give up continued existence of a species. This benefit measure is known as willingness to accept (WTA). Given that WTA is not strictly bounded by income, WTA usually exceeds willingness to pay (WTP) (Hicks 1943; Just et al. 1982; Mitchell and Carson 1989; Hanemann 1991). If the services provided by a species have poor substitutes, WTA can greatly exceed WTP (Hanemann 1991). The appropriate measure for any benefit–cost framework depends on the assignment of property rights (Hicks 1943; Mitchell and Carson 1989; Hanemann 1991). Many economists prefer to use WTP in situations in which property rights assignment is ambiguous, because WTP is more conservative (Mitchell and Carson 1989).

Measuring Nonmarket Benefits

The measurement of nonmarket benefits often entails asking survey respondents questions about their monetary valuations because appropriate market prices do not exist. The most common survey techniques used to measure nonmarket benefits can be classified into two major groups including direct and indirect measurement. Direct measurement can be further classified into direct elicitation of value or WTP, and direct elicitation of behavior. The former is called contingent value and the latter is called contingent behavior (Mitchell and Carson 1989). The Contingent Value Method (CVM) is based on survey data (Mitchell and Carson 1989). CVM studies simulate referendum outcomes for tax-based expenditures. The CVM is used to value unpriced commodities such as logging restrictions on the publicly owned Pacific Northwest forests in order to protect habitat for rare avian species (Rubin et al. 1991; Hagen et al. 1992).

The outcome of the survey referendum simulation may be punitive penalties or costs incurred by private agents (Loomis et al. 1990; Arrow et al. 1995; Welsh et al. 1995). Several economists have argued that survey-based CVM simulated referenda are not a reliable guide to policy actions in such cases (Just et al. 1982; Kahneman and Knetsch 1992; Perrings et al. 1995). Davis (1963) was perhaps the first economist to apply the CVM. Randall et al. (1974) and Carson et al. (1992, 2003) pioneered the use of CVM surveys to measure existence benefits. Many CVM-related issues are discussed in Mitchell and Carson (1989), Arrow et al. (1995), and Bateman and Willis (1999).

The primary indirect survey method is known as the travel cost method (TCM) and is based on respondents' reports of their behavior, or revealed preferences. Survey participants are drawn only from people who use the amenity, such as park visitors or on-site fishermen. Actual expenditures incurred to participate in the activity reflect individual willingness to pay. Because these expenses vary by distance from the site, some users gain consumer's surplus by living nearby. Assuming the utility of all visitors to the site is the same, the sum of consumer's surplus (i.e., benefits above costs incurred) reflects the net willingness to pay for using the site. Hotelling (1949) sketched the TCM concept in a letter. TCM approaches and issues are discussed in Clawson and Knetsch (1966), Just et al. (1982), Creel and Loomis (1990), and Ward and Beal (2000). When direct survey is precluded by institutional constraints such as time or money limitations, a benefit transfer technique may be applied. When benefit transfer is done using meta-analysis (Loomis and White 1996a; Loomis 1999), the researcher finds values from existing studies of similar nonmarket goods with similar market conditions and then adjusts the estimate for a set of parameters that can be readily quantified, such as population, income, education, employment status, gender distribution, age distribution, and ethnic group (Welsh et al. 1995).

Differences in Valuing Existing and Hypothetical Projects

Each method has its merits, and best applications depend on the attributes, including the time and budget for economic analysis, of the needed decision. The TCM method is limited to existing projects, whereas the CVM can be applied to both existing and proposed projects. However, the required sample size and cost of implementation are smaller for the TCM because the survey is relevant to all the interviewees and the response rate is higher. The TCM observes actual behavior (revealed preferences), whereas the direct measurement methods inquire as to hypothetical willingness to pay for future consumption. The direct methods, especially CVM, are more susceptible to strategic behavior where respondents may guide their answers toward a particular personal agenda. The dichotomous choice version of CVM reduces this opportunity for bias. Loomis (1999) has obtained meta-analysis benefits estimates based on the application of the CVM. This study provided benefits estimates of proposed dam removal habitat enhancement for threatened Columbia River system anadromous fish species (Loomis 1999).

Because existence and bequest values do not involve visits to a recreation site, there is no relevant travel cost for the amenities. Thus existence benefits CVM studies are used if offsite or nonuser values are critical components of the social value of a nonmarket amenity. Existence benefits CVM studies have been used by several economists to value rare species. Summaries can be found in Pearce and Moran (1994), Loomis and White (1996a, 1996b), and Jakobsson and Dragun (1996). In Table 10.1, we present estimates of annual household CVM WTP for terrestrial species. These estimates generally represent total economic value (e.g., use, existence, and bequest values). The few studies that have partitioned total value into use and existence value generally find that the bulk of the WTP total values are existence values (Mitchell and Carson 1989).

Table 10.1. *Estimates of annual household CVM willingness to pay for preservation of rare species (1993 dollars)*

Species	Annual Value per Household	Low Estimate	High Estimate
Northern spotted owl (a)	$70.00	$44.00	$95.00
Grizzly bear (b)	$46.00	—	—
Red-cockaded woodpecker (c)	$13.00	$10.00	$15.00
Bald eagle (d)	$24.00	$15.00	$33.00
Whooping crane (e)	$35.00	—	—
Bighorn sheep (f)	$21.00	$12.00	$30.00
Gray wolf—one time payment (g)	$67.00	$16.00	$118.00

Source: Table based on (a) Boyle and Bishop (1987), (a) Hagen et al. (1992), (a) Rubin et al. (1991), (b) Brookshire et al. (1983), (c) Loomis and White (1996a), (d) Stevens (1991), (e) Bowker and Stoll (1988), (e) Stoll and Johnson (1984), (f) King et al. (1988), (g) Duffield (1991).

Note: Dashes mean no data.

The amenity in most of these estimates is a set of actions that would prevent extinction and thus represents an upper bound for evaluating policies that increase the probability a species will persist (Loomis and White 1996a). Although the per household preservation values for some species may be small, the total value may be large because amenities that provide existence benefits are pure public goods and can be extended to large populations.

Questions of survey design, survey administration, and survey elicitation format have been widely discussed in the literature (Mitchell and Carson 1989; Arrow et al. 1995; Welsh et al. 1995). Written mail-back surveys have lower response rates, higher mean bids, and noisier data than CVM survey data gathered from surveys administered by telephone (Mitchell and Carson 1989). Thus administration in person is usually recommended (Arrow et al. 1995). The National Oceanic and Atmospheric Administration CVM panel (Arrow et al. 1995) recommends the use of the dichotomous choice format as the bid elicitation format. In a dichotomous choice format survey the respondents are given a specific WTP sum for an amenity and asked if they will pay the sum. This response is a "yes" or "no"; hence the elicitation format is said to be a "dichotomous choice." In older, open-ended CVM surveys respondents are asked to write down their maximum WTP amount. Often a payment card with a table of plausible WTP sums is listed in the survey to facilitate the solicitation of the WTP sum (Mitchell and Carson 1989). Although dichotomous choice is the preferred format, it does generate higher estimates than open-ended surveys (Loomis et al. 1990; Welsh et al. 1995).

For a considerable period after the first CVM and TCM studies were published, many resource economists noted that the two approaches generated similar benefits estimates for use values (Carson et al. 1992). By contemporary standards, the benefits estimates were modest and often less than the cost of the amenity improvement or preservation in question, hence they seemed reasonable. The debate on the range of amenity values that are appropriate for each of the nonmarket valuation methods will continue to provide interesting dialogue in the journals and is not likely to result in definitive recommendations for a while (Kahneman and Knetsch 1992; Arrow et al. 1995; Perrings et al. 1995; Douglas and Taylor 1999; Carson et al. 1992, 2003; Douglas and Harpman 2004; Douglas and Johnson 2004).

Empirical Estimates of Economic Values

Basic research has developed important uses for rare species and will continue to do so in the future (Phillips and Meilleur 1998). Some species, especially plants, have proven valuable for direct medicinal purposes or as aids in diagnoses (Simpson and Reid 1996). Perhaps 25% of all medical drugs were originally derived from plants (Pearce and Moran 1994). Recent examples include taxol, originally derived from the Pacific yew tree, for treating ovarian cancer, and several other highly effective cancer drugs (Janni 2003). The rosy periwinkle has provided compounds that are the basis for cancer-fighting drugs with global sales that have exceeded $100 million (Janni 2003).

Valuation of medicinal benefits is difficult. One approach has been to use the market values of the bioprospecting contracts by pharmaceutical companies. These range from $175,000 for the enzymes and bacteria at Yellowstone National Park to as much as $3.2 million for the right to take 30,000 samples of Brazilian biota (Nunes and van den Bergh 2001). Another valuation venue is to infer value from the end result of using the drugs that provide a longer life expectancy and improved health. This procedure allows the analyst to tap into a well-developed literature on the value of marginal increases in life expectancy and health.

However, it is difficult to determine the probability that a given species will yield any medicinal value. Pearce and Moran (1994) suggest that this probability is between 0.1 and 0.01%. The small probability makes the expected value of preserving any single species for medicinal applications quite low, despite the fact that the value of a statistical human life ranges from $4 to $7 million (Pearce and Moran 1994; Viscusi 2004). Nonetheless, it is possible, or perhaps likely, that wild plant species make a multibillion-dollar contribution to the medical well-being of humans (Pearce and Moran 1994). Given the present lack of knowledge regarding which of the millions of species are critical to treating current and future diseases, there is a substantial quasi option value to maintaining the species in situ (see also chaps. 3, 4, and 5 for values for RLK species).

Wildlife resources in general can have high direct use values that may be captured for local communities through visitor fees or recreation-associated economic activities (Perrings et al. 1995). For example, rare species may be global tourist attractions. Whooping cranes attract bird-watchers from around the world (Stoll and Johnson 1984). Costa Rica has protected nearly 10% of their country from development, and concern for the environment has made the country a major ecotourism destination (Perrings et al. 1995). Tourism provides two types of benefits. First, it provides a consumer surplus value to the consumer. This consumer surplus can be measured as the visitors' net WTP above their travel costs. Second, it provides employment opportunities and tax revenues to the host country.

Wildlife viewing values for whooping cranes and bald eagles provide some indication of the value of rare species as ecotourism attractions. Stoll and Johnson (1984) studied the Arkansas National Wildlife Refuge, where nearly 100,000 people visit during the whooping crane migration there. Bowker and Stoll (1988) calculated WTP for a visit with and without whooping cranes, and found a net WTP of $2.50 per person viewing the cranes. With nearly 100,000 visitors each year, the aggregate benefits conferred are $250,000 per annum. Swanson (1996) estimated the mean WTP per trip for bald eagle viewing on the Skagit River Bald Eagle Natural Area in Northwest Washington to be about $55.69. With more than 4000 visitors per year, the net aggregate WTP was $223,000. Nonmarket resources have been valued on a regional basis as Peterson et al. (1992) did for Alaska wildlife.

There is some documentation of sizeable values for species with declining populations that are endemic to a region. Stevens et al. (1991) estimated that New England survey respondents have annual CVM WTP values of $19, $12, and $8 for bald eagle, wild turkeys, and Atlantic salmon, respectively. The study provided critical assessments of CVM benefits estimates for wildlife preservation.

The activity of bird-watching is linked to participants' desire to observe, photograph, and/or provide feed to regionally rare avian species. Birders in the United States spend over $20 billion per annum on seed, equipment, and travel (Kerlinger 1993). This expenditure estimate has robust policy implications because it is much larger than CVM wildlife preservation-linked benefits estimates.

Green National Accounts

Increased awareness of the importance of natural resources and the environment has led to a movement to supplement U.S. national economic accounts by including changes in the abundance of natural resources and the environment (Nordhaus and Kokkelenberg 1999). Modifying the national accounts to include physical and environmental assets reflects the desire to construct indicators that incorporate sustainable development constraints. Existing market-based accounts are not realistic because both economic and social welfare depend upon many excluded nonmarket activities. Traditional national income accounts report the value of the final output of goods and services. Limiting the accounts to market transactions produces estimate bias. This artificial boundary distorts our measures of economic activity. For example, GDP captures the value of recreating in a commercial facility, while some of the value(s) of recreating on the public lands and waters are excluded.

Other green goods and services that are omitted include changes in the quality of physical assets such as air and water (U.S. Environmental Protection Agency 1997). Increasing GDP often entails a decrement in the quality and quantity of renewable and nonrenewable resources (Hartwick 1997). Allowances are made in the national accounts for depreciation of structures and equipment but not for the depletion of petroleum pumped from reservoirs. Adjustments are made for damage to physical capital but not for the deterioration of an urban air shed. Economic techniques have been developed and are under constant refinement for measuring the money value of these types of nonmarket goods on a regional basis. However, these techniques have not been applied on a national scale (Nordhaus and Kokkelenberg 1999).

Creation of environmental accounts can provide valuable information on the interaction between the environment and the economy. This will, in turn, help economists to determine if environmental capital is being used in a sustainable manner by providing data on the effects of different public management policies, regulations, and taxes on the growth or degradation of environmental capital. Augmented national economic accounts would provide better measures of all outputs that provide social utilities.

The current market-based GDP could be maintained and satellite accounts developed that account for environmental services (Cobb et al. 1995). Providing information on the structure and interactions of the economy and the environment is increasingly recognized as an essential function of government (U.S. Environmental Protection Agency 1997).

The Bureau of Economic Analysis of the U.S. Department of Commerce has explored the issues involved in adjusting the GDP to include activities associated with the environment and natural resources (Nordhaus and Kokkelenberg 1999; Nordhaus 2005). Congress subsequently suspended the Bureau's work pending an independent review by the National Research Council's Committee on National Statistics. That process was completed and the committee's findings are available (Nordhaus and Kokkelenberg 1999).

From an economic standpoint, an ideal model would be where price varies when environmental effects result during production. As such, economic actions taken by a firm would have direct and measurable environmental consequences that are reflected by changes in price. The model could incorporate profit optimization, relate nonmarket costs to market costs, and minimize the output of negative effects related to the output of positive outcomes for a given price structure by industry (including tax, subsidy, etc.). The model would be complex and at first costly, because limits on expected environmental effects would be defined uniquely over many ecosystems rather than only as a national total. Relevant ecological information has generally been inadequate to specify this type of comprehensive model, but new thrusts at the federal level to develop environmental indices may soon remove that shortfall.

Conclusion

Economists have observed that the market system can help protect endangered species but the result can be uneven and sporadic. For example, tradable pollution rights may result in excess protection in one region and insufficient protection in another, thereby causing irreversible species loss in the underprotected region (Just et al. 1982). The tradable pollution rights method can fail because it does not incorporate a safe minimum standard (Perrings et al. 1995). Thus it is critically important to study the effects of alternate methods used to protect species.

If minimum physical thresholds are required to retain viability of populations of species, these limits must be set by regulatory mechanisms such as the Endangered Species Act (Hartwick 1997). Actions by individuals or organizations contrary to the maintenance of environmental thresholds must often be precluded by vigorously enforced regulations.

In this chapter we discussed a wide array of nonmarket valuation techniques for valuing RLK species. For public and private policy to fully reflect the wants and desires of individuals and society, it is critical that preservation benefits as well as costs be considered in natural resource allocation decisions. Thus a wide assortment of alternative decision rules based on criteria linked to the valuation criteria are currently available. However, economics should not be the sole criterion by which society decides to preserve RLK species. Other social criteria such as equity or ethics play a central role (chap. 9). Still, one advantage of explicitly evaluating the economics of species preservation is that benefits are often larger and costs lower than one might expect (Just et al. 1982). Society must make decisions in this arena that incorporate information from various sources and considerations including economics, ecology, politics, and environmental justice.

References

Arrow, K. J. 1965. *Aspects of the theory of risk bearing*. Finland: Academic Bookstore.

Arrow, K. J., and A. C. Fisher. 1974. Environmental preservation, uncertainty, and irreversibility. *Quarterly Journal of Economics* 88:312–19.

Arrow, K. J., and F. H. Hahn. 1986. *General competitive analysis*. Amsterdam: Elsevier Sciences.

Arrow, K. J., R. Solow, P. R. Portnoy, E. E. Leamer, R. Radner, and H. Schuman. 1995. *Report of the NOAA panel on contingent valuation*. Washington, DC: National Oceanic and Atmospheric Administration.

Bateman, I. J., and K. G. Willis, eds. 1999. *Valuing environmental preferences: Theory and practice of the contingent valuation method in the US, EU, and developing countries*. New York: Oxford University Press.

Bator, F. M. 1958. The anatomy of market failure. *Quarterly Journal of Economics* 72:351–79.

Bishop, R. C. 1978. Endangered species and uncertainty: The economics of a safe minimum standard. *American Journal of Agricultural Economics* 60:10–18.

———. 1982. Option value: An exposition and extension. *Land Economics* 58:1–15.

Bowker, J. M., and J. R. Stoll. 1988. Use of dichotomous choice nonmarket methods to value the whooping crane resource. *American Journal of Agricultural Economics* 70:372–81.

Boyle, K., and R. Bishop. 1987. Valuing wildlife in benefit–cost analyses: A case study involving endangered species. *Water Resources Research* 23:943–50.

Brookshire, D. S., L. S. Eubanks, and A. Randall. 1983. Estimating option prices and existence values of wildlife resources. *Land Economics* 59:1–15.

Carson, R. T., R. C. Mitchell, W. M. Hanemann, R. J. Kopp, S. Presser, and P. Ruud. 1992. A contingent valuation study of lost passive use values resulting from the *Valdez* oil spill. Report to the attorney general of the State of Alaska. www.evostc.state.ak.us/Publications/Downloadables/Economic/Econ_Passive.pdf.

———. 2003. Contingent valuation and lost passive use: Damages from the Exxon *Valdez* oil spill. *Environmental and Resource Economics* 25:257–86.

Caufield, C. 1986. *In the rainforest: Report from a strange, beautiful, imperiled world.* Chicago: University of Chicago Press.
Clawson, M., and J. L. Knetsch. 1966. *Economics of outdoor recreation.* Baltimore: Johns Hopkins University Press.
Clinton, W. J., and A. Gore Jr. 1993. The forest plan: For a sustainable economy and a sustainable environment. Appendix A in *The Northwest Forest Plan: A Report to the President and Congress,* ed. E. T. Tuchmann, K. E. Connaughton, L. E. Freedman, and C. B. Moriwaki. Portland, OR: U.S. Department of Agriculture, Forest Service, Pacific Northwest Research Station.
Coase, R. H. 1960. The problem of social cost. *Journal of Law and Economics* 3:1–44.
Cobb, C., T. Halstead, and J. Rowe. 1995. *The genuine progress indicator: Summary of data and methodology.* San Francisco: Redefining Progress.
Costanza, R., R. d'Arge, R. deGroot, S. Farber, M. Grassot, B. Hannon, K. Limburg, et al. 1997. The value of the world's ecosystem services and natural capital. *Nature* 387:253–80.
Creel, M. D., and J. B. Loomis. 1990. Theoretical and empirical advantages of truncated count data estimators for analysis of commercial deer hunting in California. *American Journal of Agricultural Economics* 72:434–41.
Daily, G. 1997. *Nature's services: Societal dependence on natural ecosystems.* Washington, DC: Island Press.
Davis, R. K. 1963. The value of outdoor recreation: An economic study of the Maine woods. PhD diss., Harvard Univ.
Douglas, A. J. 1990. Renewable resources. Pp 512–16 in *Encyclopedia of environmental science,* ed. D. E. Alexander and R. W. Fairbridge. Dordrecht: Kluwer Academic.
Douglas, A. J., and J. G. Taylor. 1999. A new model for the travel cost method: The total expenses approach. *Environmental Modeling and Software* 14:81–92.
Douglas, A. J., and D. A. Harpman. 2004. Lake Powell management alternatives and values: CVM estimates of recreation benefits. *Water International* 29 (3): 375–83.
Douglas, A. J., and R. L. Johnson. 2004. Empirical evidence for large nonmarket values for water resources: TCM benefits estimates for Lake Powell. *International Journal of Water* 2:229–45.
Duffield, J. 1991. Existence and nonconsumptive values for wildlife: Application to wolf recovery in Yellowstone National Park. Western Regional Research Project W-133/Western Regional Science Assn. Joint Session, Measuring Nonmarket and Nonuse Values. Pp. 1–39 in *Benefits and costs in natural resources planning,* comp. C. Kling. Davis: University of California.
Ehrlich, P. and A. Ehrlich. 1996. *Of science and reason: How anti-environmental rhetoric threatens our future.* Washington, DC: Island Press.
Fisher, A. C., and W. M. Hanemann. 1986. Environmental damages and option values. *Natural Resources Modeling* 1:111–24.
Freeman, A. M. 2003. *The measurement of environmental and resource values.* Washington, DC: Resources for the Future Press.
Hagen. D. A., J. W. Vincent, and P. G. Welle. 1992. Benefits of preserving old-growth forests and the spotted owl. *Contemporary Policy Issues* 10:13–26.
Hanemann, W. M. 1991. Willingness to pay versus willingness to accept: How much can they differ? *American Economic Review* 81:635–47.

Hardin, G. 1968. The tragedy of the commons. *Science* 162 (13): 1243–48.
Hartwick, J. 1997. National wealth, constant consumption and sustainable development. Pp. 55–81 in *The international yearbook of environmental and resource economics 1997/1998*, ed. H. Folmer and T. Tietenberg. Cheltenham, UK: Edward Elgar.
Henderson, J. M., and R. E. Quandt. 1980. *Microeconomic theory: A mathematical approach*. 3rd ed. New York: McGraw-Hill.
Hendrickson, H., S. Joshi, and L. B. Lave. 1998. Economic input–output models for environmental life-cycle assessment. *Environmental Science and Technology* 32:184A–91A.
Hicks, J. R. 1943. The four consumer's surpluses. *Review of Economic Studies* 11:31–41.
Hof, J. G., and M. G. Raphael. 1997. Optimization of habitat placement: A case study of the northern spotted owl in the Olympic Peninsula. *Ecological Applications* 7:1160–69.
Jakobsson, K. M., and A. K. Dragun. 1996. *Contingent valuation and endangered species: Methodological issues and applications*. Northampton, MA: Edward Elgar.
Janni, O. 2003. *Biotechnology and biodiversity conservation: Economic viability and international institutional framework*. Milan: Instituto di Ricerca Sulla Dinamica dei Sistemi Economici.
Just, R. E., D. L. Hueth, and A. Schmitz. 1982. *Applied welfare economics and public policy*. Englewood Cliffs, NJ: Prentice-Hall.
Kahneman, D., and J. L. Knetsch. 1992. Valuing public goods: The purchase of moral satisfaction. *Journal of Environmental Economics and Management* 22:50–70.
Kerlinger, P. 1993. Birding economics and birder demographics studies as conservation tools. Pp. 32–38 in *Status and Management of Neotropical Birds*, ed. D. M. Finch and P. W. Stangel. General Technical Report GTR-RM-229. U.S. Department of Agriculture, Forest Service, Fort Collins, CO, Rocky Mountain Forest and Range Experiment Station.
King, D. A., D. J. Flynn, and W. W. Shaw. 1988. Total and existence values of a herd of desert bighorn sheep. Western Regional Research Project W-133, Benefits and Costs.
Krutilla, J. V. 1967. Conservation reconsidered. *American Economic Review* 57:777–86.
Lave, L. B., E. Cobasflores, C. T. Hendrickson, and F. C. McMichael. 1995. Using input–output analysis to estimate economy-wide discharges. *Environmental Science and Technology* 29:420A–26A.
Leontief, W. 1936. Quantitative input–output relations in the economic system of the United States. *Review of Economics and Statistics* 18:105–25.
———. 1941. *The structure of the American economy: 1919–29*. New York: Oxford University Press.
———. 1966. *Input–output economics*. New York: Oxford University Press.
———. 1974. The structure of the American economy. *American Economic Review* 64:823–34.
Loomis, J. 1999. Recreation and passive use values from removing the dams on the Lower Snake River to increase salmon. Walla Walla, WA: Department of the Army Corps of Engineers.

Loomis, J., W. M. Hanemann, and T. C. Wegge. 1990. *Environmental benefits study of San Joaquin Valley's fish and wildlife resources*. Sacramento: Jones and Stokes Associates.

Loomis, J., and D. S. White. 1996a. Economic benefits of rare and endangered species: Summary and meta analysis. *Ecological Economics* 18:197–206.

———. 1996b. Economic values of increasingly rare and endangered fish. *Fisheries* 21:6–11.

Meadows, D. H., D. Meadows, R. Jorgen, and W. Behrens. 1972. *The limits to growth*. New York: Universe Books.

Miller, R. E., and P. D. Blair. 1985. *Input–output analysis: Foundations and extensions*. Englewood Cliffs, NJ: Prentice-Hall.

Mishan, E. J. 1976. *Cost–benefit analysis*. New York: Praeger.

Mitchell, R. C., and R. T. Carson. 1989. *Using surveys to value public goods*. Washington, DC: Resources for the Future.

Montgomery, C., and G. M. Brown. 1992. Economics of species preservation: The spotted owl case. *Contemporary Policy Issues* 10:1–112.

Montgomery, C., G. M. Brown, and D. M. Adams. 1994. The marginal cost of species preservation: The northern spotted owl. *Journal of Environmental Economics and Management* 26:111–28.

Nordhaus, W. D. 2005. Principles of national accounting for nonmarket accounts. Prepared for the Conference on Research in Income and Wealth. http://nordhaus.econ.yale.edu/CRIW_0120305.doc.

Nordhaus, W. D., and E. C. Kokkelenberg. 1999. *Nature's numbers: Expanding the national economic accounts to include the environment*. Washington, DC: National Research Council, Commission on Behavioral and Social Sciences and Education.

Nunes, P. A., and J. C. van den Bergh. 2001. Economic valuation of biodiversity: Sense or nonsense. *Ecological Economics* 39:203–22.

Olsen, D., J. Richards, and D. Scott. 1991. Existence and sport values for doubling the size of Columbia River basin salmon and steelhead runs. *Rivers* 2:44–56.

Pearce, D., and D. Moran. 1994. *The economic value of biodiversity*. London: Earthscan, Publications in Association with the International Union for the Conservation of Nature.

Perrings, C., E. B. Barbier, G. Brown, S. Dalmazzone, C. Folke, C. Gadgil, N. Hanley, et al. 1995. The economic value of biodiversity. Pp. 872–76 in *Global Biodiversity Assessment*, ed. V. H. Heywoods. Cambridge: Cambridge University Press.

Peterson, G. L., C. S. Swanson, D. W. McCollum, and M. H. Thomas. 1992. *Valuing wildlife resources in Alaska*. Boulder: Westview Press.

Prewitt, R. A. 1949. *The Economics of Public Recreation: An Economic Study of the Monetary Evaluation of Recreation in the National Parks*. Hotelling's letter is on p. 9. Washington, DC: U.S. Department of the Interior, National Park Service.

Phillips, O. L., and B. A. Meilleur. 1998. Usefulness and economic potential of the rare plants of the United States: A statistical survey. *Economic Botany* 52:57–67.

Randall, A. 1991. Total and nonuse values. Pp. 303–22 in *Measuring the Demand for Environmental Quality*, ed. J. B. Braden and C. D. Kolstad. Amsterdam: Elsevier Science.

Randall, A., B. Ives, and E. Eastman. 1974. *Benefits of abating aesthetic environmental damage from the four corners power plant, Fruitland, New Mexico*. University

Agricultural Experiment Station Bulletin no. 618. Las Cruces: New Mexico State University.
Randall, A., and J. R. Stoll. 1983. Existence value in a total valuation framework. Pp. 265–74 in *Managing air quality and scenic resources at national parks and wilderness areas*, ed. R. D. Rowe and G. Row. Boulder: Westview Press.
Ready, R., and R. C. Bishop. 1991. Endangered species and the safe minimum standard. *American Journal of Agricultural Economics* 73:309–12.
Rubin, J., G. Helfand, and J. Loomis. 1991. A benefit–cost analysis of the northern spotted owl: Results from a contingent valuation survey. *Journal of Forestry* 89:25–30.
Samuelson, P. A. 1954. The pure theory of public expenditure. *Review of Economics and Statistics* 36:387–89.
Schamberger, M. L., R. L. Johnson, J. J. Charbonneau, and M. J. Hay. 1992. *Economic analysis of critical habitat designation effects for the northern spotted owl.* Washington, DC: U.S. Department of Agriculture, Fish and Wildlife Service.
Schumpeter, J. A. 1942. *Capitalism, socialism, and democracy*. New York: Harper and Row.
Simpson, R. R., and J. Reid. 1996. Valuing biodiversity for use in pharmaceutical research. *Journal of Political Economy* 104:163–85.
Spurr, S. H., and B. V. Barnes. 1980. *Forest ecology*. New York: John Wiley and Sons.
Stevens, T. 1991. CVM wildlife existence value estimates: Altruism, ambivalence, and ambiguity. Western Regional Research Project W-133, Fourth Interim Report. Corvallis: Oregon State University.
Stevens, T., J. Echeverria, R. J. Glass, T. Hagar, and T. A. Morel. 1991. Measuring the existence value of wildlife: What do CVM estimates really show? *Land Economics* 67:390–400.
Stoll, J., and L. A. Johnson. 1984. Concepts of value, nonmarket valuation and the case of the whooping crane. *Transactions of the North American Wildlife and Natural Resources Conference* 49:382–93.
Swanson, C. S. 1996. Economics of endangered species: Bald eagles on the Skagit River Bald Eagle Natural Area, Washington. *Transactions of the North American Wildlife and Natural Resources Conference* 61:293–300.
Swanson, C. S., D. W. McCollum, and M. Maj. 1994. Insights into the economic value of grizzly bears in the Yellowstone recovery zone. *International Conference on Bear Research and Management* 9:575–82.
Tuchmann, E. T., K. P. Connaughton, L. E. Freedman, and C. B. Moriwaki. 1996. *The Northwest Forest Plan: A report to the president and Congress*. Portland, OR: U.S. Department of Agriculture, Forest Service, Pacific Northwest Research Station.
U.S. Bureau of Census. 1992. *Statistical abstract of the U.S.* 103rd ed. Washington, DC: U.S. Department of Commerce.
U.S. Environmental Protection Agency. 1997. *Final report to Congress on benefits and costs of the Clean Air Act, 1970–1990*. EPA 410-R-97-002. Washington, DC: U.S. Government Printing Office.
Viscusi, W. K. 2004. The value of life: Estimates with risks by occupation and industry. *Economic Inquiry* 42:29–48.
Walsh, R. G., L. D. Sanders, and J. B. Loomis. 1984. Measuring the economic benefits of proposed wild and scenic rivers. Pp. 260–71 in *National River Recreation Symposium Proceedings*, ed. J. S. Popadic, D. I. Butterfield, D. H. Anderson, and

M. R. Popadic. Baton Rouge: Louisiana State University, U.S. Forest Service, U.S. National Park Service, and Bureau of Land Management.

Ward, F. A., and D. Beal. 2000. *Valuing nature with travel cost model: A manual.* New Horizons in Environmental Economics. Northampton, MA: Edward Elgar.

Welsh, M. P., R. D. Bjonback, R. D. Rosenthal, and R. A. Aiken. 1995. *GCES Nonuse Values Study Summary Report.* Madison, WI: Hagler Bailly Consulting.

11

Implementation Considerations

Deanna H. Olson

During implementation of any resource management program, numerous considerations can alternatively facilitate or constrain program development. Such considerations may have socioeconomic, political, or scientific foundations and can functionally serve as program "drivers," "pivots," or "barriers" to implementation (Olson et al. 2002). Drivers are the impetus factors that instigate programs, pivots are issues that cause the path of a program to change, and barriers stall program development. They may occur anytime, during program initiation or later. These implementation considerations are especially critical to address during the planning phases of conservation programs for rare or little-known (RLK) species. Many can interact directly with the conservation approach selected, whether it is a species- or systems-based approach. Implementation considerations can ultimately determine program or approach effectiveness.

Eight programmatic considerations are particularly crucial for successful conservation of RLK species (table 11.1). These include authority, priorities, risk and uncertainty attitudes, capability, accountability, adaptive management, monitoring, and communication. Key considerations are the authority, capability, and accountability of programs. Like three legs of a stool, these elements stabilize successful conservation programs, which may falter or collapse if they have not been well developed during design and initiation stages, or effectively monitored and adapted later. Interestingly, these eight considerations are not mutually exclusive categories, and intricate interdependencies may develop. Program priorities tie authority to capability. Without sufficient authority and capability and clear priori-

Table 11.1. *Eight implementation considerations may function to facilitate or constrain rare or little known (RLK) species conservation programs*

Consideration	Key Elements
Authority	• Legal foundation of RLK species conservation (e.g., laws, policies, statutes) • Enforcement may vary despite legal framework • Leadership and advocacy can be a socially authoritative force • Agency or institutional ownership of program can enhance program authority
Priorities	• Clear goals and measurable objectives are needed • Targets may vary with different RLK species • Objectives may vary with time, spatial scale, or taxonomic group • Science-based objectives are needed.
Risk and uncertainty	• Risk to species may be relative to abundance levels or persistence • Risk attitudes vary among stakeholders and should be acknowledged, some may have no-risk philosophies and some may have an acceptable-risk level • Uncertainty captures both unknowns and errors
Capability	• Components include funding, time, effort, species recovery potential, and personnel infrastructure (persons with biological, technical, and administrative skills provide five functions: design, management, monitoring, analysis, and communication)
Accountability	• Measurement of successful program implementation and RLK conservation, which is tied to adaptive management, monitoring and communication
Adaptive Management	• A dynamic program design can include a cyclic learning environment, where all program aspects are revisited and changed to improve the program as knowledge accrues over time
Monitoring	• Implementation, validation and effectiveness monitoring may be implemented • Effectiveness monitoring objectives should match overall program objectives, and these should be measurable, science-based, and short- and long-term • RLK species may introduce monitoring difficulties due to abundance or distribution issues and should be considered during objective-setting procedures • A multipronged monitoring approach may be needed
Communication	• Stakeholders need updates on design, implementation, and effectiveness of RLK species conservation programs • Reports, meetings, and electronic venues for information should be used, with each likely being received by different audiences • Clarity of communications is paramount

ties, species conservation efforts can be immediately futile, whereas accountability may be a factor with delayed impact upon a program. Integral to capability is adaptive management, an element that is also critical to accountability. Accountability further relies on the considerations of monitoring and communication. A consideration of particular importance for RLK species are the assessments of and responses to risk and uncertainty. Each of these eight implementation considerations is described further in the following sections.

Authority

The authority of species conservation programs for RLK taxa is often tied directly to the legal foundation upon which it is based, the clarity and specificity of policies, and their enforcement (see also chaps. 2 and 10). If RLK species protection guidelines do not have a legal basis, they may have reduced authority and strength to achieve the stated objectives. A conceptually effective species- or systems-based approach for RLK species conservation (chaps. 6 and 7) may falter if it garners little authoritative enforcement.

Legal issues are often drivers of species conservation programs and may have several forms, including statutory laws, jurisdictional issues relative to landownership, and certification incentives for private landowners (chap. 2). Each has a particular scope relative to species management, either limiting the species or the lands included for protection. Laws, regulations, or guidelines address species directly or indirectly by addressing geographic areas, ecosystems and habitats, and threats to species or systems. Thus, depending upon the conservation approach (species, systems) selected for RLK species, different regulations or a mix of regulations may apply. RLK species considered under several policies, such as species, habitat, and threat laws, may result in more effective piggy-backed protections. Some aquatic riparian-dependent species may exemplify this if they are addressed by species laws and other clean water or habitat-based protections.

Despite the potential for successful rare species management, gradients in authority resulting in variable effectiveness are evident across U.S. state and federal agencies. Federal species laws such as the Endangered Species Act (ESA), Marine Mammal Protection Act, and Migratory Bird Treaty Act carry established legal weight and have strong track records of enforcement. The ESA is the most powerful statute to directly prevent rare species from extinction on all U.S. lands. For target species to be listed as federally threatened or endangered, they must not only fulfill criteria for rarity, they must also demonstrate losses or threats and lack of existing mitigations to stem losses. Thus this statute does not apply to all species that may be considered rare, only a subset in the direst situations such that extinction is imminent and a listing package has been developed. ESA listing has been considered a "last resort to species conservation" (Restani and Marzluff 2002). Existing knowledge is used to base an ESA listing decision. Such decisions may exclude little-known species for which insufficiency of information may be cited. Consequently, the authority of the ESA will apply to only a small subset of RLK species. Once a species is ESA listed, recovery is usually addressed with species-based approaches, in part due to the direct accountability back to the species of concern. Of the spectrum of species approaches, ESA recovery plans are always species-specific rather than surrogate species, geographically based, or systems-based approaches.

Similar to the ESA, federal species management provisions tied to a Record of Decision, such as the former Survey and Manage provision of the Northwest Forest Plan (USDA and USDI 1994; Molina et al. 2006), have greater authority than other more internal policy statements or planning guidelines. Public oversight and threat of litigation on federal threatened and endangered species issues and rare species issues tied to Records of Decision have undoubtedly played key roles in their enforcement. Regulations of the 1976 National Forest Management Act (NFMA) carried almost 20 years of case law that served to clarify its interpretation and greatly aided specificity and hence authority of those policies, particularly as they relate to biodiversity issues. Depending upon the provision considered, such Record of Decision provisions may apply to species- or systems-based approaches.

In contrast, other policies, such as the federal agency sensitive species programs, have historically been tied to documents that carry considerably less authority and are less prescriptive. Subsequently, they have not been as readily subject to litigation and have been more flexible in implementation. The flexibility in federal sensitive-species programs is demonstrated both among and within regions, which vary dramatically in the interpretation and implementation of common national guidelines. Criteria for rare or concern-species inclusion within sensitive-species programs may differ across administrative units. For example, during the evaluation of mollusks for inclusion in federal agency sensitive-species programs in the U.S. Pacific Northwest, some taxa were afforded listing within some administrative units and not in other neighboring units due to the different evaluation criteria applied in each (USDA and USDI 2004a). Little-known species may again fail to meet information sufficiency criteria to be included in such programs. Although it may seem that sensitive species programs would logically be species-specific approaches to conservation, many may rely on surrogate species, geographically based, or habitat-based protections instead that may be less effective for RLK species (chaps. 6 and 7).

Furthermore, the lands to which different statutes and policies apply may vary. For example, federal policies may differ among lands often regarded as types of reserves, such as national parks, federal wilderness areas, and resource natural areas. Also, city or county ordinances may carry considerable authority within local communities and differ from direction on adjacent lands. In particular, how different policies interact with property rights laws requires consideration on privately owned lands (chap. 10). Thus species- or systems-based approaches to species conservation that cross lands with different statutes may have varying authority and enforcement. This may be a critical implementation consideration for species with broad potential ranges, or systems approaches with large spatial extents. Varying authority among stakeholders could result in a mosaic of program effectiveness that would need consideration in conservation program design phases.

State boundaries may be particularly troublesome in this regard, relative to both federal and state statutes that differ with state lines. Some U.S. state governments have strong programs for threatened, endangered and sensitive species, and biodiversity generally, whereas others do not. A comparison of state laws, policies, and programs for biodiversity recently was compiled by the Defenders of Wildlife (http://www.defenders.org/bio-st00.html). They found only four states had formal biodiversity policies, and one state endangered species act provided as much basic protection as the federal act. These state policies carried various authorities, including an interagency memorandum of understanding and initiatives (California); a governor executive order (Kentucky); and laws (New York, Michigan). Despite biodiversity or rare species policy development, implementation has generally lagged or lacked enforcement. For example, in New York, a recent survey of amphibians and reptiles sold in state pet stores detected 15 of 29 (52%) state species of concern, including 4 of 16 (25%) protected species that required a license for possession (Hohn 2003); Hohn thus proposed a review of state regulations. This situation is likely echoed in many other states, as regulations relative to many rare, little-known, or somewhat low-profile species are not well understood or enforced.

Authority is not strictly a legal issue but can also be a social force (chap. 8). Rare species do not become recipients of either concern or law without advocacy. Bioethics and aesthetics are involved in social forces, as biota are considered critical resources and values to many individuals and may be significant drivers of conservation practices. The role of key individuals can be critical drivers of biodiversity conservation. In particular, individuals in leadership roles may effectively drive a program. In Brazil, a large number of conservation advocates have played pivotal roles in the development of the nation's protected areas (e.g., Mittermeier et al. 2005; see also other papers in *Conservation Biology* 19[3]). In northern Italy, Bani et al. (2002) listed both legislation and land management leadership among circumstances that can lead to effective implementation of ecological networks for conservation. In the United States, ESA listing of some species has resulted from the actions of primarily one or a few concerned professional biologists (e.g., Houston toad [*Bufo houstonensis*], Brown and Mesrobian 2005).

Leadership may not require key individuals but instead may result from agency representation or cooperative partnerships forged during the planning and decision-making process. If a rare species conservation program has been developed by an agency, institution, or partners such that they have direct ownership in its policies, are stakeholders in its success, and control lands over which it is implemented, then sufficient authority may exist for program success. The Natural Community Conservation Plans program in the State of California (e.g., Pollack 2001a, 2001b) provides a model in this regard. It was initiated to preserve biodiversity via cooperation among landowners, conservation organizations, the public, and government agencies at the local, state, and federal levels. Social authority applies equally to conservation programs with species or systems approaches, and conceptually to both little-known and rare species.

Problems arise, however, if biodiversity conservation priorities (legally, ecologically, or socially driven) are not paramount to conflicting land management objectives, particularly those with socioeconomic bearing (e.g., conflicting social or economic values from an area considered for conservation; chaps. 9 and 10). Stankey and Shindler (2006) address this concept in their assertion that rare species conservation is a fundamentally sociopolitical issue. They maintain that success follows accepted social goals and the stability of long-term social institutions. Likewise, Meffe (2001) called conservation biology the "'science of engagement' in addition to the traditional 'science of discovery'" (see also Robertson and Hull 2001; Song and M'Gonigle 2001). Without social understanding or support for rare species, conflicts arise, and the authority of a conservation program can be challenged and compromised if policies are vaguely outlined or overly discretionary in their implementation. As time and budgets always tend to be at a premium, perceived optional programs or programs described with indistinct language may not be wholly implemented in the spirit originally intended. Rare species policies with specific requirements, prescriptive standards, and guidelines are more enforceable than recommendations with conditional (e.g., "should") language. The Natural Community Conservation Plans of California mentioned earlier were initiated in 1991 via multiple partnerships, and within a decade resulted in the development of 2001 legislation to formalize the process, enhancing its authority. A legal framework for rare species conservation programs will serve to ameliorate the potential swing of a sociopolitical pendulum.

In summary, authority issues are inherently multifaceted. Enforceable regulations, leadership, advocacy, and sociopolitical tendencies may determine the authority or acceptance of a conservation approach for rare species. Current species laws promote effective protection of primarily a small subset of those RLK species we have considered in this book, using dominantly species-based approaches. Little-known species appear particularly vulnerable to a lack of authority and enforcement of protections. Lack of knowledge can be used to divert attention from these taxa and may promote an unawareness or lack of focus on such species.

Priorities

Clear direction is essential for any program, and priorities need to be set at several levels for RLK species conservation programs (chaps. 2, 3, and 5). These priorities will directly reflect the balance of multiple resources and the conservation approach selected, whether species or system based. Priorities may serve as pivots or barriers constraining full implementation of conservation ideals. At the highest levels, how do rare species programs rank with other conflicting programs that may be implemented on common landscapes such as those for timber harvest, hydroelectric power, urban development, recreation, and fire suppression? Are biodiversity goals on a par with socioeconomic objectives? Are conservation objectives policy-driven or evidence-based targets (Svancara et al. 2005)? As conflicting objectives are addressed, are there established orders of importance or mechanisms to determine trade-offs (see also chap. 9)? Are species priorities subject to established processes of adaptive management as system capability is matched with multiple land and resource objectives? Or do priorities swing with the sociopolitical pendulum, as implied earlier? A strong and stable commitment by the leadership of a region will be needed to support rare species goals (Molina et al. 2006), tying priorities to both authority and capability.

In the United States, a complex web of stakeholder interactions affects rare species management under the ESA, including priorities enacted by Congress, the public, the states, nongovernmental organizations, and federal agencies (Restani and Marzluff 2002). The interplay of stakeholders can fluctuate with geography, time, and rare species management programs. In different circumstances, diverse entities may play dominant roles in this network, with the outcome of priorities being a function of advocacy, legal authority, and funding capability.

Once RLK species conservation programs are established, in addition to the need for overarching vision, mission, and goals statements, concise language for species objectives and priorities is needed. Clarification of potentially vague terms such as "rare," "conservation," "protection," "viability," "persistence," "diversity," "at risk," or "restoration" are essential (chaps. 3 and 4). Although this may be considered primarily a semantic matter, it is by no means trivial. Furthermore, measurable objectives are paramount (Tear et al. 2005). Later, accountability of a program will be tied directly to the levels to which target objectives are met, and hence those targets need to be explicitly defined and measured. Communication avenues will need to use precise language during program development and later during accountability reporting.

In addition to setting clear and measurable targets, Tear et al. (2005) provided three additional fundamental principles and six standards for setting conservation objectives. Their principle to "separate science from feasibility" emphasized use of science-based information to set objectives, which can later be considered relative to practicality or capability constraints. Similarly, the evidence-based approach described by Svancara et al. (2005) favors use of science to set conservation objectives. Sanderson (2006) further describes various approaches to setting population-based targets for animals. The principle proposed by Tear et al. (2005) to "follow the scientific method" speaks to developing a transparent and repeatable process during goal setting. Their last principle, "anticipate change," is an up-front acknowledgment during conservation plan development that objectives are expected to change as knowledge is accrued. This principle ties the objective to the adaptive management process, discussed later. The six standards given by Tear et al. (2005) provide additional guidance for creating conservation priorities that can lead to successful programs. These include (1) use the best available science; (2) provide and evaluate multiple alternatives; (3) set short- and long-term objectives; (4) as possible, incorporate conservation concepts of representation, redundancy, and resiliency; (5) tailor objectives to the biological system, rather than adopting generic standards used elsewhere; and (6) evaluate errors and uncertainties. Tear et al. (2005) provide case studies to show how these guidelines have been applied, including a species- and systems-based approach.

An important consideration for RLK species, in particular, is that species conservation objectives can differ in the relative rarity status threshold they use to differentiate program emphasis, as well as the spatial scale (global to local planning area, species range to specific site locality), or biological scale of organization (sites, populations, species, biodiversity) at which they focus efforts. Programs may place emphasis on preventing extinction; conserving biodiversity; precluding federal listing as threatened and endangered; ensuring against future regulatory changes (i.e., "no surprises" agreements); identifying species with a downward population trend, population declines or disappearances; restoring species to a secure status; maintaining stable, well-distributed populations across species' ranges; and/or gathering basic species' information to better assess species risks and effectively address conservation goals. Although many programs blend these concepts, for those using species approaches there is a gradient in species rarity targets from maintaining a species inventory list for biodiversity conservation (simplistically, only "species presence" or representation addressed per area), through precluding federal listing as threatened and endangered (e.g., maintaining taxa above a minimal threshold for likely persistence), to maintaining well-distributed populations across a species' range. On one hand, a minimalist approach to biodiversity maintenance can be the target, and at the other extreme, assurance of thriving populations across the range may be the goal (see also Sanderson 2006).

Hierarchical objectives across scales may be most effective, with site-specific information gathering and conservation targets for local populations balanced with landscape-level goals for information gathering and conservation of metapopulations, larger scales of biological organization, and systems or processes. For example, Semlitsch (2002) includes maintenance of both local habitats and connectivity among them as critical elements of effective recovery and restoration plans for aquatic breeding amphibians. This melding of species- and systems-based approaches is likely the most holistic and subsequently successful in meeting conservation goals because it not only provides for a species-population but buffers it within a resilient ecological network.

Furthermore, if a program includes diverse taxa, multiple species-specific issues and priorities need to be addressed (chap. 4). Given the commonness of rarity and taxonomic diversity, different criteria may be imposed to set priorities (i.e., tailor objectives to the biological system, Tear et al. 2005). Known rare species might be distinguished from those that are little known for a two-pronged approach addressing relevant information and conservation needs of the different categories. If taxonomic bias exists, it should be recognized. Some programs differentiate objectives among taxa, others set criteria with inherent taxonomic bias, and some result in bias due to priorities set during program implementation. A few examples illustrate this. Clark et al (2002) found, "taxonomic bias was pervasive in [ESA] recovery plans," with vertebrates receiving more attention than other taxa. They cite two legal constraints contributing to this bias: U.S. congressional earmarking of funds for select species and ESA language giving greater protection to animals than to plants. Restani and Marzluff (2001) found that, among avifauna, charismatic species (e.g., eagles, owls, woodpeckers) received the majority of recovery funds and efforts, rather than taxa with more highly ranked conservation priority. They also cite the role of sociopolitical efforts in this process: congressional earmarking and litigation by special interest groups for select high-profile species. Similarly, Restani and Marzluff (2002) reported that ESA recovery expenditures did not track species priority ranks and attributed this to a variety of social, political, and institutional barriers. In particular, they found recovery funds limited for species with low public appeal and little economic impact.

In the U.S. federal Northwest Forest Plan (USDA and USDI 1994), species conservation priorities were met with a combination of systems and species approaches to land management. Species protections were stratified by the bulk of taxa being addressed by surrogate species or habitat mitigations, and a minority of taxa addressed individually. For those RLK species addressed individually, standards for vertebrate species under the 1976 National Forest Management Act viability provision were applied to nonvertebrate species "to the extent practicable" (USDA and USDI 1994, 2000, 2001). This practicality clause was somewhat open to interpretation, but certainly tied to programmatic capability (discussed later). Practicality was further included in differentiating categories of species protection under the Survey and Manage provision of the Northwest Forest Plan; two categories included the criterion that surveys were not practical (USDA and USDI 2001). Survey practicality referred to the "ability to reasonably and consistently locate occupied sites during surveys prior to habitat disturbing activities" (see also chap. 4). Two main issues led to predisturbance surveys being not practical. First, there may have been a low detection probability because the species do not produce identifying structures annually or predictably, or they may have identifying parts but they are visible for only a very short time. This was an issue dominant in many forest fungi with life histories including infrequent fruiting. Second, there may have been an inability to readily identify a species due to its minuscule size, identification methods requiring efforts beyond simple field or laboratory procedures, or species knowledge restricted to a limited number of experts. This practicality constraint had initial taxonomic bias relative to fungi, lichens, bryophytes, and mollusks due to the poorly known status of these taxa relative to others and to the lack of general knowledge or expertise about the taxa (see chap. 2). With time, as expertise was gained in the Survey and Manage program, solutions to some practicality constraints were developed for several taxa (e.g., species identification constraints became resolved).

Additional examples of taxonomic bias in rare species protections are evident. In 2002, the U.S. Forest Service Pacific Southwest and Pacific Northwest lists of sensitive species did not include fungi, lichens, bryophytes, or many invertebrates. Their standards for inclusion had eligibility criteria that may have preempted these taxa from consideration.

As already mentioned, information insufficiency is one of these criteria used to prioritize among taxa for inclusion in rare species programs. Thus little-known species may be left out of rare species programs entirely, leaving them in a "catch-22" rarity pseudostatus without recognition or a mechanism to accrue information. Information insufficiency also applies to the next example: when the Pacific Northwest federal Northwest Forest Plan was developed, a species-specific approach was adopted for all taxa considered, except arthropods, for which functional groups were assessed. The rationale to treat arthropods differently included lack of information on an estimated 70 to 80% of known/suspected species, inadequate surveys, and suspected greater diversity than any other class in Pacific Northwest forests (USDA and USDI 1993). Thus information insufficiency in addition to possibly practicality and expected programmatic capability, were key in this priority-setting process. In retrospect, relative to other taxa in the Survey and Manage program, the result has been almost a stasis in forest arthropod species information since the onset of this precedent-setting RLK species program. However, exceptions include new information on management effects on southwestern Oregon arthropod assemblages (e.g., Niwa and Peck 2002). Rather than dismissal of little-known species groups as infeasible to consider in conservation efforts, biodiversity and community structure metrics (e.g., McCune and Grace 2002) offer an initial means to include such enigmatic RLK assemblages (e.g., soil biota, Wall 2004) into ecological studies and potentially conservation programs.

In many cases, different rare species programs exist for plants and animals, which can easily lead to distinct standards and priorities being applied to each. For example, the U.S. federal agency sensitive species programs are not always implemented equally between plants and animals. Although this is difficult to quantify, botanical programs sometimes appear better developed, having more effective multispecies surveys and more straightforward and simple conservation practices across taxa with common life histories (e.g., plants are sessile, facilitating common approaches); faunal programs may rely less on field survey data and more on habitat assessments. Some of these flora–fauna program differences may have roots in the differing educational systems for botany and zoology, with training for botanists being more evenhanded across taxa. In contrast, many agency animal specialists come from college or university fisheries and wildlife programs that split training between aquatic and land fauna, and further may focus on game species: fishes, birds, and mammals. As a legacy of this educational bias, invertebrates and herpetofauna become the lesser-known taxa by default and are possibly neglected or deprioritized taxa in agency conservation programs. If programmatic funding is tied to fishing or hunting permits, as many state conservation programs are, then a taxonomic bias for species management may result if personnel are predominantly trained in fish and game taxa. Thus priorities may also result due to legacies of training, funding, or past program emphasis, rather than strict application of conservation criteria.

Additionally, all taxa are not easily or equally distinguished into discrete units such as sites, populations, or species, thus requiring reconsideration of program objectives and priorities in terms of differing levels of biological organization per taxon. A site may indicate suitable habitat, an individual occurrence, or a population. Multiple site locations may be the same individual fungus, or may represent separate individuals within a population of small mammals, or may be discrete populations or species of mollusks. Site-scale management thus may have minimal to considerable meaning for species persistence. Some species have highly divergent genetic structures, demonstrating complex evolutionary histories. Discrete population segments may be highly relevant for management of such species complexes. They have been used to distinguish fish stocks into categories of concern (e.g., Waples 1991; Nielsen 1995) and some other vertebrate populations for listing under the ESA (Pennock and Dimmick 1997), but the application of this conceptual framework across all taxa has lagged. Species designations have been difficult for some taxa such as fungi and mollusks (USDA and USDI 1993, 2000), providing challenges to development of species management guidance. Integration of biological scales of organization and spatial scales into species management priorities per taxon may be required.

Overall, many guidelines are available when considering priorities for RLK conservation programs. Setting goals and objectives is a critical and dynamic process. In particular, objectives should be clear, measurable, science-based, short- and long-term, hierarchical across spatial scales, and specific to different biological contexts such as rare species, little-known species, taxonomic group, and rarity status.

Risk and Uncertainty

In the development of any species conservation program, risk and uncertainty attitudes need up-front consideration. These are separate but related concepts. First, in conservation contexts, risk may mean a species' risk to persistence or risk of extinction. This is often expressed as a likelihood or probability estimate. The converse, likelihood of persistence, may also be used. Risk ratings may be developed subjectively or quantitatively, and might also be applied to habitats, habitat elements, processes, or functions. A risk "attitude" recognizes that different individuals assessing species' status or management may be comfortable taking different levels of risk. For some people or agencies, a conservation goal may to be to achieve a near-100% likelihood of species persistence over a set time frame, whereas for others, the goal may be lower. For example, during the development of the U.S. federal Northwest Forest Plan, species that were rated as having less than an 80% likelihood of being well distributed in the planning area were considered for additional protections (USDA and USDI 1993, 1994). In this regard, an 80% level may be interpreted as the acceptable risk level.

Risk attitudes are particularly variable among diverse conservation stakeholders, including biologists, land and natural resource managers, and the public (see chap. 9). A no-risk conservation approach may be paramount for some individuals and a priority for some conservation scenarios. This could highlight use of the precautionary principle to preserve species' or systems' potentialities and eliminate threats. In contrast, other individuals may view species and systems with an eye to their resilience, and consequently do not see all disturbances as such dire events; some risks might be acceptable because they are likely to affect species' abundances but not persistence. An approach with such risk may be warranted to balance conflicting priorities, such as the socioeconomics of managed lands, could be applied if a species' or system's situation were not necessarily on the brink of elimination, or may be used to test approach efficacies to advance knowledge. Programmatic risk attitudes need clarification so that various stakeholders with their own divergent risk philosophies can work together for a common goal (see consensus building, chap. 9).

Uncertainty is the expression of unknowns or sources of error (chap. 5). For RLK species, unknowns pertinent to conservation may include their distribution, abundance patterns, habitat associations, dispersal ability, reproductive rate, threats, and responses to disturbances. Uncertainty shrouds little-known species conservation, in particular. Questions of their basic biology and ecology, and how to approach effective conservation practices, outnumber known facts in many instances (chap. 4). For example, as aforementioned, species designations (i.e., scientific names) of many little-known fungi, mollusks, and arthropods have not been determined. How can they be fully considered in a conservation program given this most basic unknown? Unknowns may lead to conservative estimates of species' population, ecology, or life history components. In some circumstances, uncertainty can be linked to a perception or assessment of risk to persistence. For example, a little-known species with poorly known distribution might be assumed to associate with particular habitats in an area, and consequently considered vulnerable to a variety of disturbances to those habitats. Risk to persistence from disturbances might be rated conservatively high to account for such uncertainties. Although risk assessments may attempt to acknowledge uncertainty and separate it from a risk rating, such separation is difficult to achieve because the concepts are highly interwoven.

Programmatically, uncertainty attitudes can be clarified. For example, many programs impose criteria to limit inclusion of little-known taxa, making them lower-priority considerations, and focus limited resources on better-known rare taxa. The federal Survey and Manage program had few exclusionary criteria and developed a programmatic process called "strategic surveys" to bolster information for these lesser-known species (USDA and USDI 2001, 2004b; Rittenhouse 2002, 2003; Molina et al. 2003, 2006). Strategic surveys included a diverse array of information-gathering approaches tiered to the need per taxon: library or museum searches and syntheses of information on species, field searches of known species' sites to collect additional habitat information, searches of specific areas for species occurrences such as near known sites or at current species range boundaries, and random survey designs to allow inference to larger landscapes (Olson et al. 2007). Over 10 years of such information gathering, at a sum of about $13 million, a tremendous amount of uncertainty was reduced, advancing species' management approaches significantly (multispecies random surveys, ~$8 million, Molina et al. 2006; other strategic surveys, ~$5 million, Olson et al. 2007).

Tear et al. (2005) addressed uncertainty in two ways in the guidelines for setting conservation objectives. First, they recognized that information gaps should be identified as a step in a transparent, repeatable process that assesses program targets. Second, they incorporated an evaluation of errors and uncertainties relevant to conservation program development. These included three main types, errors in (1) species occurrence; (2) responses to conservation actions; and (3) species persistence. Hence, not only could uncertainty affect status assessments and development of conservation program targets, it may also affect a clear understanding of program effectiveness. Conservation itself is an imprecise discipline, due to these interacting uncertainties in addition to the nature of often unpredictable sociopolitical and science interactions. Key points to consider, however, are clear statements of known errors, known uncertainties, assumptions when "in-lieu" data are being applied, and risk and uncertainty attitudes.

Capability

Capability includes the time, effort, resources, expertise, and recovery potential relative to RLK species conservation. These elements directly interact with authority and priority issues to determine program effectiveness. Capability is the engine that will move the species management "train" once it has an engineer and a track. The capability of a program is more often a barrier or pivot to conservation than a driver.

At the most basic level, given authority, a conservation program cannot proceed without funds (chap. 10). Economic considerations are dominant drivers of any conservation program. Given funding, a management infrastructure can be built, and necessary expertise and tools acquired for program implementation. Both funding adequacy and funding stability over time are needed. Miller et al. (2002) reported that increased ESA recovery plan funding was associated with improved species status, but concluded that the funding level was less than 20% of the amount they estimated was needed "to get the job done." In contrast, Restani and Marzluff (2001) examined recovery spending for ESA-listed birds; for 85 species, total annual expenditures ranged from $75 million to $141 million. They identified the 10 bird species receiving the most recovery funds from 1992 to 1995, ranging from $2.2 million to $24.9 million annually per species, and of these only 5 showed increasing population trends. These were not species with the highest priority ranks, and expenditures on them likely affected the recovery potential of species with higher-priority ranking (Restani and Marzluff 2001). A combination of sociopolitical and economic forces constrain full implementation of the ESA and explain common complaints that the listing process and development of recovery plans are slow, and the perceived overuse of the "warranted but precluded" category for species in apparent peril.

The concept of recovery potential is an important capability issue, although complex to quantify due to its socioeconomic and ecologic contexts. Miller et al. (2002) suggested that species recoverability was affected by the specific threats the species faced. Generally, they found recoverability was high for ESA-listed taxa examined that were affected by resource extraction, direct human-caused mortality, natural threats, and development, and was low if the threats were exotic species, dams/drainage/diversion, and altered disturbance regimes. Efforts placed on taxa that have restricted distributions or are vulnerable to extant threats but are not rare enough to be ESA listed may have particularly high conservation capability in this regard. Where threats can be managed, Restani and Marzluff (2002) suggested full implementation of the spectrum of legal and political frameworks to maintain healthy ecosystems (e.g., Clean Water Act, National Environmental Policy Act, National Forest Management Act) as measures to avoid ESA listing for such species. They pointed out that ESA provisions are "meant as a last resort of species conservation." As an example of this approach that focuses on recoverability of unlisted concern-taxa, the annual budget of the Survey and Manage program ranged from less than $5 million to $11 million (USDA and USDI 2004b; budget declined from $11 million in fiscal year 2001 to less than $5 million in fiscal year 2004) and addressed over 300 RLK species. This program was comparatively cost-effective in terms of dollars per species (e.g., relative to ESA-listed species' recovery plan costs noted earlier), and successful overall in reducing threats (primarily from timber resource extraction) and precluding listing, at least in the first 12 years since its implementation.

Capability of rare species conservation can depend upon the programmatic infrastructure, particularly the personnel with biological, technical, and administrative skills who implement the program. Infrastructure can vary widely with goals and priorities. A single-species conservation plan may require only a relatively small panel of consulting species experts working in conjunction with existing land managers and planners. This would be a relatively simple program to implement. Several state-level conservation plans for threatened species have developed recovery plans using this blend of personnel. These usually involve one or a few taxa and small spatial scales. Large-scale single-species programs or a regional multispecies effort may require its own infrastructure of managers and planners, taxon specialists, field crews, data managers, and analysts. In particular, if little-known taxa are included in such a multispecies program, scientific expertise to identify species and develop species management guidelines may be limiting. If several little-known taxa are included, the program becomes much more complex to initiate and operate. If an effort crosses political or landownership boundaries, then the infrastructure may include a mix of individuals, organizations, agencies, and partnerships.

Salafsky et al. (2002) categorized conservation practitioners by three skill sets (knowledge, programmatic, administrative) and five functional roles (design, management, monitoring, analysis, and communication). The resulting 15 areas of organizational expertise have some overlap, and single individuals may be able to fill multiple roles. Multiple stakeholders such as landowners may add a third dimension or subdivide "cells" of this 3 by 5 matrix. Additionally, a comprehensive, large-scale, cross-taxonomic program may add an additional dimension or further subdivide skill set categories (e.g., knowledge per taxon, administration per landownership). Multiple organizations and taxa effectively transform this matrix of conservation expertise into a much more complex infrastructure.

Figure 11.1a shows this 3 by 5 matrix of conservation practitioner skills and roles, with two federal stakeholders splitting each of the 15 cells. This flat diagram shows the potential personnel infrastructure with a single-species conservation program, which could also apply to surrogate, flagship, indicator, or focal species approaches where a single-species focus is used. An infrastructure with 30 cells potentially needs to be populated. Of course, individuals may take on multiple roles or skills, and each stakeholder may not contribute personnel to every skill-role combination, such that many fewer than 30 people could run the program. Multiple flat-figure 11.1a models could result if conservation programs are not coordinated across taxa, as might occur for plants, fish, and nonfish fauna.

Figures 11.1b–d show how such an infrastructure might vary with conservation approach. Increasing the number of taxa involved in a single approach could only elevate the last row (fig. 11.1b), if programmatic and administrative duties are shared. This infrastructure might result from hot spot biodiversity or reserve systems approaches. The *z*-axis need not rep-

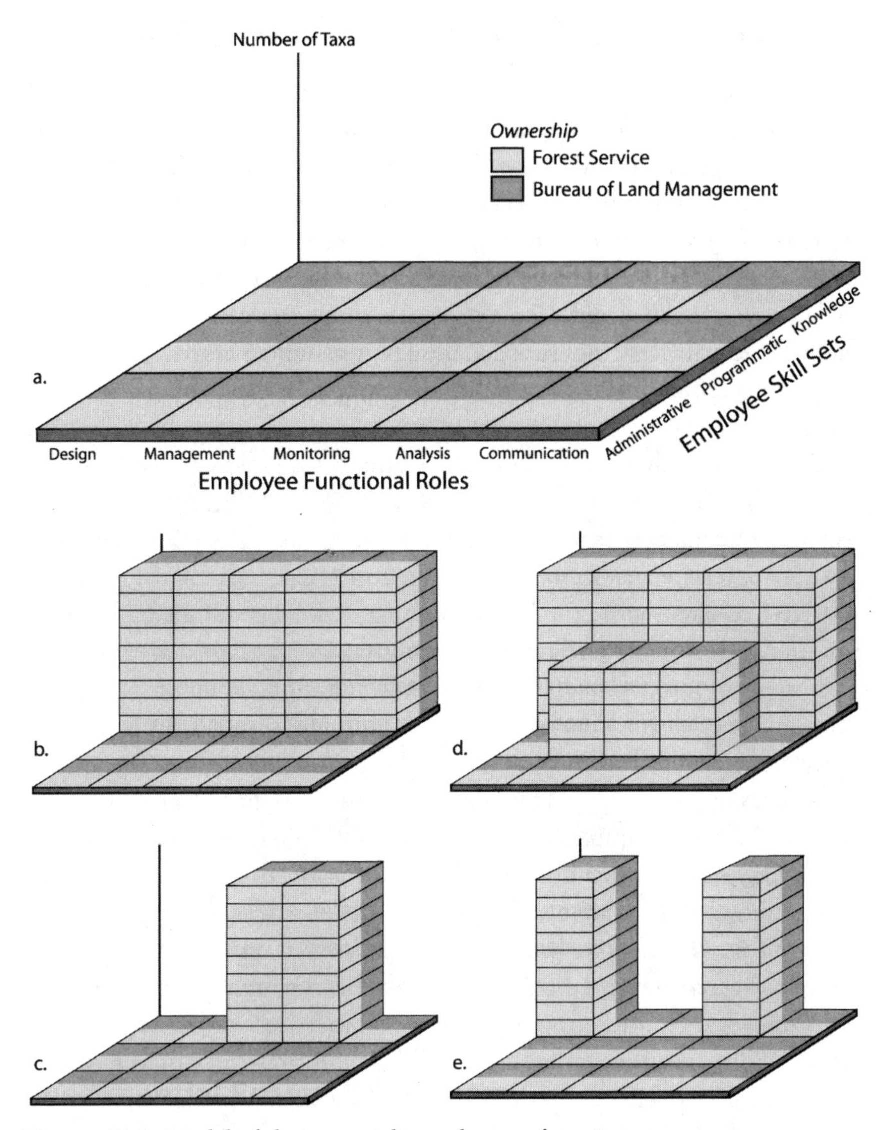

Figure 11.1. Model of the potential complexity of species conservation program personnel infrastructures, using a foundation matrix of three employee skill sets (y-axis: administrative, programmatic, and knowledge) and five employee functional roles (x-axis: design, management, monitoring, analysis, and communication; from Salafsky et al. 2002). These conservation practitioners may be further split among stakeholders, in this case federal land ownerships (cell shading). (a) A simple conservation program such as a single-species program in a small area is shown by a flat model, with no cells raised along the z-axis (Number of Taxa). Single employees could take on roles of several cells in this model to reduce infrastructure for small programs. (b–e) Species conservation program infrastructures increase in complexity as multiple taxa are addressed through different approaches, by personnel with different skills and functions. See text for further explanation.

resent the number of taxa for systems or habitat approaches, but rather the knowledge complexity these alternative methods entail. Different approaches require different types and levels of information. As such this last row could vary tremendously depending on the sophistication and details needed to implement a specific program. For example, acquiring new knowledge about little-known species might increase personnel in the knowledge skill set across functional roles. Managing habitats or landscape conditions similarly would require different skill sets than site management.

Some approaches may be oriented around single-species or relatively simple systems relative to design and management functions, but monitored for incidental benefits to multiple taxa or system functions (fig. 11.1c). Other scenarios can be conceptualized (figs. 11.1d and 11.1e). Figure 11.1d shows the sharing of programmatic skills in design and communication across personnel with differing taxa or system expertise. This might occur, for example, if management and monitoring of localities of multiple species is intended. Figure 11.1e might result from a focus on development of species-habitat models for conservation planning.

This example demonstrates the multidimensionality of program capability. The more complex infrastructures rely on greater resources and integrated management to retain program effectiveness. However, cooperation among stakeholders and cross-training of practitioners among skill sets and functional roles would reduce personnel and costs. In this way, systems approaches integrating multiple taxa priorities may be more cost-effective than numerous independently organized single-species programs. Balancing the conflicting priorities of optimizing cost versus conservation effectiveness is necessarily a management decision made when the capability of the program is determined. Conservation effectiveness might be maximized by a more extensive species-based approach.

Capability Examples

Capability issues are repeatedly addressed in the design and implementation of most conservation programs. In Brazil, retrospective analysis suggests that successful development of widespread conservation efforts are partially attributed to several "capacity building" measures (Mittermeier et al. 2005). These included (1) development of "catalysts" (i.e., nongovernmental organizations) that effectively served as intermediaries between researchers and land management interests; (2) primate conservation, which successfully leveraged funds to develop model conservation programs, facilitating the capability for many additional conservation endeavors; and (3) establishment of professional courses and opportunities that have developed expertise to populate the personnel infrastructure of conservation programs.

Implementation of recovery plans for ESA-listed species have met many of the capability issues mentioned here. For example, delay in recovery plan development and implementation, and lack of cooperation or coordination among stakeholders may have their roots in both insufficient funding and inadequate personnel infrastructure. These issues were cited as affecting two ESA-listed amphibians, the Wyoming toad (*Bufo baxteri*; Dreitz 2006) and the Houston toad (*Bufo houstonensis*; Brown and Mesrobian 2005).

In the Pacific Northwest, a host of capability issues can be identified from the federal Survey and Manage program for RLK species. Upon creation of the Survey and Manage provision in 1994, existing federal agency personnel were assigned Survey and Manage duties to oversee implementation relative to about 400 RLK species over 9.7 million ha. Usually, these individuals were assigned to interagency working groups to fill the new need, and most balanced these new tasks with the continuance of their previous jobs. Several problems became evident with this original framework. The extent of the workload eventually exceeded the capacity of the infrastructure; work could not be accomplished in a timely and effective manner by part-time committee members, with new members constantly coming and going. This framework also proved ineffective because individuals had outside work priorities and supervision, setting up complex conflicts at individual to programmatic levels. Taxonomic expertise for some groups (e.g., mollusks) was lacking within the agencies, stalling the implementation of survey and management procedures. The role of agency scientists in the development of management guidelines and policy was questioned, and the need to partner science experts with manager liaisons per taxonomic group was identified. Additionally, there was a need for clarification of the new standards and guidelines, and development of better cross-taxonomic criteria to assess species status.

The program was restructured in 2001 (USDA and USDI 2000, 2001), clarifying direction and reassigning species to newly derived rarity categories, and a distinct, funded "program" was created. Through this adaptive management "ratchet," 72 species were removed from the program, largely due to accrual of new information regarding their abundance, distribution, or habitat such that they no longer met the criteria for Survey and Manage. Program budgets ranging from almost $4 to $11 million per year (USDA and USDI 2004b) supported regional personnel at several levels to manage and implement tasks. The program became a self-contained entity, and when fully staffed, funded 95 persons with a total of 35 full-time equivalents (R. Huff, pers. comm.), fulfilling many if not all of the 3 by 5 areas of expertise outlined by Salafsky et al (2002), but with the "third dimension" added for several cells. Personnel included a program manager, coordinators of key areas such as conservation planning, annual species review, and strategic surveys, taxon-specific "experts" and "leads" to address the science and management aspects of the program, respectively, database managers, a biometrician, and federal agency representatives providing oversight. The number of federal agency stakeholders was greater than represented in figure 11.1 and included separate roles of U.S. Forest Service management and research arms for each of two Forest Service regions, the Bureau of Land Management, and the U.S. Fish and Wildlife Service. These six stakeholders had roles within the program as team members on taxon-specific teams, liaisons between field-research-management personnel, information synthesizers, intermediate managers, and high-level managers with decision authority. The more complex programmatic infrastructure resulted in a highly effective conservation program, although adaptive management (discussed later) could likely improve efficiencies (e.g., Olson et al. 2007).

In 2002, the program was challenged in court by the timber industry due to its effect on timber resource management in the region (USDA and USDI 2004a). The federal agencies developed a "settlement agreement" that would include an alternative to remove the Survey and Manage provision from the Northwest Forest Plan, with a court-imposed deadline to enforce its completion. In 2004, this alternative was chosen and the Survey and Manage provision was eliminated. Rare species management for eligible taxa was turned over to the agency special-status and sensitive-species programs. This decision realigned resource management priorities and capabilities on federal lands and led to an infusion of "capability" (funding, trained personnel, infrastructure categories) to the newly coordinated federal agency sensitive-species programs in the region. In 2006, the Survey and Manage program was reinstated by court order. Retention of aspects of its earlier capability (again, funds, people, infrastructure) has allowed this recent transition to be relatively smooth. Although the future path of this situation is uncertain, the established capability of the agencies to implement both a complex and more streamlined version of rare species conservation programs has enabled its resiliency to respond to change as directed by legal, political, social, and/or science-based needs.

Accountability

Program accountability is the measurement of its success at meeting goals and priorities. Without accountability, renewal of funds or infrastructure for a program capability may be lost. The Nature Conservancy measures success by regular evaluations of its conservation targets and threats to those targets (http://nature.org), assessing its institutional conservation impact, activity, and capacity. Accountability is necessary for continued program support regardless of approach, species- or systems-based. Implementation of an adaptive management framework for species conservation has become an effective mechanism to ensure program accountability, whereas monitoring and communication of findings are central to establishing accountable conservation programs.

Adaptive Management

Conservation science is not a static system, such that a program can be established and set on its course without checks and balances. Species populations are dynamic entities, as is our knowledge base concerning them and their interactions with their environments. Our toolbox for successful species conservation or restoration is similarly under development. Thus our capacity to implement a conservation program will in part depend on our resilience to respond to changes in species status and knowledge. Adaptive management provides a process to implement dynamic change in the program over time as it responds to new situations and knowledge. To effectively create a dynamic program, however, subject to change, is not easily done. Salafsky et al. (2002) called this dynamic program structure a "learning portfolio" in which a conservation program develops a formal learning framework, a cycle of collecting data, developing knowledge and tools, testing species management strategies, and using this new information to improve programmatic capacity and effectiveness. This changes the concept of "cells" in figure 11.1 from static, closed, independent entities to dynamic, interacting functions.

Within such a cyclic learning environment, goals and priorities are regularly revisited, new information is acquired and integrated with existing knowledge, and altered management practices can refine and improve the program's effectiveness. As in any organization, installation of checks and balances improves performance. When done as part of the programmatic structure, iterative changes are expected. With timely reviews, drastic changes are less likely and can be smoothly implemented. In contrast, irregular reviews to static systems may uncover unexpected imperfections that impact the program, possibly compromising its integrity. This accountability and adaptive management process requires the establishment of timelines, benchmarks, standardized tools for performance assessment, and new information acquisition through census, research, and/or monitoring. Furthermore, a synthesis process is needed to integrate new information and assess success relative to established targets.

For RLK species conservation, in particular, accountability and adaptive management are critical. For the rarest taxa with declining trends, species management practices must readily stem losses or alleviate threats for species' persistence. If we discover that management actions are not well applied, are ineffective, or pose risks to species, that knowledge needs to feed back into a reassessment of the program so new approaches can be developed. Lack of an iterative review to hone management procedures could result in irreversible losses. For many RLK species, due to our lack of baseline ecological information on their life history, tolerance limits to environmental parameters, and habitat requirements, conservation actions may be trial-and-error in nature such that adaptive management is key for mitigating losses or concern. New knowledge accrual can be rapid for never-studied taxa, and such information may realign species priorities for program attention. Furthermore, the science of conservation biology is a relatively young discipline; direct links between biology, ecology, and management tools are not well developed (Clark et al. 2002). Thus adaptive feedback loops relative to both program implementation and effectiveness are needed due to our limitations in both taxonomic knowledge and the application of a relatively new science.

Monitoring

Such feedback implies development of a monitoring network as an iterative review process of progress and new developments. Three types of monitoring are needed (implementation, validation, and effectiveness monitoring). Implementation monitoring provides short-term reporting of the progress in establishing actions. Validation monitoring establishes links between the causes and effects under consideration (e.g., cause: threat to species; effects: declining species numbers due to threat). Effectiveness monitoring evaluates the long-term outcomes of imposed conservation actions.

It is important to tie effectiveness monitoring objectives directly to the conservation plan objectives set forth at the outset of plan design. Measurable plan objectives (Tear et al. 2005) are those elements quantified during monitoring. Criteria for measuring success at both short-term and longer intervals are needed (Semlitsch 2002; Tear et al. 2005). Beever (2006) reviewed monitoring targets and methods for both species- and systems-based approaches to conservation. For RLK species, specifically, careful consideration of monitoring elements is needed because species population numbers may be inherently low or difficult to assess. For example, trend analyses may not be possible with such data. Nevertheless, concrete comparisons may be possible for elements such as occupancy rates, habitat condition, and threat abatement. Additionally, monitoring techniques selected may affect data quality (Semlitsch 2002) and potentially the inference of the results. Many lessons have been learned from past monitoring efforts so new monitoring programs do not need to be developed from "scratch" (Stem et al. 2005). In particular, with multiple plan objectives, a single-pronged monitoring approach may not suffice. One of the more common lessons learned in establishing monitoring programs is to match different monitoring approaches to different objectives.

There are several examples of the need for accountability and monitoring in rare species management programs. Reviews of ESA recovery plans have uncovered numerous implementation and effectiveness issues, as well as program successes. The review by Clark et al. (2002) listed numerous recommendations for recovery plan improvement. Some of these included (1) make amelioration of threats a primary focus; (2) specify adequate monitoring tasks; (3) ensure current, quantitative, and documented data on species status; and (4) improve and standardize the revision (adaptive management) process. Standardization in program implementation across spatial and time scales, or taxonomic boundaries is an emerging consideration affecting accountability. Comparison of findings across taxa or over time/space is difficult if different methods were used to collect or evaluate data. Standard protocols such as those followed for the National Environmental Policy Act or environmental assessments are useful in this regard. Without consistently applied measures of assessment or species management, a mosaic of approaches results, potentially compromising interpretations of conservation effectiveness.

Communication

Accountability has a strong socioeconomic component, as program advocates often need assurance that preset goals or their concerns are being met and cost-effectiveness is being achieved. Communication is imperative, to the extent that marketing of the conservation program's development, successes, and progress may be needed, especially if multiple resource conflicts result from conservation decisions. Annual reporting may provide this review as part of the learning cycle. However, annual reports may reach only those persons on the "mailing list," whereas rare species or biodiversity issues may have a much broader constituency among the various biodiversity stakeholders, including public, private, and governmental entities. Internal and external communication networks may need to be established to ensure program visibility and, subsequently, viability. Internally, reporting should reach the hierarchy of directly and indirectly involved persons and stakeholders. Externally, a more active outreach or "marketing" of the program and its primary annual accomplishments may be key in maintaining or enhancing program advocacy. Conservation International uses internal and external reviews and reports as ways to document progress (e.g., Conservation International annual report, http://www.conservation.org/xp/CIWEB/about/annualreport.xml). Because RLK species conservation programs are likely to be acquiring new species knowledge and developing new tools to conduct ecological management, external communication to both science and management professionals is important.

Several communication outlets are available to meet the need for external program accountability. Communication venues such as presentations at regional, national, and international professional meetings allow for reporting new findings, gaining external review of procedures and results from peers that may not otherwise be aware of program existence, and developing external liaisons for potential group resolution of common problems. This method of technology transfer is effective and moderately far-reaching. Development of hard-copy reports, handouts, fact sheets, or brochures is similarly effective, as made available at conventions of interested personnel or in educational contexts. Electronic venues of information sharing are fast becoming the most far-reaching mechanism of technology transfer. Such communication may include electronic mail and mailing lists, topical list-servers, program Web sites, and conference Web sites. Emerging interactive online Web technologies of interest include various Internet "portals" for biodiversity data (Kagan 2006) and collaborative Web sites, termed "wikis." Classic peer-reviewed publications should be included in this regard as the external communication mechanism with professional stature and permanency.

Furthermore, clarity of communications should not be overlooked. Conservation terminology is inconsistent among researchers, managers, and the public, which may result in confusion among the diversity of conservation stakeholders (Stem et al. 2005). Conservation is a fast-developing discipline and new terms are quickly incorporated into concepts and application. RLK species conservation is a particularly complex problem because it involves new innovative techniques and designs and introduces new and often undefined terms. Improving the understanding of conservation issues among all stakeholders and the public is an underlying consideration for overall conservation program development.

Conclusion

Implementation considerations may facilitate or constrain species conservation programs or approaches (species, systems). These considerations may be socioeconomic, political, or scientific in nature. Primary categories of implementation considerations include authority, capability relative to funding and program infrastructure, and accountability. Secondary categories that integrate these include clear priorities, adaptive management, monitoring, and communication. Authority can be legally founded, or a social or sociopolitical force. Legal authority may not extend to little-known species or some systems approaches to conservation. Capability is multidimensional, including time, effort, resources (funds), and expertise (program infrastructure). It is tied to the complexity of the approach selected. Capability particularly increases with number and type of taxa, systems, and stakeholders involved. Conservation priorities functionally link authority and capability, whereas adaptive management links capability to accountability. Incorporation of learning objectives and monitoring improve accountability for RLK species conservation. Communication of successes to the extent of marketing the program is needed when multiple resource conflicts result from conservation decisions.

Relatively simple single-taxon programs may require little management and scientific expertise, whereas broad-ranging multitaxa programs can be complex, involving personnel from multiple agencies or institutions. If little-known taxa are included, expertise may be limiting and practicality issues may limit capability, feeding back into constraints on conservation priorities. By nature, RLK species conservation programs are information poor relative to both species' biology and management knowledge. Thus the adaptive management process requires a well-designed monitoring component that in itself can serve as part of the mitigation measure for RLK species conservation.

Acknowledgments

I thank Rob Huff for comments on an earlier draft of this chapter, and Kathryn Ronnenberg for graphic design. Funding was provided by the Aquatic and Land Interactions research program of the USDA Forest Service Pacific Northwest Research Station.

References

Bani, L., M. Baietto, L. Bottoni, and R. Massa. 2002. The use of focal species in designing a habitat network for a lowland area of Lombardy, Italy. *Conservation Biology* 16:826–31.

Beever, E. A. 2006. Monitoring biological diversity: Strategies, tools, limitations, and challenges. *Northwestern Naturalist* 87:66–79.

Brown, L. E., and A. Mesrobian. 2005. Houston toads and Texas politics. Pp. 150–67 in *Amphibian declines: The conservation status of United States species,* ed. M. Lannoo. Berkeley: University of California Press.

Clark, J. A., J. M. Koekstra, P. D. Boersma, and P. Karieva. 2002. Improving U.S.

endangered species act recovery plans: Key findings and recommendations of the SCB Recovery Plan Project. *Conservation Biology* 16:1510–19.

Dreitz, V. J. 2006. Issues in species recovery: An example based on the Wyoming toad. *BioScience* 56:765–71.

Hohn, S. M. 2003. A survey of New York State pet stores to investigate trends in native herpetofauna. *Herpetological Review* 34:23–27.

Kagan, J. S. 2006. Biodiversity informatics: Challenges and opportunities for applying biodiversity information to management and conservation. *Northwestern Naturalist* 87:80–85.

McCune, B., and J. B. Grace. 2002. *Analysis of ecological communities.* Gleneden Beach, OR: MjM.

Meffe, G. K. 2001. The context of Conservation Biology. *Conservation Biology* 15:815–16.

Miller, J. K., J. M. Scott, C. R. Miller, and L. P. Waits. 2002. The endangered species act: Dollars and sense? *BioScience* 52:163–68.

Mittermeier, R. A., G. A. B. Da Fonseca, A. B. Rylands, and K. Brandon. 2005. A brief history of biodiversity conservation in Brazil. *Conservation Biology* 19:601–7.

Molina, R., D. McKenzie, R. Lesher, J. Ford, J. Alegria, and R. Cutler. 2003. *Strategic survey framework for the Northwest Forest Plan survey and manage program.* General Technical Report PNW-GTR-573. Portland, OR: U.S. Department of Agriculture, Forest Service, Pacific Northwest Research Station. http://www.or.blm.gov/surveyandmanage/USFS/PNW-Pubs/gtr573.pdf.

Molina, R., B. G. Marcot, and R. Lesher. 2006. Protecting rare, old-growth, forest-associated species under the Survey and Manage Program guidelines of the Northwest Forest Plan. *Conservation Biology* 20:306–18.

Neilsen, J. L., ed. 1995. *Evolution and the aquatic ecosystem: Defining unique units in population conservation.* Symposium 17. Bethesda, MD: American Fisheries Society.

Niwa, C. G., and R. W. Peck. 2002. Influence of prescribed fire on carabid beetle (Carabidae) and spider (Araneae) assemblages in forest litter in southwestern Oregon. *Environmental Entomology* 31:785–96.

Olson, D. H., S. S. Chan, and C. R. Thompson. 2002. Riparian buffers and thinning designs in western Oregon headwaters accomplish multiple resource objectives. Pp. 81–91 in *Congruent management of multiple resources: Proceedings from the Wood Compatibility Workshop*, ed. A. C. Johnson, R. W. Haynes, and R. A. Monserud. General Technical Report PNW-GTR-563. Portland, OR: U.S. Department of Agriculture, Forest Service, Pacific Northwest Research Station.

Olson, D. H., K. J. Van Norman, and R. D. Huff. 2007. *The utility of strategic surveys for rare and little known species under the US federal Northwest Forest Plan.* General Technical Report PNW-GTR-708, Portland, OR: U.S. Department of Agriculture, Forest Service, Pacific Northwest Research Station.

Pennock, D. S., and W. W. Dimmick. 1997. Critique of the evolutionarily significant unit as a definition for "distinct population segments" under the U.S. Endangered Species Act. *Conservation Biology* 11:611–19.

Pollack, D. 2001a. *Natural community conservation planning (NCCP): The origins of an ambitious experiment to protect ecosytems.* Part 1 of a series. CRB-01-002, March 2001. Sacramento: California Research Bureau, California State Library. http://www.library.ca.gov/crb/01/02/01-002.pdf.

———. 2001b. *The future of habitat conservation?: The NCCP experience in southern California*. Part 2 of a series. CRB-01-009, June 2001. Sacramento: California Research Bureau, California State Library. http://www.library.ca.gov/crb/01/09/01-009.pdf.
Restani, M., and J. M. Marzluff. 2001. Funding extinction? Biological needs and political realities in the allocation of resources to endangered species recovery. *BioScience* 52:169–77.
———. 2002. Avian conservation under the endangered species act: Expenditures versus recovery priorities. *Conservation Biology* 15:1292–99.
Rittenhouse B. 2002. *2002 Strategic survey implementation guide*. Version 2.4. Portland, OR: U.S. Department of Agriculture, Forest Service, Region 6. http://www.or.blm.gov/surveyandmanage/StrategicSurveyGuides/2002/SS_Imp_Guide.pdf.
———. 2003. *2003–2004 Strategic survey implementation guide*. Version 1.2. Portland, OR: U.S. Department of Agriculture, Forest Service, Region 6. http://www.or.blm.gov/surveyandmanage/StrategicSurveyGuides/2003/2003_SS_Implementation_Guide.pdf.
Robertson, D. P., and R. B. Hull. 2001. Beyond biology: toward a more public ecology for conservation. *Conservation Biology* 15:970–79.
Salafsky, N., R. Margoluis, K. H. Redford, and J. G. Robinson. 2002. Improving the practice of conservation: A conceptual framework and research agenda for conservation science. *Conservation Biology* 16:1469–79.
Sanderson, E. W. 2006. How many animals do we want to save? The many ways of setting population target levels for conservation. *BioScience* 56:911–22.
Semlitsch, R. D. 2002. Critical elements for biologically based recovery plans of aquatic-breeding amphibians. *Conservation Biology* 16:619–29.
Song, S. J., and R. M. M'Gonigle. 2001. Science, power, and system dynamics: The political economy of conservation biology. *Conservation Biology* 15:980–89.
Stankey, G. H., and B. Shindler. 2006. Formation of social acceptability judgments and their implications for management of rare and little-known species. *Conservation Biology* 20:28–37.
Stem, C., R. Margoluis, N. Salafsky, and M. Brown. 2005. Monitoring and evaluation in conservation: A review of trends and approaches. *Conservation Biology* 19:295–309.
Svancara, L. K., R. Brannon, J. M. Scott, C. R. Groves, R. F. Noss, and R. L. Pressey. 2005. Policy-driven versus evidence-based conservation: A review of political targets and biological needs. *BioScience* 55:989–95.
Tear, T. H., P. Kareiva, P. L. Angermeier, P. Comer, B. Czech, R. Kautz, L. Landon, et al. 2005. How much is enough? The recurrent problem of setting measurable objectives in conservation. *BioScience* 55:835–49.
USDA and USDI. 1993. *Forest ecosystem management: an ecological, economic, and social assessment*. Report of the Forest Ecosystem Management Assessment Team. Portland, OR: U.S. Department of Agriculture, Forest Service, and U.S. Department of Interior, Bureau of Land Management.
———. 1994. *Record of decision on management of habitat for late-successional and old-growth forest related species within the range of the northern spotted owl* [Northwest Forest Plan]. Portland, OR: US Department of Agriculture, Forest Service, and US Department of Interior, Bureau of Land Management.
———. 2000. *Final supplemental environmental impact statement for amendment*

to the survey and manage, protection buffer, and other mitigation measures standards and guidelines. 2 vols. Portland, OR: U.S. Department of Agriculture, Forest Service, and U.S. Department of the Interior, Bureau of Land Management.

———. 2001. *Record of decision and standards and guidelines for amendments to the survey and manage, protection buffer, and other mitigation measures standards and guidelines.* Portland, OR: U.S. Department of Agriculture, Forest Service, and U.S. Department of the Interior, Bureau of Land Management.

———. 2004a. *Record of decision to remove or modify the survey and manage mitigation measure standards and guidelines in Forest Service and Bureau of Land Management planning documents within the range of the northern spotted owl.* Portland, OR: U.S. Department of Agriculture, Forest Service, and U.S. Department of the Interior, Bureau of Land Management.

———. 2004b. *Survey and manage: fiscal year 2003 annual status report.* Portland, OR: U.S. Department of Agriculture, Forest Service, and U.S. Department of the Interior, Bureau of Land Management. http://www.or.blm.gov/surveyandmanage/AnnualStatusReport/2003/S_and_M-2003.pdf.

Wall, D. H., ed. 2004. *Sustaining biodiversity and ecosystem services in soils and sediments.* Washington, DC: Island Press.

Waples, R. S. 1991. Pacific salmon, *Oncorhynchus* spp., and the definition of "species" under the Endangered Species Act. *Marine Fisheries Review* 53:11–22.

12

A Process for Selection and Implementation of Conservation Approaches

Martin G. Raphael, Randy Molina, Curtis H. Flather, Richard S. Holthausen, Richard L. Johnson, Bruce G. Marcot, Deanna H. Olson, John D. Peine, Carolyn Hull Sieg, and Cindy S. Swanson

In this book, we have described a variety of approaches designed to achieve biological conservation objectives for both rare and little-known species. Resource managers are usually most concerned, for practical reasons, with approaches that conserve large-bodied and better-known taxa. Because we know so little about population status and life history requirements of rare or little-known (RLK) species, conservation managers have been forced to assume that the needs of these species are being met via proxy through our management of vegetation communities and well-studied species. Because so much of the world's biological diversity consists of rarer and more poorly known species (see chaps. 3 and 4), our focus on these species is warranted. In this chapter, we propose a process by which a land manager might evaluate the conservation issues on a particular planning area and select a set of approaches to address those issues. We organize this discussion around a set of steps, as illustrated in figure 12.1:

1. Identify geographic area and scope
2. Identify social, economic, and ecological requirements and considerations

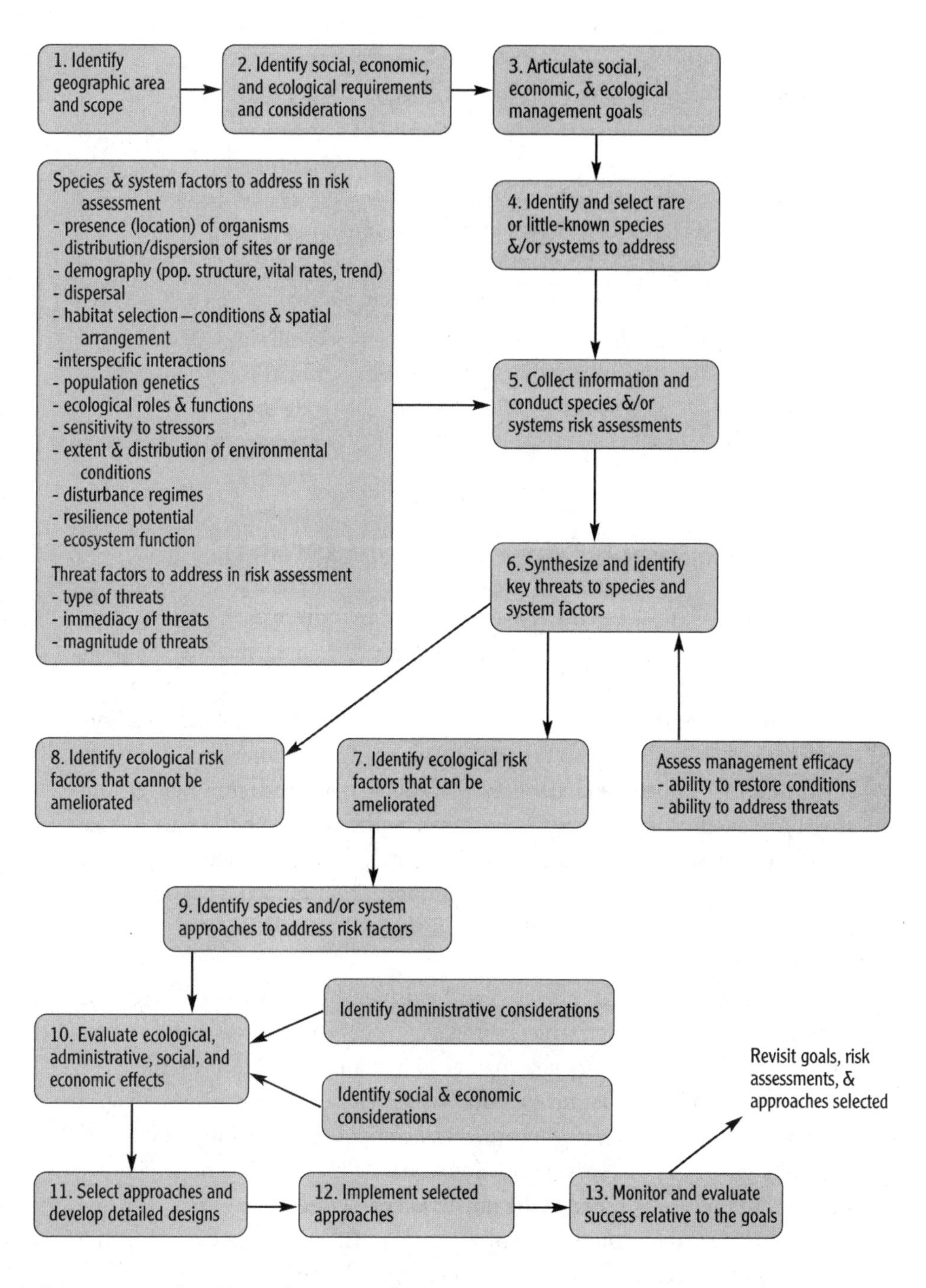

Figure 12.1. Flow chart of a process for selecting and testing alternatives to conservation of rare or little-known species as discussed in this book.

3. Articulate social, economic, and ecological management goals
4. Identify and select rare or little-known species and systems to address
5. Collect information and conduct species and/or systems risk assessments
6. Synthesize and identify key threats to species and system factors
7. Identify ecological risk factors that can be ameliorated
8. Identify ecological risk factors that cannot be ameliorated
9. Identify species and/or system approaches to address risk factors
10. Evaluate ecological, administrative, social, and economic effects
11. Select approaches and develop detailed designs
12. Implement selected approaches
13. Monitor and evaluate success relative to the goals

Identify Geographic Area and Scope (Step 1)

The first step is to delineate the geographic area of interest. It is important to identify not only the overall extent of the area but the range of ecosystems or subareas within the extent that will be the focus of planning. For example, one may be interested only in grassland systems within a larger landscape, so the specific planning area may be a set of disjunct units within some larger context. The size of the planning area and the number of ecosystem types within it will have an obvious influence on the complexity of the conservation problem. As area and variety of ecosystems increase, so will the number of potential species to be addressed, following the familiar species–area curve (e.g., Fisher et al. 1943).

Delineation of a planning area can be driven by a variety of factors or considerations. Administrative boundaries, such as a ranger district or project area, may be a primary consideration. Landownership will likely be important as well. The boundary may be determined by the range of a critical species or ecosystem type or it may encompass the combined ranges of several key species or ecosystems. Boundary delineation is a management decision that must be made at the outset of the planning process. The planning area needs to encompass an area that is sufficiently large to affect the species and processes of greatest concern, and at the same time allow assessment of the planning area's context at larger scales.

Identify Social, Economic, and Ecological Requirements and Considerations (Step 2)

Once the planning area is identified, the next step is to articulate the various social, economic, and ecological issues and considerations that apply. First, there may be constraints imposed by the ownership pattern in the area. There may be multiple jurisdictions (e.g., counties, towns) and there may be a mix of federal, state, or private lands, each with their attendant regulations and requirements. In addition, it will be important to identify and list the interests and priorities of the various stakeholders within the planning area. Any existing plans that cover the area may need to be modified or replaced.

Part of this step also includes characterizing the primary ecological issues. For example, on the Sheyenne National Grassland, the presence of the western prairie fringed orchid (*Platanthera praeclara*) is a major consideration because this species is federally listed as threatened under the Endangered Species Act (ESA) and is ranked as globally imperiled under the NatureServe ranking of species conservation status (see chap. 3). Note, however, that a full evaluation of the various species and systems in the planning area, as well as any threats to these species or systems, occurs in steps 5 and 6. At this point in the process the intent is to bring forward the ecological, social, legal, administrative, and political considerations that will influence objectives for the conservation plan.

The initial identification of social and economic considerations occurs in this step. Social considerations may include values of affected communities, the inclusiveness and balance of the decision-making process that will be used, and political and institutional dynamics that may affect the decision. Economic considerations should include economic efficiency and economic impacts. Interrelationships between economic and social effects should also be considered.

Articulate Social, Economic, and Ecological Management Goals (Step 3)

After identifying the major issues to be addressed, the next step is to specify the ecological, social, and economic goals and objectives for the plan (see chap. 2). Some of these goals will be quite broad (e.g., reintroduce fire as a natural disturbance process) whereas others may be quite specific and measurable to tie into tangible monitoring elements to assess program success later (e.g., provide conditions to support a viable population of the western prairie fringed orchid). Both short- and long-term objectives will facilitate the adaptive management of the program and help assure its success. Setting goals and objectives will involve feedback from other steps in this process. Goals may be revised or refined as information develops in subsequent steps. Consequently, we anticipate a fluid process between succeeding steps.

For example, if a land management plan is already in effect for the planning area of interest, any new goals or objectives will need to be reconciled with it. There may be additional mandates that will influence the specific goals and objectives for the area; these will have been brought forward in Step 2. These various goals and objectives, as well as existing mandates, will be blended to set priorities for management action. Setting priorities for the area should be an inclusive process that involves collaboration among affected parties and recognition of social, economic, and ecological requirements and considerations for the planning area.

Identify and Select RLK Species and Systems to Address (Step 4)

This is a critical step that will significantly impact future resource needs. The process of selecting species and systems to address is strongly influenced by the overall scope, complexity, and goals noted previously.

Identify and Select RLK "Species"

In this step, RLK species may include taxonomic species and also selected subspecies, populations, or other entities below the species level (hereafter referred to collectively as species). Before selecting individual species, the process begins by addressing which major taxonomic groups (viz., taxonomic classes) will be considered for risk analyses. Considerations are influenced by a combination of factors that relate to the initial goals as well as to the practical science issues and resource needs. Legal considerations are often defining factors. Species listed under the ESA, for example, are given high priority in conservation programs. Broad goals such as conserving total biodiversity may include many more species and other taxa than simply protecting ESA-listed species and may therefore require comprehensive processes for dealing with large numbers of species. Such was the case in developing the Northwest Forest Plan, where eight taxonomic groups, reflecting a broad array of biodiversity considerations, were selected for risk assessments (Meslow et al. 1994).

Several science- and resource-related issues must be weighed when considering which taxonomic groups to include, because each taxon has unique attributes. Some practical considerations for selecting species and taxa include number of species; availability of information on natural history, ecology, habitat requirements, dispersal, and threats; issues of detectability and sampling difficulty; availability of taxon experts; and overall practical experience in developing feasible management strategies. The size of the planning area and scope of the problem (box 12.1) strongly influence which taxa are selected.

After taxonomic groups are selected, the level of species analysis is determined. Most assessments have traditionally focused on individual species, but depending on the scope of the issue, the focus could be on evolutionary significant units (ESUs), individual populations, or other entities below the species level. For some taxa or ecosystems, multiple species assemblages may be identified for consideration rather than individual species. This could be the case where taxonomic groups include hundreds or thousands of poorly known species. For example, given the tens of thousands of arthropods in forest ecosystems, the Forest Ecosystem Management Assessment Team (FEMAT) analysis addressed only 15 functional groups of arthropods (Meslow et al 1994).

A detailed and preferably streamlined process is then developed for the actual selection of species (or species units) for risk assessment. Examples of methods for generally identifying species at risk, which may be extended to RLK species, have been suggested or tested by Wright et al. (2001) for plants and Lehmkuhl et al. (2001) for nonfish vertebrates. In a further consideration of risk factors, being both rare and specialized can mean higher risk of local extinction in forest fragments—as reported by Davies et al. (2004), who suggested that synergistic characteristics of some rare forest beetle species prone to extinction include the interplay between their low abundance and their high degree of specialization. Several major ranking systems such as those by NatureServe and the International

Box 12.1. Selecting approaches to rare or little-known species conservation: a multiobjective assessment approach to risk assessment and management.

Essentially, the general procedure suggested in this chapter for selecting approaches to conservation of rare or little-known species is a problem of multiobjective management. There are a number of useful methods for conducting multiobjective risk management. For example, Mendoza and Prabhu (2000) applied multiple criteria analysis (MCA) to assess criteria and indicators for forest management by ranking and rating alternative management decisions in a participatory setting with stakeholders and experts.

Other approaches include multiattribute utility theory (MAUT), goal hierarchy, analytic hierarchy process (AHP), multiple criteria decision making (MCDM), quantitative risk analysis (QRA), and others (e.g., Varis 1980; Basak and Saaty 1993; Helles et al. 1999; Mendoza and Prabhu 2000). All of these approaches entail identifying potentially desirable conditions, such as high confidence of persistence of rare or little-known species; identifying sources and degrees of risk; identifying the expected influence of potential management actions on those causes; exploring potential costs, benefits, risk attitudes, and acceptable levels of influence; and eventually selecting the best course of action that will minimize risks and costs or maximize likelihoods of desired outcomes. More formally, the performance level of each objective is quantified, the preferences among different objectives are weighted, the alternative management approaches are ranked, and an interim decision is made on the optimal course of action. Sensitivity testing can be used to help identify the relative influence of risk factors on desired species and system conditions and the relative influence of alternative management decisions on those risk factors.

Union for Conservation of Nature and Natural Resources (IUCN) provide detailed processes to assess risk (see chap. 3). Local species analyses should take advantage of these and other ranking systems in developing a final species list for consideration.

Carefully considering information already available (e.g., species lists developed for the area) is advantageous. Information can range from private records, collections, and databases to federal and heritage listings. Taxon experts can be engaged at this point to collect and synthesize information. Selection criteria are chosen and applied to develop a list of species that require risk assessments. The entire process of species selection, including specific criteria used, should be well documented.

The early analyses leading to the development of the Northwest Forest Plan provide a large-scale example of this initial species selection. Using a combination of agency databases and knowledgeable taxa experts, the FEMAT science team selected approximately 1100 species, 21 groups of fish, and 15 groups of arthropods for risk analysis (FEMAT 1993). Efforts to develop the Sierra Nevada Framework (http://www.fs.fed.us/r5/snfpa) provide another large-scale example in selecting species. They used a carefully designed process to identify species for analysis that also explicitly recognized that the level of risk used as a cutoff in these evaluations was itself an important decision.

Identify Systems to Address

Selecting systems to address should be guided by the project size, scope, and objectives. Legal considerations such as protecting critical habitat of threatened and endangered species can play an initial role in the selection. Some systems themselves carry legal protection (e.g., wetland protection) which often influences their inclusion in the risk assessment.

Two areas need careful consideration when selecting systems. The first is availability of information (literature, databases, maps, etc.) and resources (particularly people) to define and characterize the systems. The second is the classification or characterization of the systems.

Definitions and delineations of systems can be a far more slippery task than that of species. There is no single taxonomy of ecosystems, ecological communities, species assemblages, or other representations of systems. Multiple classifications exist for these types of systems and for vegetation types, wildlife habitat types, ecoprovinces, and other systems-level entities (e.g., O'Neil et al. 1995; Rieman et al. 2000; Treitz and Howarth 2000; Kintsch and Urban 2002). Even the term "ecosystem" can mean vastly different things (Corn 1993; DeLeo and Levin 1997; Watson 1997). And, in addition, all such classifications are artificial abstractions whose attributes (other than geography) are often impossible to assign definitively at actual locations in the field, where a particular location may display properties of multiple classes.

Classifications can also include existing vegetation communities, potential natural vegetation types, and designations specific to particular environments such as riparian ecosystems or old-growth forest ecosystems, or even special landscape features such as caves (Sieg et al. 1999). Some systems are delineated by administrative boundaries such as planning units or reserve lands. Similar to selecting taxa groups, different system designations have unique attributes that influence the complexity of systems risk analyses (e.g., availability and quality of existing information). These unique attributes should be carefully weighed in selecting the final systems for risk analyses. As with taxon selection, all criteria used should be clearly documented and all assumptions and uncertainties described.

Collect Information and Conduct Species and Systems Risk Assessments (Step 5)

These analytical steps begin with collecting pertinent information. Sources vary widely and differ in quality and accessibility. They might include government and nongovernment databases, maps of species locations and distribution, maps of vegetation or ecosystems types, unit boundaries and land designations (e.g., reserves), and published scientific literature. Next, a list of risk criteria and threat factors is developed for the assessment process. The large box in the flow diagram (see fig. 12.1) provides a range of examples. Risk analyses should focus on the factors that put the species or system at risk, currently or historically. The main stressors and potential effects of threats should be thoroughly characterized (e.g., type, immediacy, magnitude), the main assumptions clarified, and the main uncertainties documented. The final desired products and outcomes for the assessment (e.g., relative importance of different threats) also influence the selection of factors.

The selection of factors is also influenced by the availability of information to address a factor, and the utility of that information to accurately reflect the condition or threat under consideration. There is often limited mapped information for the actual threats considered important, and the magnitude of threats often has to be estimated from surrogate data and coarse modeling efforts. An example might be sedimentation in streams, which may be treated in a risk assessment by looking at rainfall, slope, soil and vegetation types, grazing and fire regime, and so forth, rather than by directly measuring the transport of silt.

Developing and using standardized processes to summarize information and assess risk is the most critical step in the risk analysis. Standardized processes are especially important when evaluating many taxonomic groups or systems. The literature on decision analysis and risk management offers a plethora of procedures (e.g., Hope and Peterson 2000; Williams 2000; McDonald and McDonald 2003, and many others). Panels of experts are often used to conduct portions of such evaluations, such as gauging effects of potential threats (e.g., von Winterfeldt 1992), and this introduces the chance of inconsistent interpretations across taxa or systems. Careful standardization of the process ultimately increases the likelihood of developing a scientifically sound assessment for decision makers (e.g., Shaw 1999).

As an example of the successful completion of this process, Thomas et al. (1993) analyzed risks to 667 species associated with old-growth forests in the Pacific Northwest. Risks included broad-scale declines in habitat, loss of fine-scale habitat features, and threats from various forms of human disturbance. The risk assessment was used to develop potential management strategies that could be applied by federal agencies.

The outcome of Step 5 is a clear description of the factors that cause species and systems to be at risk. These may include past reduction in abundance or distribution of habitat, changes over time in level of threat from various human activities, or other parameters. Results of the risk assessment are carried into the next step and provide the foundation for designing management approaches.

Synthesize and Identify Key Threats to Species and System Factors (Step 6)

This step entails combining the results of the species and system risk assessments. Doing this requires summarizing the expected types, degrees, and effects of risk factors on systems and on the persistence of RLK species. As an example of this step, Cane and Tepedino (2001) summarized results from a workshop (expert panel) to identify causes and degrees of declines of native invertebrate pollinators in North America, which include some RLK species of bees, flies, and other taxa.

Identify Ecological Risk Factors That Can Be Ameliorated (Step 7)

After identifying the common causes of risks to species and systems, the next step is to identify which causes can be ameliorated by management activities, and determine the degree to which management could reduce or eliminate each risk factor and thereby restore desired conditions. For example, Feldman et al. (1999) used a comparative risk assessment approach to help set priorities for environmental conservation. Their approach entailed identifying the relative efficacy and benefits of environmental policy decisions. They specifically examined how projects were administered; how they involved the public; how they characterized, ranked, and prioritized risks; whether and how they implemented projects based on results of rankings; and whether and how they evaluated project results.

Identify Ecological Risk Factors That Cannot Be Ameliorated (Step 8)

This step requires identifying which risk factors likely cannot be fixed by management, and outlining their expected effect on species or systems of interest. This will provide an understanding of the degrees to which management can and cannot be expected to solve specific problems. In so doing, it will help provide realistic expectations for the effects of management and identify any critical causes of risk that may need to be addressed by other concerned parties, ownerships, or organizations outside the immediate management focus.

An example is the decline of pendant arboreal lichens from poor air quality caused by point and nonpoint air pollution sources, such as urban industrial centers and automobiles. This may be a major cause of decline in some RLK species of lichens (see Stolte et al. 1993), although there may be little to nothing that forest management activities can do to slow or reverse the degradation. Similar off-site causes of risks to RLK species may be adverse regional climate change, spread of invasive species, noise pollution, and other risk factors.

In general, Steps 6 through 8 can be conducted as part of a structured decision-making procedure. Such a procedure serves to synthesize findings on risk levels and causes, and, eventually, on potential management actions that can ameliorate, restore, or mitigate for adverse effects. This is essentially a problem in *multiobjective risk management,* to which the preceding three steps and the next two steps pertain.

Identify Species and/or System Approaches to Address Risk Factors (Step 9)

The objective of this step is to make a preliminary determination of approaches that could be useful in conserving species (chap. 6) and systems (chap. 7) that were identified in Step 4, and that will address those threats identified in Step 7. At this point in the process, the identification of approaches should not be heavily constrained by economic, political, or administrative considerations. Those factors should be considered when implications of the approaches are reviewed in the next step (10). Final selection and refinement of the approaches takes place in Step 11.

There is no accepted, comprehensive guide to the selection and design of conservation strategies for various situations. However, consideration of the following factors can help inform the preliminary selection of conservation approaches:

Identified Ecological Risk Factors That Can Be Dealt with by Management

The risk factors identified in step 7 may provide important clues to the general type of conservation approach that would be helpful, as well as clues to some specific design considerations for that approach. As shown in chapter 8 (see table 8.2), each of the approaches addresses a subset of the factors that are important to species conservation. So if we know the factors that are causing a species to be at risk, this table can help guide the selection of a conservation approach. For example, if the loss of current species locations is a significant risk factor, and the species has strong co-occurrence associations with other more easily monitored species, then a useful strategy might combine the use of a biodiversity indicator species, a geographic approach, and some applications of individual viability strategies. Other surrogate approaches (e.g., umbrella or flagship species) and system approaches would be less applicable. Conversely, if the dynamics of a species' habitat was a key risk factor, the system approaches would tend to be useful, and the species approaches would be less useful. Of course, it will often be the case that there are a variety of risk factors that would best be dealt with through some combination of approaches.

Level of Our Knowledge about the Species/Systems

The level of our knowledge about a species or system may also be critical in the selection of a conservation approach. Although level of knowledge should not be considered an absolute long-term constraint to the selection of a conservation approach (i.e., with the proper investment it is possible to gain additional knowledge), it may be a very important consideration in the choice of a conservation approach that can be applied in the short term. Table 12.1 shows information that is needed for application of the various conservation approaches and may be helpful in determining which approach can be applied in a given situation. For example, if the only information available for an RLK species is information on its current locations, then the most useful approach over the short term may be management for specific locations of the species. If additional information is available, there may be a broader selection of possible approaches.

Basic Natural History of the Species/Systems

Natural history of a species may influence the choice of conservation approach. If a species tends to occur in small, isolated populations, then a focus on management of the locations of those populations may be most appropriate. On the other hand, if a species tends to occur more broadly, is relatively mobile, and has individuals that disperse among populations, then an approach based on viability of individual species, which would focus on the current location of populations, suitable but unoccupied habitat, and dispersal habitat, would be more appropriate.

Underlying Management Goal

As discussed in chapters 2 and 8, managers may operate under a variety of different management goals, and different conservation approaches may best

Table 12.1. *Information needed to implement species- and system-based approaches for conservation of rare or little-known species.*

Approach	Presence (locations) of organisms	Distribution/dispersion of sites or range	Demography (pop. structure, vital rates, trend)	Dispersal	Habitat selection—conditions and spatial arrangement	Interspecific interactions	Population genetics	Ecological roles and functions	Sensitivity to stressors	Extent and distribution of environmental conditions	Disturbance regimes	Resilience potential	Ecosystem function
Species-Based Approaches													
Viability of Individual species													
Viability of individual species	2	1	2	2	1	2	2	2	2	—	—	—	
Surrogate species[a]													
Focal species	2	1	—	1	1	—	—	—	1	—	—	—	
Umbrella species	—	1[c]	—	—	1	—	—	—	—	—	—	—	—
Guilds	—	2	—	—	1	2	—	2	—	—	—	—	—
Habitat assemblages	—	2	—	—	1	—	—	—	—	—	—	—	—
Biodiversity indicator species	2	1[c]	—	—	2	—	—	—	—	—	—	—	—
Flagship species	—	—	—	—	—	—	—	—	—	—	—	—	—
Geographic approaches													
Management for locations of target species	1	2	—	—	2	—	—	—	—	—	—	—	—
Hot spots (of RLK spp.)[b]	2	1	—	—	2	—	—	—	—	—	—	—	—
Reserves or protected areas	2	1	—	2	2	—	—	—	—	—	—	—	—

(continues)

Table 12.1. *Continued*

Approach	Presence (locations) of organisms	Distribution/dispersion of sites or range	Demography (pop. structure, vital rates, trend)	Dispersal	Habitat selection—conditions and spatial arrangement	Interspecific interactions	Population genetics	Ecological roles and functions	Sensitivity to stressors	Extent and distribution of environmental conditions	Disturbance regimes	Resilience potential	Ecosystem function
System-based Approaches													
Maintaining system structure and composition													
Managing for RNV	—	—	—	—	—	—	—	—	—	1	1	—	—
Managing for diversity of habitats using concept other than RNV	—	—	—	—	—	—	—	—	—	1	2	—	—
Strongly interacting species[c]	—	1	—	—	1	—	—	1	—	—	—	—	—
Maintaining system function													
Maintaining disturbance regimes	—	—	—	—	—	—	—	—	—	2	1	2	—
Maintaining other ecosystem functions	—	—	—	—	—	—	—	—	—	2	2	2	1

This table assumes that approaches would be crafted anew based on the central objective of conserving rare or little-known species.
RNV = range of natural variability
1 = minimum required information
2 = desired but not essential information
— = neither minimum required nor desired (i.e., n/a)
[a] may include non-RLK species
[b] information needed on > 1 RLK species
[c] includes information on co-occurrence with other species

contribute to each of those goals. Although the goals may not be mutually exclusive, emphasis on particular goals could influence the choice of conservation approach. For example, if the primary goal is to maintain RLK species over at least the short term, focus on approaches that maintain the current locations of the species is warranted. On the other hand, if the primary goal is to restore system resiliency, system approaches would be appropriate. Again, it is likely that a variety of goals could best be achieved through some combined approach. Many overall strategies also have a variety of specific implementation scenarios that may perform similarly for species conservation, for example, in establishing a network of protected sites totaling to some particular total acreage or budgetary target. In this case, the land manager would want to know relative costs and benefits for each proposed scenario.

Evaluate Ecological, Administrative, Social, and Economic Effects (Step 10)

The conservation approaches identified in Step 9 may have profound implications for social, economic, and ecological systems. They may also differ greatly in requirements for knowledge, trained personnel, budget, and administrative structures. These implications must be assessed before managers can make an informed decision about adopting an approach or set of approaches. Effects evaluation should tell the land manager whether the approach protects the species, is socially supported, and is worth adopting based on its cost–benefit analysis.

Evaluation of Ecological Effects

The following guidelines should be consulted in the evaluation of ecological effects:

- If the conservation approaches to be evaluated include the use of surrogate species or species assemblages, appropriate analysis should be completed to identify them. They then become the focus of species evaluations.
- Evaluation of effects should be framed as a risk and uncertainty assessment (e.g., Meslow et al. 1994; Raphael et al. 2001).
- The evaluation must include assessment of both short-term

and long-term risks. The timeframe over which long-term risks are projected should be determined based both on biology of the species (e.g., generation time, response time to changed conditions, recolonization capability) and on the time needed for the overall ecosystem to respond to proposed management. Assessment over such a timeframe is important to a full understanding of the long-term effects of management on ecosystems and species, but it must be understood that confidence in the accuracy of risk evaluations decreases rapidly as the timeframe of projections increases.

- The spatial scale of the evaluation should reflect the scale at which biological populations of the species operate and the scale at which ecological processes occur within the system, including its ecological context.
- In addition to the projected future, the analysis should also address the current condition and, where possible, the historical condition of the species and/or system.
- For most species, the only practical quantitative analysis is an assessment of habitat conditions. It is, however, essential that we connect habitat conditions with population consequences, even if this connection is estimated using general ecological principles due to lack of population data or knowledge of species population processes.
- The evaluation should be logical and consistent, consider relevant information, and disclose both risks and levels of uncertainty. It is important to document important sources of uncertainty, including uncertainty due to environmental stochasticity.
- Where surrogate species or species assemblages are used, the evaluations of the surrogates or assemblages should be applied to all of the species that they represent.

Evaluation of Administrative Considerations

Careful evaluation of administrative issues is key to determining the practical implications of implementing the conservation approaches identified in step 9. In particular, the various approaches will require different information sets, expertise, administrative structures, and costs. First, different approaches require different types and levels of information. Specific information needs of approaches are shown in table 12.1. Figure 12.1 provides a general view of the level of information required to implement a generalized set of conservation approaches.

Second, availability of specialists with the appropriate expertise will vary with approach chosen. Some approaches require the availability of highly specialized skills. For example, approaches that focus on protection of extant locations of RLK species require the availability of personnel who can identify the species. In some cases, identification of RLK species is an extremely specialized skill, and there may be few qualified personnel available. On the other hand, approaches that rely on conservation of habitats rather than conservation of species occurrences generally require less-specialized personnel assuming there is an accepted classification of vegetation or ecosystems within the planning area, and an understanding of the role of disturbances in maintaining the systems.

Third, administrative structures must be able to accommodate the conservation approach chosen and number of stakeholders involved. Well-designed administrative structures are needed for consistent implementation of conservation approaches over large geographic areas incorporating multiple landowners. For some approaches, these administrative structures may deal with facets of management that are familiar to the agencies involved and so may not be much of a departure from day-to-day management. For example, managing systems based on the range of natural variability involves forms of management that are similar to historic practices of the land management agencies. Conversely, managing for sites of individual species may involve issues that are much less familiar to the agencies and require coordinating structures that are new to the agencies.

However, perhaps the most important administrative consideration for implementing an approach is its cost. The fundamental capability of administratively implementing an approach involves its cost and available funding source. In general, the cost per species is often higher for individual species approaches and lower for systems approaches, surrogate approaches, and other approaches that address multiple species.

Evaluation of Social Considerations

The social sciences provide a mechanism to measure the social/cultural human context of the biological requirements of species conservation strategies (see chap. 9). The following guidelines should be consulted in evaluating the social consequences of alternative species conservation strategies.

Values

The foundation for building consensus begins with an understanding of the social values at play at the personal, family, business, and community levels. Values range across a wide spectrum from personal environmental stewardship ethics, to family values, including their traditional use of natural resources, to economic prosperity.

Decision-Making Process

Inclusiveness and balance of perspectives are keys to credible and successful consensus building. It is imperative to understand the perspectives of the stakeholders, their priorities, and how they are manifested in the process.

Institutional Dynamics

Traditional approaches and policy can create a precedent that narrows perspectives and stifles unbiased assessments of alternative conservation strategies. An open decision-making process should foster new ideas and consideration of alternative strategies.

Political Dynamics

Social values are manifested in political positions on issues of concern. More often than not, political influence drives the decision-making process and can detract from an open-minded assessment of alternatives. The political perspective concerning federal land stewardship and use is shifting away from national public opinion toward local special interests. These dynamics influence all aspects of the social dimensions of evaluating alternative species conservation.

Evaluation of Economic Effects

Economic considerations should be included early and continuously throughout the RLK species conservation process. This reduces surprise and distrust when biologically and technically feasible conservation outcomes may not be achievable based on economic efficiency grounds later in the decision process. The evaluation of economic efficiency should be framed as a comprehensive benefit to cost analysis. Costs should include the opportunity costs of alternative resource uses when conservation requirements preclude existing or next-best uses, and benefits should include nonmarket benefits accruing not only to the conserved species or systems, but when possible, also to ancillary positive externalities, such as cultural and biological diversity. Evaluation of risk and uncertainty should be framed as risk to benefit analysis so that biological and ecological risks can be compared with potential economic benefits.

Economic impacts such as jobs, taxes, and income must be estimated as equity or distributional effects, separate from efficiency benefits. In particular, both analysts and managers must avoid counting potential increases or decreases in regional employment as net benefits or costs in the benefit–cost analysis. The level of economic analysis effort should vary depending on the expected economic and social consequences of alternative management practices and policies. For example, new willingness-to-pay studies may be well worth the cost when expected nonmarket benefits are large, but they may not be cost-effective when expected nonmarket benefits are small. In the latter case, benefits-transfer methods that estimate RLK species benefits from existing similar studies may be used. Likewise, new economic impact analysis may be needed if expected employment effects are large, otherwise not.

Economic effects should be closely linked to other social effects, including cultural values and institutional dynamics. Because connected ecological and economic systems are complex, variable, and changing, they should be linked in common computer simulation models that can test and monitor the effects of alternative management actions meant to conserve RLK species.

Develop Final List of Approaches for Consideration

The final component of this step is the articulation of a reasonable set of approaches to be considered by managers in their selection process in Step 11. This may include only a subset of the approaches originally identified in Step 9. Some of the approaches identified in Step 9 may not be fully developed for consideration by management because the evaluation in Step 10 indicates that they would be ineffective, inefficient, or unfeasible.

Select Approaches and Develop Detailed Designs (Step 11)

The final selection of the conservation approach or approaches is a management decision and will rarely be determined entirely by the ecological, social, or economic data. Evaluation of information from Step 10 should be clearly articulated and summarized for use by managers in making this decision. It is extremely important that the evaluations be conducted in a consistent way across the alternative approaches, and that the summaries developed for managers are also consistent and focus on key differences. It is important to be clear about sources and levels of uncertainty associated with the results, and to emphasize true differences among the alternative approaches rather than apparent differences. Selection of approaches to be implemented should be based on a judgment about how well the approaches will accomplish the full set of goals described in Step 3. Managers should carefully document the basis for their choices, particularly where accomplishment of one goal has to be weighed against accomplishment of other goals.

The approaches developed and evaluated in the foregoing steps may still be somewhat conceptual and require further refinement before they can be implemented. The following may be important steps in the development of a final design:

- Determine data sources (e.g., species point locations, remotely sensed imagery, historical studies of disturbance patterns) and models (e.g., species habitat affinity, vegetation dynamics) that will be used in various parts of the conservation approach.
- Reconcile existing inventory standards or modeling approaches that might vary across the plan area.
- Develop new survey/inventory protocols and models if current data and existing models are found to be inadequate.
- Develop a framework to receive and/or harvest information from citizen monitoring and other nonexpert field observations, attach source/reliability metadata, and make it available for review, annotation, and analysis.

- Define a set of species or system elements that will be monitored to judge the success or failure of the plan, determine an appropriate monitoring schedule for the elements selected, define the monitoring protocols, specify how monitoring data will feed species or system models, and develop a strategy for how the results from monitoring will be used to revise the plan.
- Finalize the boundaries for specific types of management areas such as reserves or hotspots.
- Finalize management standards for specific types of management areas (e.g., the level of salvage that might be allowed within reserves, or standards designed to provide for persistence of RLK species on sites they currently occupy).
- Develop standards for management that are designed to emulate natural disturbance or move the system toward range of natural variability (RNV) conditions.

Implement Selected Approaches (Step 12)

Implementation of the selected approach or approaches begins as a planning process. With a focus on getting the job done, the scope and complexity of the program can be adjusted to fit the circumstances (see chap. 11). Approaches with objectives addressing large spatial scales, multiple habitat types, numerous species in multiple taxonomic groups, and the priorities of many stakeholders may easily lead to a complex program structure. Approaches with the broadest scopes may have management tiers that include decision makers, program managers, leaders of primary objectives, a variety of species specialists or resource specialists, data managers and analysts, communication specialists, and the necessary support staff and services. Depending upon the level of cooperation among the landowners involved, some similar roles may need to be filled for each of the primary stakeholders so that integrated teams are developed. The plan should also clearly define criteria for success, that is, measurable outcomes by which rare or little-known species can be judged to be sufficiently conserved or restored. The number of taxa included in the plan and the level of accountability desired to achieve the conservation objectives are key in determining the workforce needed. Simpler approaches and those monitoring only surrogate response parameters would require less staff, although the utility of such approaches for meeting resource objectives needs to be tested.

During this planning process, the capability of the resource managers to implement the selected approach should be assessed. Capability is largely a function of the level of involvement needed to conduct the program, funding, available expertise, time, and implementation constraints. If the approach requires only delineating a protected zone around a specific remote area, then the program may have some start-up effort, but subsequent maintenance would involve only required monitoring programs. Higher maintenance would be needed for an actively managed area, or an approach that requires intensive species- or resource-level monitoring, or for other approaches such as reserve boundaries that might be adjusted in response to disturbance or environmental changes.

Many constraints can arise during planning or actual program implementation. Funding, in particular, will limit the extent to which a complex program can be implemented and the extent to which specific design elements can be effectively managed and monitored. For example, many little-known taxa belong to taxonomic groups for which there may be little scientific expertise, or for which techniques for detecting their presence and managing sites or populations are not straightforward. Funding can help acquire or develop expertise and techniques, provided that there is time available for such new knowledge to be gathered. In addition to funding, authority may arise as an issue if the scope of the selected approach results in serious natural resource conflicts, such as reduced economic outputs. In particular, the approach may be challenged by some stakeholders if it does not appear to have a legal basis or opposes other policies. However, authority need not rely solely on a legal foundation but also on a mix of social advocacy, sociopolitical management decisions, and conservation leadership. And in the case of those RLK species that provide ecosystem services, act as architects, serve as potential biotechnology products, or otherwise enhance economic well-being, the policy may be undertaken because it produces economic benefits or reduces economic risks.

The implementation phase involves reiteration of many previous steps. Considerations from social, economic, and ecological arenas may be reevaluated. The initial objectives may be reassessed—and importantly, reprioritized. Implementation constraints may require the selection and design of the approach to be fine-tuned or outright changed. Implementation also requires looking forward to the next steps, because the monitoring plan and adaptive management process will need to be developed and incorporated into the program structure. Once the planning process has proceeded and resolved the major hurdles, and the resulting conservation plan has been approved for actual implementation, the infrastructure can be built and the approach(es) put to the test on the ground.

Monitor and Evaluate Success Relative to the Goals (Step 13)

Developing and implementing conservation approaches for RLK species is not a linear process; it is cyclical (see fig. 12.1). Once plans are implemented, the final step should include monitoring, adapting management, and being accountable to stakeholders. Selected approaches should be viewed as experiments, because rarely in our history have we undertaken species or systems management to the extent that is now being explored. As with any innovation, to ensure our best intentions do not go awry, the "test" needs to be checked, and, as necessary, adjustments made to the management plan.

Two monitoring elements are critical for conservation: implementation monitoring and effectiveness monitoring. Implementation monitoring is designed to assess if specific actions were taken, whereas effectiveness monitoring accounts for how well these actions are meeting the initial goals. Administrative progress reports may be sufficient to monitor the process of implementation, but data collection and analyses will likely be needed to assess whether ecological and socioeconomic goals are being met. With objectives involving species or ecological systems, repeated inventories of species or systems elements will likely be needed to assess species or systems status and trends, and to ensure that desired environmental conditions are being maintained or restored. However, conducting effective inventories to assess status and trends is difficult when dealing with low sample sizes of system responses, such as occurs when monitoring RLK species. This area warrants technique advancement. But there are a number of existing methods that may help solve this problem (e.g., sequential Bayesian analysis), and as new knowledge is synthesized, our ability to reassess and apply effective conservation approaches will be enhanced.

Adaptive management allows the conservation approach to be revised based on new knowledge. With the new knowledge obtained through both the experience of program implementation and monitoring for status and trends, conservation approach effectiveness can be addressed. More broadly, the initial goals and considerations may be readdressed, and with the new knowledge from implementation, the conservation plan may be redesigned with fine-tuned priorities and a new balance of considerations. Time intervals for adaptive management may be contingent upon implementation schedules and specific tasks being conducted. Nevertheless, more regular and frequent intervals might be needed, especially at program initiation, when changes can be anticipated. Frequent, iterative fine-tuning adjustments to a conservation plan may be preferred over irregular or infrequent, larger-scale changes. In some cases, iterative smaller changes in a conservation plan may be less dramatic to implement and may result in a lessened impact both on the program and on risks to RLK species.

Accountability blends both the monitoring and the adaptive management aspects of this step. Stakeholders will be interested in how the approach is implemented and how effective it is and will be interested in developing plans to change the approach or implementation process. Accountability may reveal the perceived problems of the original design and implementation process but will also demonstrate successes and how problems are being resolved. Accountability keeps the stakeholders engaged in the program and should improve partnership trust. Initial conflicts may resurface, but accountability will likely reinforce the importance of the goals and priorities and may enrich the program by an infusion of stakeholder feedback from their different perspectives. Accountability of the conservation approach to the public, beyond primary stakeholders, may similarly result in a recounting of conflicts, but may also increase advocacy.

Accountability can be achieved by a variety of communication avenues to diverse audiences. Oral and visual communication routes may reach different people. Oral routes, including stakeholder meetings, workshops, conferences, and tours, may be the most effective means to achieve one-on-one communication, forging new alliances between individuals. More remote visual means of communication could include publications, reports, brochures, videos, and Web sites. Involvement of communication specialists to help reach target audiences, such as funding agencies and advocacy groups, may be a key step for program longevity.

Conclusion

In any given collection of species that inhabit a particular planning area, most species will be represented by very few individuals (i.e., most species will be rare), and an even greater number of species will be nameless to science. The sheer number of species that are either rare or little known presents difficult challenges to biodiversity conservation. Inclusion of these RLK species in conservation programs adds inherent complexity due to the potentially large number of species to consider and uncertainty due to the fundamental lack of knowledge on their ecologies and natural histories. Careful attention is thus needed in prioritizing the selection of species based on perceived importance (ecological, social, economic, legal) of the species or species group, the probability of support of stakeholders and the public for the effort, the availability of expertise to guide information gathering and evaluation, and practicality from a programmatic point of view. Similar arguments pertain to selection of systems, with an emphasis on our ability to characterize and measure the selected system variables. When selecting priority species or systems, it is critical to develop a standardized process that is transparent to all stakeholders and documents all assumptions and uncertainties.

Although our text and associated figure (see fig. 12.1) discuss our suggested implementation process as a linear sequence of steps, we emphasize that our process is both cyclical (the last step informs a new first step) and contains internal feedbacks among all steps such that the process adapts continuously. We believe it is critically important to (1) set clear goals, (2) identify measurable short- and long-term objectives, and (3) include learning objectives to increase knowledge for little-known species.

We have described a variety of conservation approaches, some focused on species and others focused on systems. Each approach has utility and each has limitations. It is our hope that resource managers can use information we have summarized, evaluate potential approaches considering the context of their planning area, and select approaches that will best meet their goals and objectives. As shown in chapter 8, no single approach is highly effective in providing for species diversity, genetic diversity, and ecosystem diversity conservation objectives. Consequently, in most cases, a combination of approaches will best meet the full set of ecological goals and objectives that managers must meet. For example, approaches designed to meet objectives for RLK species will generally have to be combined with approaches for other species and additional approaches that are intended to meet ecosystem diversity objectives.

We have shown that selection of conservation approaches is not solely based on biological considerations. Regional economies and communities associated with the conservation of RLK may also be complex systems characterized by interacting private ownership and public agency responsibilities. The social effects of alternative RLK conservation actions must be considered, and these actions should also be separately evaluated for local economic impacts and efficiency cost–benefits, nonmarket benefits included. Conservation works best when social and economic considerations are integrated early with biological/ecological considerations in search of management alternatives that can enhance full measures of social, economic, and ecological well-being while conserving RLK species.

References

Basak, I., and T. Saaty. 1993. Group decision making using the analytic hierarchy process. *Mathematical Computer Modeling* 17:101–9.

Cane, J. H., and V. J. Tepedino. 2001. Causes and extent of declines among native North American invertebrate pollinators: Detection, evidence, and consequences. *Conservation Ecology* 5:1. http://www.consecol.org/vol5/iss1/art1.

Corn, M. L. 1993. Ecosystems, biomes, and watersheds: Definitions and use. Report for Congress. Washington, DC: Committee for the National Institute for the Environment, Congressional Research Service.

Davies, K. F., C. R. Margules, and J. F. Lawrence. 2004. A synergistic effect puts rare, specialized species at greater risk of extinction. *Ecology* 85:265–71.

DeLeo, G. A., and S. Levin. 1997. The multifaceted aspects of ecosystem integrity. *Conservation Ecology* 1:3. http://www.consecol.org/vol1/iss1/art3.

Feldman, D. L., R. A. Hanahan, and R. Perhac. 1999. Environmental priority-setting through comparative risk assessment. *Environmental Management* 23:483–93.

FEMAT (Forest Ecosystem Management Assessment Team). 1993. Forest ecosystem management: An ecological, economic, and social assessment. Washington, DC: U.S. Government Printing Office 1993-793-071. Available at: Regional Ecosystem Office, P.O. Box 3623, Portland, OR 97208.

Fisher, R. A., A. S. Corbet, and D. B. Williams. 1943. The relation between the number of species and the number of individuals in a random sample of an animal population. *Journal of Animal Ecology* 12:42–58.

Helles, F., P. Holten-Andersen, and L. Wichmann, eds. 1999. *Multiple use of forests and other natural resources: Aspects of theory and application*. Dordrecht: Kluwer Academic.

Hope, B. K., and J. A. Peterson. 2000. A procedure for performing population-level ecological risk assessments. *Environmental Management* 25:281–89.

Kintsch, J. A., and D. L. Urban. 2002. Focal species, community representation, and physical proxies as conservation strategies: A case study in the Amphibolite Mountains, North Carolina, U.S.A. *Conservation Biology* 16:936–47.

Lehmkuhl, J. F., B. G. Marcot, and T. Quinn. 2001. Characterizing species at risk. Pp. 474–500 in *Wildlife–habitat relationships in Oregon and Washington*, ed. D. H. Johnson and T. A. O'Neil. Corvallis: Oregon State University Press.

McDonald, T. L., and L. L. McDonald. 2003. A new ecological risk assessment procedure using resource selection models and geographic information systems. *Wildlife Society Bulletin* 30:1015–21.

Mendoza, G. A., and R. Prabhu. 2000. Development of a methodology for selecting criteria and indicators of sustainable forest management: A case study on participatory assessment. *Environmental Management* 26:659–73.

Meslow, E. C., R. S. Holthausen, and D. A. Cleaves. 1994. Assessment of terrestrial species and ecosystems. *Journal of Forestry* 92:24–27.

O'Neil, T. A., R. J. Steidl, W. D. Edge, and B. Csuti. 1995. Using wildlife communities to improve vegetation classification for conserving biodiversity. *Conservation Biology* 9:1482–91.

Raphael, M. G., M. J. Wisdom, M. M. Rowland, R. S. Holthausen, B. C. Wales, B. G. Marcot, and T. D. Rich. 2001. Status and trends of habitats of terrestrial vertebrates in relation to land management in the interior Columbia River basin. *Forest Ecology and Management* 153:63–88.

Rieman, B. E., D. C. Lee, R. F. Thurow, P. F. Hessburg, and J. R. Sedell. 2000. Toward an integrated classification of ecosystems: Defining opportunities for managing fish and forest health. *Environmental Management* 25:425–44.

Shaw, C. G. 1999. Use of risk assessment panels during revision of the Tongass Land and Resource Management Plan. PNW-GTR-460. Portland OR: U.S. Department of Agriculture, Forest Service, Pacific Northwest Research Station. 43 pp.

Sieg, C. H., C. H. Flather, and S. McCanny. 1999. Recent biodiversity patterns in the Great Plains: Implications for restoration and management. *Great Plains Research* 9:277–313.

Stolte, K., D. Mangis, R. Doty, and K. Tonnessen. 1993. Lichens as bioindicators of air quality. USDA Forest Service General Technical Report. RM-224. Fort Collins, CO: U.S. Department of Agriculture, Forest Service, Rocky Mountain Forest and Range Experiment Station. 131 pp.

Thomas, J. W., M. G. Raphael, R. G. Anthony, E. D. Forsman, A. G. Gunderson, R. S. Holthausen, B. G. Marcot, G. H. Reeves, J. R. Sedell, and D. M. Solis. 1993. Viability assessments and management considerations for species associated with late-successional and old-growth forests of the Pacific Northwest. Portland, OR: U.S. Department of Agriculture, Forest Service. 523 pp.

Treitz, P., and P. Howarth. 2000. Integrating spectral, spatial, and terrain variables for forest ecosystem classification. *Photogrammetric Engineering and Remote Sensing* 66:305–18.

Varis, O. 1980. The analysis of preference in complex environmental judgements: A focus on the analytic hierarchy process. *Journal of Environmental Management* 28:283–94.

Von Winterfeldt, D. 1992. Expert knowledge and public values in risk management: The role of decision analysis. Pp. 321–42 in *Social theories of risk*, ed. S. Krimsky and D. Golding. Westport, CT: Praeger.

Watson, V. 1997. Working toward operational definitions in ecology: Putting the system back into ecosystem. *Bulletin of the Ecological Society of America* 78:295–97.

Williams, C. 2000. Risk and uncertainty, science and management. *Natural Areas Journal* 20:307.

Wright, R. G., J. M. Scott, S. Mann, and M. Murray. 2001. Identifying unprotected and potentially at risk plant communities in the western USA. *Biological Conservation* 98:97–106.

Contributors

Aaron J. Douglas received a BA in economics from the University of Chicago and a doctorate in economics from Stanford. He taught economics at the University of California at Berkeley and was a research associate at Harvard. He received an MS in forestry from the University of Arizona at Tucson in 1986. After leaving Tucson, he was hired by the U.S. Fish and Wildlife Service and now works for the U.S. Geological Survey as a resource economist.

Curtis H. Flather is a research wildlife biologist and landscape ecologist with the USDA Forest Service at the Rocky Mountain Research Station in Fort Collins, CO. He also holds affiliate faculty appointments within the Department of Fish, Wildlife, and Conservation Biology and the graduate degree program in ecology at Colorado State University. He received his BS from the University of Vermont, and his graduate degrees from Colorado State University.

Richard S. Holthausen retired after 28 years with the U.S. Forest Service, 15 of which were spent as the agency's national wildlife ecologist. In that position he was involved in large-scale efforts, including the Northwest Forest Plan, Interior Columbia Basin Ecosystem Management Plan, Tongass Forest Plan, and Sierra Nevada Framework. He was also involved in developing direction for implementation of the National Forest Management Act, including provisions for species viability.

Richard L. Johnson applies economic theory and private market incentives to improve public sector decisions. He has been an economist with the U.S. Interior Department for more than 30 years, developing input–output methods for estimating economic impacts, and nonmarket valuation methods for better including natural resources in public policy decisions. He led an interagency economics team to estimate the economic implications of designating critical habitat for the northern spotted owl. He worked in a multidisciplinary team to better integrate economic development and ecological conservation through adaptive environmental assessment and management. Currently, he is developing market incentives to conserve biological and cultural diversity in developing countries. He received his BS and MS from Utah State University and additional graduate work at University of Minnesota and at Michigan State University.

Bruce G. Marcot is a research wildlife biologist with the Ecosystems Processes Research Program of the USDA Forest Service in Portland, Oregon. He participates in applied science research and technology application projects dealing with older-forest management, specifically on modeling of rare and little-known species, assessment of biodiversity, and ecologically sustainable forest management. He has served on numerous regional assessment teams, including the Interior Columbia Basin Ecosystem Management Project and the Forest Ecosystem Management Assessment Team, and has worked on forest conservation and ecology in India, Congo, Canada, and many other locations. He received a BS in natural resources planning and an MS in wildlife management at Humboldt State University in Arcata, California, and a PhD in wildlife science at Oregon State University, Corvallis.

Nancy Molina is a consulting ecologist with Cascadia Ecosystems, Gresham, Oregon. She is retired from the USDA Forest Service and USDI Bureau of Land Management, where she helped guide the interagency Northwest Forest Plan's rare and little-known species conservation effort from which this book developed.

Randy Molina is a research botanist and team leader for forest mycology, USDA Forest Service, Pacific Northwest Research Station, Portland, Oregon (retired). He specializes in the ecology, ecosystem function, conservation, and management of fungi in forest ecosystems.

Deanna H. Olson is a research ecologist with the USDA Forest Service, Pacific Northwest Research Station in Corvallis, Oregon. She earned a BS from UC San Diego in 1980 and a PhD in 1988 from the Department of Zoology at Oregon State University. She maintains courtesy appointments at Oregon State University (Departments of Zoology, Fisheries and Wildlife, Forest Resources), serves as associate editor for *Herpetological Review*, and is regional chair for the Pacific Northwest section of Partners for Amphibian and Reptile Conservation and the Declining Amphibian Populations Task Force. Her research interests examine issues of conservation biology, behavioral ecology, and population and community ecology of amphibians in western Oregon. Ongoing work includes examining the effects on various fauna of forest management practices, landscape designs, and policies.

John D. Peine is a research sociologist with the U.S. Geological Survey stationed at the Southern Appalachian Field Lab on the campus of the University of Tennessee at Knoxville. He is an adjunct professor with the Department of Forestry, Wildlife and Fisheries and is a member of the Institute for a Secure and Sustainable Environment at the university. From 1982 to 1992 he was the research administrator at Great Smoky Mountains National Park. He edited a book on the principles and practices of ecosystem management and was the lead author of a book chapter on the contributions of sociology on ecosystem management. He received his BS from Purdue University and his MS and PhD from the University of Arizona.

Martin G. Raphael is a research wildlife biologist and team leader with the USDA Forest Service, Pacific Northwest Research Station, Olympia, Washington. He also holds an affiliate faculty appointment with the College of Forestry, University of Washington. He received a BA from California State University at Sacramento and BS, MS, and PhD degrees from the University of California, Berkeley. His research interests include ecology and conservation of at-risk vertebrates, especially species associated with older-forest ecosystems in the Northwest.

Carolyn Hull Sieg is a research ecologist with the USDA Forest Service, Rocky Mountain Research Station. Until 2000, she was stationed in Rapid City, South Dakota, and served on the recovery team for the federally listed threatened western prairie fringed orchid. She is coauthor of population viability exercises for a conservation biology workbook, a book on historical ecology of eastern North Dakota, and a book chapter on exotic invasive species. She is now stationed at the Flagstaff, Arizona lab. She received her BS and MS from Colorado State University and her PhD from Texas Tech University.

Cindy S. Swanson is the director of Watershed, Wildlife, Fisheries and Rare Plants for the Northern Region (Montana, Northern Idaho, and North Dakota) of the USDA Forest Service, Missoula, Montana. She previously served on the wildlife and fisheries staff in the Washington office of the USDA Forest Service. She has published extensively in the area of nonmarket valuation, with a focus on wildlife and fish values. She is a coeditor and author of several chapters in *Valuing Wildlife Resources in Alaska* (Boulder: Westview Press, 1992). She received her PhD at Ohio State University in natural resource economics in 1994 and has earlier degrees in wildlife biology and economics from the University of Wyoming.

Index

Island Press Board of Directors